678·7

**Rocket Propulsion Establishment
Library**

B. 5622

Please return this publication, or request renewal, before
the last date stamped below.

Name	Date 3 days.
Mr. Barrett	22-12-72
P. J. Sherwood.	15·1·73.
~~G.R.R~~	22·1·73
~~A. J. Buswell~~	13·2·73.
Mr. Quinton	5-2-77
E. H. Matthews	29·7·77
Mr. Plant	5-4-86
Dr. L. Mason	7.2.87
Mr. Cheshire	25.5.88.
R. Windsor	3/4/90

RPE Form 243 (revised 6/71)

739490

Polymer Science and Materials

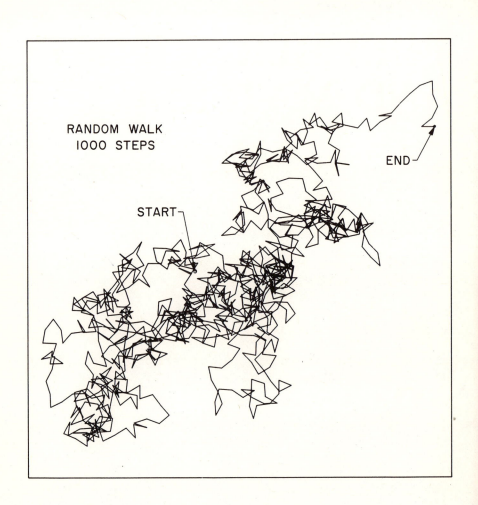

RANDOM WALK
1000 STEPS

START

END

POLYMER SCIENCE AND MATERIALS

Edited by

ARTHUR V. TOBOLSKY

Frick Chemical Laboratory
Princeton University

HERMAN F. MARK

Polymer Research Institute
Polytechnic Institute of Brooklyn

WILEY-INTERSCIENCE, a Division of John Wiley & Sons, Inc.
New York · London · Sydney · Toronto

Library of Congress Catalogue Card Number: 70-155908

ISBN 0-471-87581-3

Printed in the United States of America.

10 9 8 7 6 5 4 3 2 1

Contributors

A. A. Bondi, *Shell Development Company, Emeryville, California*

R. D. Deanin, *Department of Chemistry, Lowell Technological Institute, Lowell, Massachusetts*

D. B. DuPré, *Department of Chemistry, University of Louisville, Louisville, Kentucky*

A. N. Gent, *Institute of Polymer Science, The University of Akron, Akron, Ohio*

H. Mark, *Polytechnic Institute of Brooklyn, Brooklyn, New York*

A. Peterlin, *Camille Dreyfus Laboratory, Research Triangle Institute, Research Triangle Park, North Carolina*

L. Rebenfeld, *Textile Research Institute, Princeton University, Princeton, New Jersey*

E. T. Samulski, *Department of Chemistry, University of Texas, Austin, Texas*

S. Siggia, *Department of Chemistry, University of Massachusetts, Amherst, Massachusetts*

A. F. Stancell, *Research and Development Laboratories, Mobil Chemical Company, Edison, New Jersey*

R. S. Stein, *Department of Chemistry, Polymer Research Institute, University of Massachusetts, Amherst, Massachusetts*

A. V. Tobolsky, *Department of Chemistry, Frick Chemical Laboratory, Princeton University, Princeton, New Jersey*

I. V. Yannas, *Department of Mechanical Engineering, Fibers and Polymers Division, Massachusetts Institute of Technology, Cambridge, Massachusetts*

Preface

This book, which will appear in two volumes, is meant for the beginning student of polymer science or engineering, whether undergraduate, graduate, or employed in the polymer industry. An attempt was made to present the fundamental ideas of polymer science in as simple a manner as possible. Yet inasmuch as the two chief authors have been engaged in polymer research for a total of over seventy-five years, it is hoped that some of the adventurous and lively spirit of research thinking and also a sense of perspective will flavor the style of the book. We feel that for the beginning student this is even more important than comprehensiveness or detail, which can be obtained by wider reading once the interest and motivation are provided.

The variegated field of polymer science and materials is nurtured by many scientific disciplines. To achieve breadth of discipline the chief authors have invited several distinguished scientists to act as coauthors or authors of individual chapters. Nevertheless an attempt has been made to retain a certain uniformity in the mode of presentation. We hope that the book will serve to inspire many new students.

ARTHUR TOBOLSKY
HERMAN MARK

Princeton, New Jersey
Brooklyn, New York
April 15, 1971

Contents

Polymer Science
and Materials

Polymer Molecules

1

H. Mark and A. V. Tobolsky

1. THE LINEAR POLYMER MOLECULE

The basic idea in polymer science is the conception of the linear chain molecule, in which the atoms comprising the main chain are united by primary valence forces that could include ionic and metallic bonding but are mostly of the *covalent* type. A few examples are given in Table 1.

The expression "molecular structure" or "structure" of a molecule did not acquire substantial meaning before the second half of the last century. Previously chemists were, in general, satisfied to establish for a molecule that they had analyzed or synthesized the chemical *composition* in terms of a stoichiometric *formula* and to describe, as completely as possible, its properties—color, crystal habit, specific gravity, refractive index, melting point, boiling point, solubility, chemical character, and so on.

When, about 100 years ago, the establishment of a structure became an important, and eventually prevalent, part of a chemical publication, it was particularly Kekulé who became the protagonist of the new approach when he had the vision of carbon atoms forming chains to which other atoms, such as hydrogen, oxygen, or nitrogen, could be attached. Later he added to the concept of an open chain that of a closed ring and explained in a global way the essential differences between aliphatic and aromatic chemistry. All geometrical formulas of those days referred to the structure and the behavior of ordinary, small molecules, but when Kekulé in 1877 became rector of the

Table 1   Examples of Linear Polymer Molecules

Polymer	Formula
Polymethylene (high-density polyethylene)	$-CH_2CH_2CH_2CH_2CH_2CH_2CH_2CH_2CH_2CH_2CH_2CH_2-$

Isotactic polypropylene

$$-CH_2CHCH_2CHCH_2CHCH_2CHCH_2CHCH_2CHCH_2CH-$$
$$\quad\ |\qquad\ |\qquad\ |\qquad\ |\qquad\ |\qquad\ |\qquad\ |$$
$$\quad CH_3\ \ CH_3\ \ CH_3\ \ CH_3\ \ CH_3\ \ CH_3\quad CH_3$$

Atactic polypropylene

$$\qquad\qquad CH_3\ \ CH_3\qquad\qquad CH_3$$
$$\qquad\qquad\ |\qquad\ |\qquad\qquad\ |$$
$$-CH_2CHCH_2CHCH_2CHCH_2CHCH_2CHCH_2CH-$$
$$\quad\ |\qquad\qquad\qquad\qquad CH_3\qquad CH_3$$
$$\quad CH_3$$

Ethylene–propylene copolymer

$$\qquad\qquad\qquad\quad CH_3\qquad\qquad\qquad CH_3$$
$$\qquad\qquad\qquad\quad\ |\qquad\qquad\qquad\ |$$
$$-CH_2CH_2CH_2CHCH_2CH_2CH_2CHCH_2CH_2CH_2CH_2CH-$$
$$\qquad\qquad\ |$$
$$\qquad\qquad CH_3$$

Atactic polystyrene

$$\qquad\qquad C_6H_5\qquad\qquad\qquad\qquad C_6H_5$$
$$\qquad\qquad\ |\qquad\qquad\qquad\qquad\ |$$
$$-CH_2CHCH_2CHCH_2CHCH_2CHCH_2CHCH_2CH-$$
$$\quad\ |\qquad\qquad\ |\qquad\ |\qquad\ |$$
$$\quad C_6H_5\qquad\quad C_6H_5\ C_6H_5\ C_6H_5$$

Polycarbonate	

Synthetic polypeptide

$$-NHCHCONHCHCONHCHCONHCHCO-$$
$$\quad\ |\qquad\quad\ |\qquad\quad\ |\qquad\quad\ |$$
$$\quad R\qquad\quad R\qquad\quad R\qquad\quad R$$

Aromatic polyimide	

University in Bonn and delivered an inaugural address of general character and wider scope, he advanced the hypothesis that the natural organic substances that are most directly connected with life—proteins, starch, cellulose—may consist of very long chains and derive their special properties from this peculiar structure.

The change in emphasis from *composition* to *structure* led to the demand that any chemist present in his publications the *structural formula* of the material he was investigating. One of the greatest promoters of structural organic chemistry around the turn of the century was Emil Fischer, who, as early as 1893, had already the structure of cellulose as a polysaccharide in mind and expressed the opinion that it might be represented as a chain of glucose units; his later systematic work did not leave any doubt that he proposed with clarity and emphasis a long-chain structure for polypeptides and natural proteins.

In 1920 and 1921 there appeared three important papers that postulated long-chain structure for several synthetic and natural compounds on the basis of general considerations and offered specifically for cellulose the long-chain character as a preferred alternative in comparison with other structures. The first of these papers was published by Staudinger and proposed for polystyrene (**1**), polyoxymethylene (**2**), and rubber (**3**) formulas that represented linear long chains such as the following:

$$
\begin{array}{ccccccc}
\text{H} & & \text{H} & \text{H} & \text{H} & \text{H} & \text{H} \\
| & & | & | & | & | & | \\
-\text{C}\!-\!\!-\!\!-\!\!-&\!\!\text{C}\!-\!\text{C}&\!-\!\!-\!\!-&\!\!\text{C}\!-\!\text{C}&\!-\!\!-\!\!-&\!\!\text{C}\!- \\
| & & | & | & | & | & | \\
\text{C}_6\text{H}_5 & & \text{H} & \text{C}_6\text{H}_5 & \text{H} & \text{C}_6\text{H}_5 & \text{H} \\
& & & \mathbf{1}
\end{array}
\qquad
\begin{array}{ccccc}
\text{H} & \text{H} & \text{H} & \text{H} & \text{H} \\
| & | & | & | & | \\
-\text{C}-\text{O}-\text{C}-\text{O}-\text{C}-\text{O}-\text{C}-\text{O}-\text{C}- \\
| & | & | & | & | \\
\text{H} & \text{H} & \text{H} & \text{H} & \text{H} \\
& & \mathbf{2}
\end{array}
$$

$$
\begin{array}{ccccccc}
\text{H} & \text{CH}_3 & \text{H} & \text{H} & \text{H} & \text{CH}_3 & \text{H} & \text{H} \\
| & | & | & | & | & | & | & | \\
-\text{C}-\text{C}&=&\text{C}-\text{C}-\text{C}-\text{C}&=&\text{C}-\text{C}- \\
| & & | & | & & | \\
\text{H} & & \text{H} & \text{H} & & \text{H} \\
& & & \mathbf{3}
\end{array}
$$

These formulas are still accepted today.

The second paper was published by Freudenberg; it offered new experimental data on the yield of cellobiose during cellulose degradation and stated that the best available data were in conformity with a long-chain structure.

In the third article Polanyi came to the conclusion that the measured X-ray diffraction spots that had been obtained from native-cellulose fibers are in agreement either with long glucosidic chains or with rings consisting of two glucose anhydride units.

On the basis of all available data and with additional evidence from X-ray diagrams of various cellulose derivatives as well as the optical activity of cellulose and its degradation products Meyer and Mark accumulated convincing material for the presently accepted chain structure of cellulose and for the crystalline–amorphous character of cellulosic fibers.

In 1900 Bamberger and Tschirner reacted diazomethane with β-aryl-hydroxylamines and obtained a white, chalklike powder with a melting point of 128°C, which they considered to be polymethylene, $(CH_2)_n$. This was the first correct formulation and description of a polyhydrocarbon—linear polymethylene—which in its structure and properties is identical with linear 1,2-polyethylene.

The chain structure of native (*Hevea*) rubber, as postulated by Staudinger found substantial support by the discovery of Katz that stretched rubber gives a fiberlike X-ray diagram and by the quantitative interpretation of this diagram by Hauser and Mark and Meyer and Mark.

During many years of studies on polypeptides and proteins Fischer never suggested anything else for the structure of these products but the character of a linear chain consisting of many amino acid units that are connected with each other by the normal —CO—NH— linkage as it occurs in all amides and peptides.

Herman Leuchs, one of Fischer's associates, made an even bolder step in the direction of true synthetic polypeptides by the preparation and investigation of the α-amino acid N-carboxylic anhydrides; they decompose at elevated temperatures and in the presence of traces of moisture with the evolution of carbon dioxide into solid bodies that he considered to be "polymers" of a "cyclic monomer."

$$\left[\overline{HN-CHR-CO} \right]_x$$

Thus by analysis and synthesis there emerged with increasing vitality the concept of the linear macromolecule or the linear polymer.

Although for the most important natural substances—cellulose, rubber, proteins, and starch—the high polymeric or macromolecular character was first postulated and later more and more reliably established, there were many scientists who were unconvinced and preferred the concept that these substances consist of *small building units* that, however, are held together by exceptionally *strong forces* of aggregation or association, supposed to be of a new and still unknown character.

The fact that all materials under investigation are the products of living beings—plants or animals—provided an attractive and probably perfectly legitimate argument in favor of something new, something that we still have to learn and to clarify in order to understand the structure and properties of all these materials. However, as often in science and history, this somewhat romantic approach had to fade away gradually under the influence of more and better experimental evidence for the macromolecular theory.

There were many factors that eventually tipped the scales in favor of the concept of very large, chainlike molecules. One of them was the rapid improvement and refinement of the X-ray diffraction method, which, again and again, not only gave answers in favor of long chains but also permitted, and still permits, a progressively detailed description of every kink and twist in a macromolecule.

Another important move was the introduction of Svedberg's ultracentrifuge, which played a decisive role because it was the first method that permitted a direct and reproducible measurement of the molecular weight in the range between 40,000 and several millions; at the same time improved osmometers added significance and reliability to these data.

But probably more than any other single factor did the work of Carothers and his associates contribute to the ultimate breakthrough in favor of the long-chain concept. His efforts extended from synthesis and characterization to ultimate properties and encompassed with the same emphasis condensation and addition polymers. A careful analysis of all prior art led him early to the conclusion that the macromolecular hypothesis was correct, and all his own experiments strengthened his conviction. For him the controversy—association hypothesis versus long-chain theory—was a matter of the past, and he advanced with full scientific and industrial success on the basis of the latter.

Once the basic concepts of the new branch of chemistry were firmly established, polymer chemists settled to useful and practical work: synthesis of new monomers; quantitative study of the mechanism of polymerization processes in bulk, solution, suspension, and emulsion; characterization of macromolecules in solution on the basis of statistical thermodynamics; and fundamentals of the behavior in the solid state. The result was a basic understanding of the properties of rubbers, plastics, fibers, coatings, and adhesives.

2. THE CONCEPT OF FUNCTIONALITY

The basic and highly successful idea used by Carothers is that *equivalent* amounts of *bifunctional* monomers, such as hexamethylenediamine and adipic acid, can lead to very-high-molecular-weight linear polymers if the reaction is forced to high levels of completion.

$$NH_2(CH_2)_6NH_2 + HOOC(CH_2)_4COOH \xrightarrow{-H_2O} \text{nylon 66}$$

Addition of small amounts of monofunctional materials, such as stearic acid, leads to a lower-molecular-weight material. Addition of small amounts of trifunctional or higher amines or acids leads to network structures.

Whereas linear polymers even of very high molecular weight are generally soluble in apt solvents (formic acid for nylon 66), network polymers are not soluble in any nondegrading solvent. Furthermore, at sufficiently high temperatures linear polymers act as high-viscosity liquids and can be extruded and otherwise molded or shaped. Network polymers have infinite viscosity and must be *synthesized* in a mold that has the shape of the desired object.

To create a fiber, molten polymer must be forced through tiny holes (spinnerets) or dissolved polymer must be forced through spinnerets into a precipitating liquid. For this reason synthetic organic fibers are made from *linear* polymers, as was first accomplished by Du Pont with nylon.

Sometimes after a linear polymer has been shaped, it is desirable to introduce a small amount of *crosslinking* to produce a loose, three-dimensional network. For example, natural rubber is milled with 2% sulfur and other chemicals and forced at moderate temperatures and pressures into a tire mold. The mold is then heated for about an hour at 140°C. The sulfur produces a few crosslinkages, monosulfide and polysulfide, between neighboring polyisoprene chains. This vulcanized material (Goodyear) is now insoluble and will no longer flow under heat and pressure.

Polyethylene tape subjected to high-energy radiation becomes crosslinked through the ejection of hydrogen atoms; this new material will no longer melt and flow at high temperatures.

Clothing made from composite fabric of cellulose and polyethylene terephthalate is placed in an oven, and the cellulose is subjected to a crosslinking reaction (e.g., dimethylolurea reacting with a few of the hydroxyl groups of the cellulose). In this way permanent-press clothing is manufactured.

Very highly crosslinked networks can be made by using polyfunctional monomers; for example, Bakelite is made by reacting phenol with formaldehyde:

$$* \underset{*}{\overset{*}{\bigcirc}} - OH + CH_2O \longrightarrow Bakelite$$

The starred positions can react with formaldehyde by splitting out water and forming methylene bridges. It takes a great deal of energy to chemically "vaporize" this three-dimensional network, which is therefore used as an ablative material or heat shield for satellite nose cones.

3. STEREOSPECIFICITY OF POLYMERS

In 1955 Natta made a discovery of overwhelming importance that can best be explained in a historical context.

The polystyrene utilized by Staudinger in his studies during the 1920s was made by heating styrene, $CH_2=CH\phi$, with a small quantity of organic peroxide. The polymerization proceeds via a growing *free radical*.

The polystyrene formed in this manner was developed as a commercial plastic in Germany during the 1930s and slightly later in the United States. It is a clear, hard, amorphous glass. Its X-ray diagram shows no evidence of crystalline lattice order. The polymer dissolves readily in benzene at room temperature.

Around 1935 Imperial Chemical Industries made polyethylene by polymerizing ethylene with traces of oxygen as catalyst at high temperatures and high pressures. This product has approximately two butyl and one ethyl side branches per 100 methylene units. The polymer is also formed via growing free radicals. The polymer is approximately 50% crystalline.

In the early 1950s Ziegler showed that ethylene can be polymerized at low temperatures and low pressures by using a "Ziegler catalyst," made by mixing aluminum trialkyls with titanium salts such as the tetrachloride. The mechanism of this surprisingly smooth polymerization is not fully understood, but it is dissimilar to the free-radical mechanism. The product is almost completely linear polyethylene (no side branches). The crystallinity of this polymer is about 90%; it is therefore of higher density than the polyethylene produced by Imperial Chemical Industries. The products are frequently referred to as low-density polyethylene and high-density polyethylene.

In 1955 Natta used the Ziegler catalyst to polymerize styrene. The result shook the chemical world; a new type of linear polymer was formed. This was no longer soluble in benzene at room temperature; it required refluxing chlorobenzene to dissolve it. The new polymer was crystalline, as contrasted to free-radical-polymerized polystyrene, which is amorphous. From X-ray studies and other evidence Natta deduced that the new polystyrene (4) was *isotactic*.

$$-CH_2CHCH_2CHCH_2CHCH_2CH-$$
$$\begin{array}{cccc} | & | & | & | \\ C_6H_5 & C_6H_5 & C_6H_5 & C_6H_5 \end{array}$$
4

In this structure the phenyl groups are all "up" or all "down," as contrasted with free-radical-polymerized polystyrene, in which the phenyl groups are randomly "up" or "down."

Natta also prepared three varieties of polypropylene: isotactic, atactic, and syndiotactic. In the syndiotactic polymer the methyl groups alternate regularly as "up" and "down." Both isotactic and syndiotactic polypropylenes are crystalline; the atactic variety is amorphous. Crystalline polypropylene is now a very important plastic and film.

Isotactic polypropylene and isotactic polystyrene are not optically active because each is symmetrical around its central chain atom. However, shortly afterward isotactic polypropylene oxide and isotactic synthetic polypeptides were prepared, and these are optically active.

4. POLYMER REGULARITY

The regularity of a linear polymer molecule determines its spatial arrangement in solution and in the solid (nondissolved) state. Thus isotactic polypropylene is partially crystalline in the solid state below 175°C; above 175°C it is in an amorphous condition. Atactic polypropylene can never be obtained in a crystalline state. Above −20°C it is in a "liquid" state (actually rubbery because of the high molecular weight); below −20°C it is in a "glassy" state. Whether liquid, rubber, or glass, the overall structure is amorphous because stereoirregularity prevents an ordered packing.

Synthetically produced polymers are usually composed of molecules of different molecular weights, and this produces an element of randomness. Irregularity can also be built into a structure by random copolymerization. Thus a *random* copolymer that is based on an equimolar amount of ethylene and propylene is amorphous under all conditions.

In general copolymers produced synthetically have a nearly random sequence of the monomers employed. By contrast natural proteins, though composed of nearly twenty different amino acid moieties, have a completely ordered arrangement of the monomer units and also a definite molecular weight. Because of this the natural protein molecules can fold up in a very regular manner. Sometimes the protein molecule exists in an extended rod-like structure, as in silk, keratin, and other fibrous proteins. Most often, however, it folds into a very compact and perfectly ordered globular structure, as is true of myoglobin, hemoglobin, and most enzymes.

5. POLYMERS IN SOLUTION

Amorphous linear polymers, such as polystyrene, usually dissolve very readily in thermodynamically apt solvents (in this case benzene, toluene, etc.). Crystalline polymers, such as polyethylene, frequently have to be heated to some temperature approaching their melting point before they will dissolve in suitable solvents. Atactic polypropylene dissolves readily in heptane at room temperature; isotactic polypropylene requires refluxing n-heptane in order to dissolve.

In solution the vast majority of synthetic linear polymers have the shape of a statistically random coil (Kuhn; Guth and Mark). This can be visualized as the totality of shapes assumed by a quite flexible rope agitated by a turbulent fluid. This is because there is an internal rotation (somewhat hindered) around the bonds of the main polymer chain.

Globular proteins, such as myoglobin, hemoglobin, and various enzymes, adapt a single internally ordered arrangement in space, whether they are dissolved in water or crystallized. In solution the globular protein molecule as a whole is randomly oriented in space; in the crystalline state (first isolated by Kunitz and Northrup) it adopts a position of complete order with respect to other molecules in the lattice.

In 1952 Watson and Crick announced that DNA molecules existed as paired intertwining helical rods, a discovery that initiated the science of molecular biology. It was shortly thereafter found (Doty and Yang) that synthetic polypeptides existed as rigid helical rods in certain solvents, the form of the helix being very probably the α-helix of Pauling deduced from X-ray analysis of fibrous proteins.

The thermodynamics, hydrodynamics, and optical properties of dissolved polymer molecules obviously relate to their shape: random coil, globular spheroid, or rigid helical rod.

6. POLYMERS IN THE SOLID STATE

Polymers in the solid state (by which we mean polymers in the absence of a solvent) can exist either in a semicrystalline condition at temperatures below their melting point or in a completely amorphous condition at all temperatures. This depends on the structural regularity of the polymer, as already discussed.

It is interesting to consider the specific volume (or specific enthalpy) of a relatively-low-molecular-weight substance, such as abietic acid (rosin), shown in Fig. 1. Abietic acid exists in a crystalline form but is easily supercooled from the melt (melting point 172°C). If cooled rapidly from above its melting point, it goes into a supercooled liquid condition, and below 70°C it goes into a glassy condition.

In the crystalline state solids have a low coefficient of expansion. At the melting point T_m there is a discontinuous volume change. The coefficient of expansion in the liquid state is on the order of 6×10^{-4} per degree Celsius. At the temperature at which liquid transforms to glass (glass-transition temperature T_g) there is a fairly sharp change in the coefficient of expansion, which in the glassy state is about 2×10^{-4} per degree Celsius.

Most pure low-molecular-weight liquids can be supercooled only with difficulty; hence they exist only as crystalline solids or as true liquids. Some semicrystalline polymers, such as polyethylene, are very difficult to supercool.

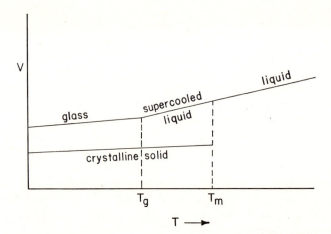

Fig. 1. Specific volume of a low-molecular-weight compound that forms a glass.

Many semicrystalline polymers can be supercooled, and so the diagram shown for abietic acid is pertinent for these materials. Moreover, many polymers with an irregular structure can never be even partly crystallized, and hence the crystalline specific-volume-versus-temperature portion of Fig. 1 is never realized.

The quantities T_g and T_m are very significant for polymers. Whereas low-molecular-weight substances are liquid above T_g, high-molecular-weight polymers are rubbery until $T_g + 100°C$, at which point they behave as viscous liquids. The T_g of natural and synthetic rubbers is about $-60°C$.

Amorphous polymers used as plastics must have a T_g above room temperature; examples are polystyrene (100°C) and polymethyl methacrylate (105°C). Crystalline polymers used as plastics must have a T_m above room temperature—for example, isotactic polypropylene (175°C) or nylon 66 (260°C). Fibers are highly oriented crystalline polymers with a T_m well above that of boiling water (e.g., nylon and polypropylene).

Bibliography

H. Mark, "Proceedings of the Robert A. Welch Foundation Conferences *10* (Polymers)," pp. 42, 43, presents a list of many of the early references discussed in this chapter.

Molecular Weight and Molecular-Weight Distribution

2

S. Siggia and H. Mark

When in the early 1920s polymer chemistry began to develop as a special branch of science, there were many distinguished chemists who seriously questioned the existence of macromolecules—that is, of chemical entities with molecular weights in the range of 100,000 or above. The only way to satisfy them was to determine the molecular weights of such representative materials as cellulose, rubber, starch, silk, polystyrene, or polyformaldehyde by reliable and strictly reproducible methods. As a consequence measurements of the molecular weights of carefully purified specimens, being in the center of the controversy on the existence of macromolecules, were of outstanding interest and importance. It was ultimately a combination of data on diffusion and sedimentation rate, osmotic pressure, viscosity, and end-group analysis that broke down the resistance of the incredulous opponents and helped establish polymer chemistry as a legitimate branch of science. But even now information on molecular weight and molecular-weight distribution still occupies a key position in polymer science. From the theoretical point of view such information is essential to proving or disproving specific concepts on the state of macromolecules in solution and, with it, on the thermo-dynamics of such systems. It is also important for the elucidation of the poly-merization mechanisms that exist under given conditions and are responsible for the start, propagation, and cessation of the individual chain molecules.

Finally the molecular weight and its distribution in many cases profoundly influence the biological and physiological properties of macromolecules and have given very useful clues to the design of effective synthetic biopolymers.

For the practical chemist molecular weight and molecular-weight distribution are of prime importance because they profoundly influence the rheology of polymer solutions and melts and are decisive for strength, elasticity, toughness, and abrasion resistance.

1. CONCEPT OF AVERAGES

When we speak of the molecular weight of a given polymer sample, we are really speaking of its *average* molecular weight since, in most cases, the material is a *polydisperse* system containing a range of molecular weights. However, a few native *monodisperse* polymers exist; they are composed of single molecular-weight species.

There are several different types of average molecular weights:

1. The *number-average* molecular weight $\overline{M}_n$ is obtained by adding the number of molecules of each species, multiplied by its molecular weight, and dividing by the total number of molecules of all species:

$$\overline{M}_n = \frac{\sum\limits_i N_i M_i}{\sum\limits_i N_i} \tag{1}$$

where N_i is the number of molecules of the species of i and M_i is its molecular weight. The boiling point, freezing point, vapor pressure, osmotic pressure, and end-group analysis give number-average molecular weights.

2. The *weight-average* molecular weight $\overline{M}_w$ is obtained by adding the product of the weight g_i and the molecular weight M_i of each fraction and dividing by the total weight:

$$\overline{M}_w = \frac{\sum\limits_i g_i M_i}{\sum\limits_i g_i} \tag{2}$$

Light scattering, diffusion, and ultracentrifugation give weight-average molecular weights.

3. The *z-average-molecular* weight $\overline{M}_z$ is defined as

$$\overline{M}_z = \frac{\sum\limits_i N_i M_i^3}{\sum\limits_i N_i M_i^2} \tag{3}$$

It is not often used and can be obtained by ultracentrifugal techniques.

The different molecular-weight averages can be compared as follows:

$$\overline{M}_n = \frac{\sum\limits_i N_i M_i}{\sum\limits_i N_i}; \qquad \overline{M}_w = \frac{\sum\limits_i N_i M_i^2}{\sum\limits_i N_i M_i}; \qquad \overline{M}_z = \frac{\sum\limits_i N_i M_i^3}{\sum\limits_i N_i M_i^2}$$

The numerator of the first is the denominator of the second and so on. If the polymer is monodisperse, then $\overline{M}_n = \overline{M}_w = \overline{M}_z$. If the polymer is polydisperse, then $\overline{M}_n < \overline{M}_w < \overline{M}_z$.

4. The *viscosity-average* molecular weight $\overline{M}_v$ can be obtained by the determination of the *intrinsic viscosity* of a given polymer in an appropriate solvent (compare Chapter 5). It was found that there exists a relation between the intrinsic viscosity and the molecular weight, and it has the form

$$[\eta] = KM^a \tag{4}$$

where K and a are empirical constants that depend on the solvent and on temperature. Using Eq. 4, one finds that the viscosity-average molecular weight is given by

$$\overline{M}_v = \frac{\sum N_i M_i^{a+1}}{\sum N_i M_i^a} \tag{5}$$

2. MEASUREMENT OF AVERAGES

a. Methods Based on Colligative Properties

The colligative properties of a solvent are those that depend on the *number* of particles dissolved therein. The measurement of these properties yields the number of dissolved particles and once the weight of the dissolved solute is known, the number-average molecular weight can be computed. The important colligative properties of a solvent are vapor pressure, freezing point, boiling point, and osmotic pressure.

Qualitatively the situation can be described as follows: The molecules of a liquid, in our case the solvent, are in a permanent random motion whose intensity depends on temperature. This movement eventually makes them escape into the gas phase and is responsible for the vapor pressure of the liquid. If the system borders on a membrane or an otherwise permeable wall, the molecules are driven into it and cause the phenomenon of osmosis. The molecules may also—at an appropriately low temperature—aggregate into a solid phase, which corresponds to the freezing of the liquid. If the solid and liquid phases are in equilibrium with each other, they evidently possess the same vapor pressure.

If one prepares a solution, a second component—the solute—is added, and its molecules interfere with the random motion of the solvent molecules, causing a reduction in its vapor pressure and diffusibility. If a membrane separates a solution from the pure solvent, the molecules of the latter diffuse more effectively through the membrane and produce the *osmotic pressure*. If one wants to bring a solution to its boiling point, one has to go to a higher temperature than that for the pure solvent because the reduced vapor pressure of the solution needs more heating to reach the external vapor pressure of 1 atm. This causes a *boiling-point elevation*. If one wants to establish equilibrium between the liquid and solid phases of a solution, one has to go to a lower temperature than that for the solvent in order to obtain a solid phase that possesses the lower vapor pressure of the solution. This causes the *freezing-point depression*.

The magnitude of the effect produced is the basis of the measurement of molecular-weight averages by these approaches.

i. Vapor-Pressure Reduction

The diminution of the vapor pressure by a dissolved solute can be measured in several ways:

1. *Direct* measurement of the *vapor pressures* of the solution and of the pure solvent. This approach is difficult and therefore not practical since usually the direct measurements are not sufficiently sensitive to permit the depression to be accurately determined.

2. The *isopiestic method*, in which solvent vapor is partitioned by evaporation between a solution of the sample and a solution of a known standard. After equilibrium the weight gain or loss of the sample solution is compared to the weight gain or loss of the standard solution. If the concentrations of both are known, the molecular weight of the solute can be obtained. This is a very accurate method, but usually a long time (several days) is needed for equilibrium to be established.

3. Measurement of the *thermal effects* produced by migration of pure solvent into a solution. One uses two thermistors in a Wheatstone bridge circuit. Pure solvent is placed on one thermistor, and a solution of the sample in that solvent is placed on the other; both units are housed in a chamber that is saturated with the vapor of the solvent. Under these conditions vapor will move from the pure solvent into the solution, causing a heating effect, due to condensation, that will unbalance the bridge. The degree of unbalance is a measure of the number-average molecular weight of the solute. This approach is widely used; it permits the measurement of molecular weights of up to 20,000 and, in some particularly favorable cases, of up to 40,000. There are several commercial instruments that function very satis-

factorily. If the solution is dilute (less than 1%) and the molecular weight is not too high, the results can be evaluated with the aid of the classical relation

$$\frac{p^0 - p}{p} = \frac{w_2}{M_2} \frac{M_1}{w_1} \tag{6}$$

where w_2 is the weight of solute of number-average molecular weight M_2, w_1 is the weight of solvent of molecular weight M_1, p^0 is the vapor pressure of pure solvent, and p is the vapor pressure of solution. Otherwise one must take care of the specific character of the polymer solution and use the extrapolation method described in Chapter 4.

ii. Boiling-Point Elevation

The boiling point is the temperature at which the vapor pressure of a liquid equals the ambient pressure on the system. There are two ways of measuring boiling-point elevation:

1. *Direct measurement* of the boiling point of the solvent and of the solution, obtaining the elevation by difference. As in the case of vapor pressure, this approach is not sufficiently sensitive for use with polymers.

2. *Differential measurement* by determining the temperature difference between condensing vapor and boiling solution. Though sensitive, this approach may suffer from superheating and other instrumental problems associated with maintaining vapor-liquid equilibrium. Furthermore, the existence of azeotropes (constant boiling mixtures) can influence the results. There exist several commercial instruments that furnish a rapid and convenient determination of the molecular weight up to about 30,000. If the solution is dilute and the molecular weight is not too high, one can apply the classical equation

$$M_2 = K_b \frac{1000w_2}{T_b w_1} \tag{7}$$

where M_2 is the number-average molecular weight of the solute, K_b is the boiling-point-elevation constant for the solvent, w_2 is the weight of the solvute, w_1 is the weight of the solvent, and T_b is the observed boiling-point elevation. Otherwise one must use the extrapolation method described in Chapter 4.

iii. Freezing-Point Depression

The simplest method for determining the freezing-point depression is the separate measurement of the freezing points of the pure solvent and the solution; the difference between these values is the depression. Since the liquid–solid equilibrium is usually more stable than the liquid–vapor equilibrium and the sensitivity of modern temperature-measuring devices is very high, the freezing-point method is quite convenient and reliable. There are several well-designed instruments on the market that allow the measurement of molecular weights of up to about 40,000.

In the case of dilute solutions the calculation is identical with the one for the boiling-point method except that K_F, the freezing-point-depression constant, is substituted for K_b and T_F for T_b.

iv. Osmotic Pressure

The osmotic method is of special interest and importance for several reasons:

1. Historically it has played an important role in ascertaining the existence of macromolecules in solution.

2. It covers conveniently the molecular-weight range of most commercially important polymers (20,000 to 500,000).

3. If carried out over a certain concentration range (from about 0.2 to 2.0%), it gives not only the molecular weight but also important information on the thermodynamics of the investigated system.

It may therefore be appropriate to discuss this method in somewhat greater detail.

There are several basic methods of osmometry; commercial instruments have been developed for all of them, each having advantages and disadvantages, but together they represent very complete and effective intrumentation for this important technique.

The *static equilibrium method* allows a solvent to diffuse through a membrane that separates it from the solution until the hydrostatic head produced just equilibrates the osmotic pressure characteristic of the system. This approach is very accurate, but it may take several days until equilibrium is actually established. Fortunately the time-consuming disadvantage of this method can be minimized as follows:

1. By using a membrane of the highest possible semipermeable porosity and of the largest possible size.

2. By adjusting the initial hydrostatic head to a value close to the anticipated final head.

These procedures, together with the minimum of attention required during the measurements and the low cost of the individual osmometer, make it possible to assemble numerous osmometers and carry out osmotic measurements in large series using different concentrations, solvents, and temperatures in the individual cells.

In the *half-sum method* of Fuoss and Mead (1) one adjusts the initial head in the osmometer somewhat (3–4 mm) above the anticipated head and plots the decreasing height of the meniscus as a function of time (curve A in Fig. 1) until one can approximate the asymptote of the curve. Then one readjusts the head so that it is now about the same distance below the equilibrium value and obtains curve *B* as a function of time. The half-sum values of *A* and *B* converge rapidly to an osmotic head that is very close to the true equilibrium value. The time scale in Fig. 1 shows that the two curves can be obtained rather rapidly so that the total test can be run within 20 min.

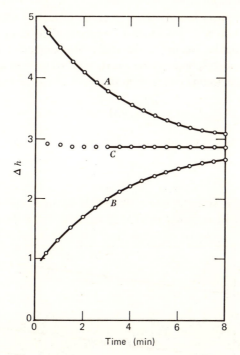

Fig. 1 Example of the half-sum method of Fuoss and Mead.

In the *dynamic method* one applies an external gas pressure to the meniscus on the solution side of the cell so that the tendency to its upward motion is exactly balanced. This pressure can be conveniently measured and is equal to the osmotic pressure of the solution. Very rapid measurements can be made by this method, but the equipment—osmometer and manostat—is somewhat more complicated and requires more attention during operation than the simpler methods.

However, several commercial instruments are available, and all are very well designed and thoroughly automated. They make osmotic tests for molecular weights of up to 500,000 at temperatures up to 130°C with non-corrosive solvents very convenient and dependable.

All early measurements with polymer solutions showed an upward curvature of the plot for osmotic pressure π versus concentration c even at very low concentrations (e.g., up to only 1 %). This contradicts van't Hoff's law, which is a simple consequence of thermodynamics and was found to be in complete agreement with all experiments on low-molecular-weight materials. Naturally this systematic discrepancy aroused the interest of all research workers in this field.

The experimentalists looked out for an empirical modification of van't Hoff's law that would permit them to evaluate their data correctly and to arrive at the molecular weight of the solute. Remembering that van't Hoff's law for solutions corresponds to the equation of state of an *ideal* gas, Ostwald suggested adding a second-order term to it and using the relation

$$\pi = \frac{RT}{M} c + Bc^2 + \dots \tag{8}$$

If one then plots π/c versus c, one obtains

$$\frac{\pi}{c} = \frac{RT}{M} + Bc \tag{9}$$

which represents a straight line with RT/M as intercept and B as slope (compare Fig. 2). In order to be able to actually make the plot indicated in

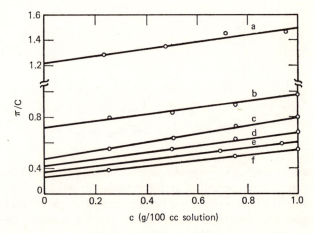

Fig. 2   Plot of π/c versus c.

Eq. 9 one must measure π at *several* concentrations. Experience has shown that one should have four or five individual measurements in the concentration range from 0.2 to 0.8 %. The researcher is rewarded for this extra effort by the fact that the intercept is the reciprocal of the number-average molecular weight $\overline{M}_n$ of the solute and that the slope provides for another interesting quantity B, which evidently is a measure of the "nonideality" of polymer solutions.

Once the necessity of using Eq. 9 instead of van't Hoff's law was firmly established, a strong incentive was given to the theoreticians to explain why even very dilute polymer solutions are nonideal and to clarify the physical meaning of the constant B. In fact the development of the thermodynamics of polymer solutions (compare Chapter 4) was essentially initiated in the 1930s by the failure of van't Hoff's law to cover osmotic data; as usual the first attempts were empirical and qualitative, but soon Huggins and Flory (2) attacked the problem at its root and laid the foundation for further improvements and refinements, which are described in more detail in Chapter 4.

b. End-Group Analysis

If the growth of a linear polymer chain is initiated or terminated by a characteristic functional group, this group can be used to determine the number-average molecular weight, since there will be only a fixed (small) number of such entities per molecule. For example, linear polyglycol ethers initiated with ethylene glycol contain *two* hydroxyl groups per molecule:

$$HO(CH_2CH_2O)_xCH_2CH_2OH$$

Polyglycol ethers initiated with glycerine result in a branched polymer, each molecule of which has *three* hydroxyl groups:

$$CH_2-(OCH_2CH_2)_xOH$$
$$|$$
$$CH-(OCH_2CH_2)_yOH$$
$$|$$
$$CH_2-(OCH_2CH_2)_zOH$$

Some polyglycol ethers that are used as synthetic detergents are initiated with substituted phenols. The terminal aromatic ring can be detected and measured via ultraviolet-absorption spectroscopy; it gives an excellent method for the molecular-weight measurement in this case.

Polyamides possess carboxyl and/or amine end groups

$$\overset{\displaystyle O}{\overset{\displaystyle \|}{HOOC(CH_2)_x(CNHCH_2)_yNH_2}}$$

Polyesters have carboxyl and/or hydroxyl groups at the ends.

In order to determine molecular weights by end-group analysis it is necessary to know the number of such groups per molecule. In some cases one does know, but in others, even though one thinks one knows, unsuspected side reactions have changed the end groups. For example, in the case of polyglycol ethers excessive heating can dehydrate one or both end groups of the molecule, resulting in a lower number of functional groups and falsifying the results.

Another problem with this analysis, as with many molecular-weight measurements of polymers, is the effect of *impurities*. For example, one reaction used to determine the hydroxyl end groups in polyethers is esterification with acetic anhydride. Evidently even a trace of aldehyde or alcohol in the polymer causes erroneous results.

Reactivity presents another problem with polymers. The different molecular-weight species present in a polymer do not react at the same rate. Hence unless enough time is given for complete reaction the molecular-weight results will be too high. Reaction rates are also decreased by the use of large samples that are required for measuring low amounts of functional groups (high-molecular-weight polymers). Occasionally samples as large as 20 grams are required. These quantities dilute the reaction mixture to such a degree that strong conditions or longer times are needed to complete the processes necessary for reacting all end groups.

Once a reliable end-group value is found, the computation of the molecular weight is not difficult. Let us assume we find 1000 ppm in a polyethylene oxide that has one hydroxyl group (MW 17) at each end the molecular weight is

$$\frac{1,000,000}{1000} = \frac{\overline{M}_n}{34}$$

$$\overline{M}_n = 34,000$$

3. LIGHT SCATTERING

When a parallel beam of light passes through a transparent system, a small part of it is scattered in directions off the primary beam. This effect can be formally interpreted as a certain turbidity τ, which is defined by

$$I = I_0 e^{-\tau l} \tag{10}$$

where I_0 denotes the intensity of the incoming beam and I represents the intensity of the beam after it has passed through the scattering medium of length l.

If it is desired to compute τ, it is necessary either to measure the transmitted intensity I or to determine the amount of light scattering in a certain direction (e.g., at an angle θ with the incident beam) and then to add up the contributions in all directions. For the computation of the intensity scattered in a given direction two factors are of importance:

1. The scattering from the individual molecules (scattering centers).
2. The manner in which the scattering centers are geometrically arranged and hence cooperate in building up the field of the scattered light.

According to classical electrodynamics the intensity scattered by a small molecule is given by

$$I(\theta) = \frac{8\pi^4}{\lambda^4 r^2}\, \alpha^2 I_0 (1 + \cos^2\theta) \tag{11}$$

where r is the distance from the scattering center and θ is the angle between the incident beam and the direction in which scattering is observed. Figure 3 gives a geometrical representation of Eq 11; it can be seen that along the circle in the YZ-plane normal to the incident beam the scattered light is equally bright and completely polarized. In the forward and backward direction it is twice as bright and completely unpolarized; for intermediate values of θ intensity and state of polarization of the scattered light are intermediate. The quantity α is the polarizability.

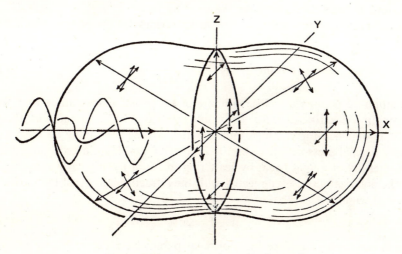

Fig. 3   Light scattering in an ideal crystal.

a. Scattering from an Ideal Gas

In reality the scattering from one small molecule can never be observed—only the effect produced by a large number of them. This is the point at which the geometrical arrangement of the scattering centers and their phase relations enter into the picture. The simplest conditions will prevail if the scattering molecules are completely independent of each other, because in this case no phase relations exist and the intensity scattered by a multitude of independent particles is obtained simply by multiplying Eq. 11 by the number of scattering centers. If we want not only the intensity scattered by all molecules in a given direction θ and observed at a given distance r but also the turbidity of the gas as defined in Eq. 10 as the fractional decrease of the primary intensity per unit length, we must consider that τ is the fraction of the incident light scattered per *unit volume* of the medium in all directions. If there are n independent molecules per cubic centimeter of the gas, each of which scatters according to Eq. 11, we must multiply $I(\theta)$ by n and integrate over all directions. The result, which was obtained by Lord Rayleigh (4), is

$$\tau = \frac{8\pi}{3}\left(\frac{2\pi}{\lambda}\right)^4 n\alpha^2 \tag{12}$$

The primary intensity I_0 has disappeared because τ is defined as its fractional decrease per unit length: r and θ have been eliminated by integrations; the polarizability α describes the optical properties of the individual scattering center, and the simple multiplication by n expresses the absence of phase relations between the secondary waves. The factor $(2\pi/\lambda)^4$ originates from the theory of electromagnetic waves and provides for the proper dimension of τ.

$$\tau(\text{cm}^{-1}) = \frac{n(\text{cm}^{-3}) \times \alpha^2(\text{cm}^6)}{\lambda^4(\text{cm}^4)} \tag{13}$$

Equation 12 permits a simple estimate of the order of magnitude of τ. By putting $n \simeq 3 \times 10^{19}$, $\lambda \simeq 5 \times 10^{-5}$, and $\alpha \simeq 1.0 \times 10^{-24}$, a value of about 10^{-7} is obtained for τ, which indicates that a tube containing an ideal gas under normal pressure has to be 10^7 cm (or 100 km) long in order to reduce the intensity of a light beam by scattering to $1/e$ of its original value.

The polarizability α is connected with the refractive index μ of the medium by the formula

$$\mu - 1 = 2\pi n\alpha \tag{14}$$

which can serve to replace the molecular parameter α Eq. 12 by the experimentally measurable quantity μ. By eliminating α from Eqs. 12 and 14, the following relation is obtained:

$$\tau = \frac{32\pi^3}{3}\frac{1}{\lambda^4}\frac{1}{n}(\mu - 1)^2 \tag{15}$$

This relation expresses the well-known fact that scattering of a gaseous system decreases sharply if a given amount of the dispersed phase is divided into a larger number of individual particles, as exemplified in the transition from a fog to a vapor; it has been used to determine Avogadro's number N_A by the scattering of light from vapors. A few recent figures resulting from such experiments are contained in Table 1; they show rather satisfactory agreement with the value of N_A obtained by other, more accurate, methods and can be considered as experimental confirmation of Eq. 15.

Table 1   Avogadro's Number N_A from Light-Scattering Measurements

Method	$N_A \times 10^{-23}$
Light scattering from a mixture of 91% argon, 8.7% nitrogen, and 0.3% oxygen	6.90
Light scattering from ethyl chloride vapor	6.50
Light scattering from water vapor	6.40
Presently accepted value	6.06

b. Scattering from a Pure Liquid

An attempt to expand the simple treatment outlined above to a liquid faces several difficulties. First, we must keep in mind that in a liquid the individual molecules, which act as scattering centers, are far from being independent from one another; in fact there exists in a liquid a considerable degree of short-range order because of the definite volume requirement of the individual particles. If a plane wave passes through a liquid, all molecules in a given part of the incident wave front will vibrate with the same phase, and the secondary waves will now possess certain phase relations because of the quasi-regular arrangement of the scattering centers. These phase relations will not affect the intensity scattered *parallel* to the primary beam, but they will sharply reduce the aggregated intensity under larger angles θ. Hence it is now not possible to simply *multiply* the intensity originating from one scattering center by the number of these centers per cubic centimeter, as was done in the derivation of Eq. 12 from Eq. 11. In adopting this procedure values of about 10^{22} would have to be used for n in liquids, and it would be concluded that the turbidity of ordinary liquids should be several hundred times larger than that of ideal gases. In reality pure liquids scatter only 10 to 50 times more light than their vapors at atmospheric pressure. A column of water has to be about 1000 meters long ($\tau \simeq 1.0 \times 10^{-5}$ cm^{-1}) in order to reduce the incident intensity by a factor of $1/e$, whereas application of Eq. 15 would lead to a value of only about 200 meters ($\tau \simeq 5 \times 10^{-5}$ cm^{-1}). This shows that the regular arrangement of the molecules in a liquid decreases the intensity of the laterally scattered light very substantially and must be considered if it is desired to arrive at the correct superposition of the secondary waves emanating from them.

Another complication arises from the fact that the relation between the polarizability of the individual molecules and the refractive index of the medium is no longer as simple as that expressed in Eq. 14 because of the fact that neighboring molecules affect and polarize each other in a condensed system.

Finally it must be considered that in a dense system, such as a liquid, any individual molecule vibrates not only under the influence of the primary wave but also under the influence of the field produced by the displaced charges in the rest of the medium.

A simple method of arriving at a useful result was originated as early as 1910 by Einstein (4), who pointed out that the scattering of a condensed system can be considered to be due to the existence of *thermal density fluctuations*, in the absence of which there would be no lateral light scattering at all. In fact an ideal crystal, in which all molecules are arranged in complete order at small distances from one another, represents a completely homogeneous optical medium and scatters no light. This can be easily understood with the aid of Fig. 4, in which *A* and *B* represent two volume elements of the

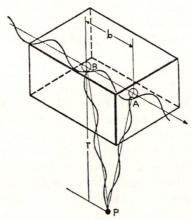

Fig. 4 Geometry of light scattering by a crystal or liquid.

medium that are small in comparison with the wavelength of light and regularly distributed throughout the systems. Because of the regular, latticelike arrangement of the molecules, each of these volume elements contains exactly the same number of scattering centers, and the secondary waves emanating from each have exactly the same amplitude. At any point of observation *P* the strength of the electric fields originating from *A* and *B* will therefore be the same. The distance between *A* and *B*, however, remains arbitrary and may be chosen in such a way that the electric vibrations coming from *A* and *B* are half a wavelength out of phase at the observation point and therefore *cancel each other*. For each volume element inside the scattering system we can find another one that just cancels its effect on *P*, and the total scattered intensity will therefore be zero. Only volume elements situated at the extreme surface of the system will remain uncanceled, but they contribute only a negligible effect.

It is thus apparent that a perfectly homogeneous and transparent medium scatters no light at all. A transparent liquid, however, is not perfectly homogeneous because of the thermal motion of its molecules. If we examine the two volume elements A and B in Fig. 4, we find that molecules are continually moving in and out of them as a result of random Brownian motion. Hence at a given instant the actual number of scattering centers in A will not be exactly the same as in B; the two waves in P will not have the same amplitude and therefore will not cancel each other completely in spite of the fact that they are half a wavelength out of phase. The *local random density fluctuations* and the corresponding fluctuations of refractivity makes a pure transparent liquid optically inhomogeneous and are therefore the reason for its turbidity. To calculate this kind of turbidity we have to know (a) the magnitude of the thermal fluctuations and (b) the optical efficiency of each of them. These two factors must then be multiplied.

In order to compute the average magnitude of the density fluctuations in a liquid, the thermal energy kT must be compared with the work that has to be supplied to accomplish a certain density change by outside pressure.

The compressibility $\chi = -1/V(\partial V/\partial p)$ indicates the relative volume decrease $-\partial V/V$ if the pressure is increased by ∂p. The larger the value of χ, the easier it is to produce a density change by outside pressure; the reciprocal compressibility $K = 1/\chi$ therefore expresses the *resistivity* of the liquid against compression and its tendency to eliminate thermal fluctuations after they have been produced by kT. The ratio kT/K will then give the average magnitude of the density fluctuations because the numerator is responsible for their creation, the denominator for their disappearance; this ratio corresponds to the number of independently scattering molecules, n, in Eq. 12.

The optical efficiency of the thermal density fluctuations is given by the refractive index of the medium, μ, multiplied by the change in the refractive index with pressure, $\partial \mu/\partial p$, and is therefore $\mu(\partial \mu/\partial p)$. This quantity corresponds to the polarizability α in Eq. 12 and must therefore appear as the second power because of the transition from the amplitudes of the electric waves to their intensities. By combining the above expressions, multiplying them by a numerical factor resulting from various integrations, and considering the wavelength dependence, we find that the turbidity of a pure transparent liquid can be put in the form

$$\tau = \frac{32\pi^3}{3} \frac{1}{\lambda^4} \frac{kT}{K} \left(\mu \frac{\partial \mu}{\partial p} \right)^2 \tag{16}$$

By comparing Eq. 16 with Eq. 15 it can be seen that kT/K replaces $1/n$, whereas $\mu(\partial \mu/\partial p)$ now stands for $(\mu - 1)$. If the proper values for λ, K, μ, and $\partial \mu/\partial p$ are introduced into Eq. 16, the correct order of magnitude for the turbidity of liquids ($\tau \simeq 10^{-5}$ cm^{-1}) is obtained; this indicates that a liquid column about 1000 meters long would scatter enough light laterally to reduce the light beam $1/e$ of its original intensity.

c. Scattering from Solutions of Small Molecules

Next we consider the turbidity of a solution and assume that the molecules of the solute are small in comparison with the wavelength of the scattered light, which is usually in the neighborhood of 5000 Å (green light). "Small" means that the molecules of the solute should have no extension larger than about one-fifteenth of λ, which is about 350 Å. A solid sphere of this diameter made out of a polymeric substance like polystyrene, rubber, or cellulose acetate would have a molecular weight between 20 and 30 million. A rod-shaped particle of this molecule weight and even a randomly kinked coil would of course have a much larger dimension along its axis and therefore could certainly not be considered to be small in comparison with the wavelength of the scattered light. In fact chainlike molecules with molecular weights as low as 150,000 are similar to the wavelength of 5000 Å, parallel to their longest extension, and can therefore not be treated as centers of simply spherical secondary waves. The intensity scattered from such molecules can no longer be represented by Eq. 11 and the turbidity of a gas consisting of such particles would not be given by a relation of the form of Eq. 12. We discuss the influence of the molecular size and shape on the scattering of light in the next paragraph and presently assume that the solute molecules are smaller in all directions than the wavelength of the scattered light.

In a dilute solution of this kind there will be not only local *density* fluctuations of the solvent but also local fluctuations of the *concentration* of the solute. In small volumes of the solution we shall find that solute molecules move randomly into a certain volume element and out of it again, so that the solute concentration within the element fluctuates about an average value. If the refractive index of the solute differs from that of the solvent, the solution will contain volume elements of different refractivity and the medium will be rendered turbid by the presence of minute local refracting heterogeneities. In order to compute the turbidity caused by these fluctuations in solute concentration we must calculate their *average magnitude* and multiply it by their *optical efficiency*.

We must now compare the average thermal energy kT available in the system with the work that is necessary to establish a certain change in concentration. This work is closely connected with the osmotic pressure π of the solution and can be expressed by $c(\partial \pi/\partial c)$. Hence the magnitude of the fluctuations in solute concentration is given by the ratio $kT/[c(\partial \pi/\partial c)]$. The optical efficiency of these fluctuations is expressed by $\mu^2(\mu_s - \mu)^2$, where μ

is the refractive index of the solvent and μ_s is the refractive index of the solution. For dilute solutions $(\mu_s - \mu)$ is very nearly proportional to the concentration of the solute and can be put equal to γc. Hence the turbidity of a solution, as far as fluctuations of solute concentration are concerned, is

$$\tau = \frac{32\pi^3}{3} \frac{1}{\lambda^4} \frac{kT}{c(\partial\pi/\partial c)} (\gamma\mu c)^2 \tag{17}$$

It can be seen that this equation, like Eqs. 11 and 12, consists of four factors: (a) a numerical factor $32\pi^3/3$ ($\simeq 350$), (b) a factor $1/\lambda^4$ that expresses the wavelength dependence of dipole radiation, (c) a factor $kT/[c(\partial\pi/\partial c)]$ that considers the *average magnitude* of the concentration fluctuations, and (d) a factor $(\gamma\mu c)^2$ that accounts for their capacity to *render the system turbid*.

For solutions of small solute molecules the osmotic pressure can be expressed by van't Hoff's law, $\pi = nkT$, in a concentration range up to a few percent (n is the number of kinetically independent solute molecules per cubic centimeter). By introducing this into Eq. 17 we obtain

$$\tau = \frac{32\pi^3}{3} \frac{(\gamma\mu c)^2}{\lambda^4} \frac{1}{n} \tag{18}$$

In comparing this with Eq. 15 it can be seen that a solution in the range of van't Hoff's law produces the same turbidity as an ideal gas, if one replaces $(\mu - 1)^2$ for the gas by $(\gamma\mu c)^2$ for the solution. Both systems—gas and solution—become less and less turbid as a given amount of matter is dispersed in a larger and larger number of independently moving particles.

If we want to introduce the molecular weight M_2 of the solute, we have to write van't Hoff's law in the following form:

$$\pi = \frac{N_A kT}{M_2} c \tag{19}$$

where N_A is Avogadro's number and c is the solute concentration. If Eq. 19 is introduced into Eq. 17, we get

$$\tau = \frac{32\pi^3}{3} \frac{\gamma^2\mu^2}{\lambda^4} \frac{M_2 c}{N_A} \tag{20}$$

which shows again that, at a given solute concentration, the turbidity increases with M_2. The coarser the distribution of the solute, the more turbid the system (1 gram of cellobiose in 100 cc of water is twice as turbid as 1 gram of glucose).

Equation 20 can be used to compute the molecular weight M_2 of the solute if the wavelength λ of the scattered light and the refractive index μ of the pure solvent are known and if *two* quantities are measured: (a) the increase γ in the refractive index with solute concentration and (b) the turbidity

τ at a given concentration c. For a given light source (λ) and a given solvent–solute system (λ and μ) the factor

$$\frac{32\pi^3\gamma^2\mu^2}{3N_A\lambda^4}$$

assumes a certain constant value of the order of magnitude of 10^{-6} to 10^{-7}, which has been denoted by H. Equation 20 can then be written as

$$\tau = HM_2c \tag{21}$$

In passing to the discussion of solutions containing small polymer molecules it must be remembered that van't Hoff's law does not hold even at very low concentrations. For such solutions Eq. 19 must be replaced by (compare, for example, reference 7)

$$\pi = \frac{N_AkT}{M_2}c + Bc^2 \tag{22}$$

where B is an empirical constant which expresses the interaction between the solvent and the individual segments of the dissolved polymer molecules and which, in first approximation, is independent of the molecular weight of the solute. If Eq. 22 is introduced into Eq. 17, the following relation for the turbidity is obtained:

$$\tau = H\frac{M_2c}{1 + (2B/RT)M_2c} \tag{23}$$

This relation reduces to Eq. 21 whenever M_2c is so small that the second term in the denominator can be neglected as compared with unity. This will be the case for very small concentrations if M_2 is large or for larger concentrations if M_2 is not too large, just as would be expected from our knowledge about the validity of van't Hoff's law in the domain of polymer solutions.

For the actual determination of M_2 from osmotic measurements it is common practice to use the equation:

$$\frac{\pi}{c} = \frac{RT}{M_2} + Bc \tag{24}$$

By plotting the *reduced* osmotic pressure π/c versus c a straight line is obtained, the intercept of which is RT/M_2, whereas the slope permits the determination of B.

Correspondingly, it is advantageous to rearrange Eq. 23 into the following form:

$$\frac{Hc}{\tau} = \frac{1}{M_2} + \frac{2B}{RT}c \tag{25}$$

If the reduced transparency Hc/τ is plotted versus c, a straight line is also obtained; its intercept is $1/M_2$, and the slope can serve for the determination of B.

Comparison of Eqs. 24 and 25 shows that osmotic pressure and turbidity measurements can be used to determine experimentally two important properties of a polymer solution: (a) the molecular weight M_2 of the solute and (b) the thermodynamic interaction as expressed by the constant B. It was Debye (5) who first emphasized the value of light scattering measurements for characterization of polymer solutions.

In order to determine these quantities from direct measurements of π it is necessary to carry out four or five individual osmotic measurements and to draw the best straight line through the experimental points. In using the light-scattering method, it is necessary to make the following measurements:

1. To measure the refractive index of the solution at several (four or five) concentrations below 1 % and to determine γ.

2. To measure the turbidity of the solution at several (four or five) concentrations below 1 %, to plot Hc/τ versus c, and to draw the best straight line through the experimental points.

An example is shown in Fig. 5, which shows plots of Hc/τ versus c for three polystyrene fractions in toluene at room temperature; the experimental points are arranged along straight lines, the intercepts of which are distinctly

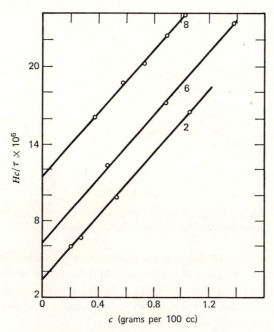

Fig. 5 Values of Hc/τ for three polystyrene fractions in toluene versus solute concentration.

different, as should be expected from the fact that the three fractions are supposed to have different molecular weights. Figure 5 also shows that the slopes of the three straight lines are very nearly the same, indicating that the thermodynamic interaction between the segments or submolecules of each of these fractions with toluene is about the same. This is in line with expectation, because all three fractions were obtained from the same polystyrene sample by fractional precipitation.

Figure 6 represents plots of Hc/τ versus c for seven polystyrene fractions in methyl ethyl ketone at room temperature. Again it becomes apparent that the points for each individual fraction are distributed fairly well along a straight line, whose intercepts on the ordinate determine, according to Eq. 25,

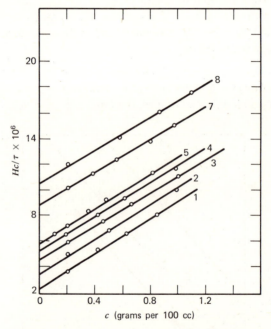

Fig. 6 Values of Hc/τ for seven polystyrene fractions in methyl ethyl ketone versus solute concentration.

the molecular weights of the various fractions. It may also be pointed out that the molecular weights of fractions 2 and 8, which were measured in toluene and methyl ethyl ketone, agree reasonably well (285,000 and 85,000 in toluene, 250,000 and 95,000 in methyl ethyl ketone), whereas the B values in various solvents show distinct differences.

The scale on the ordinates in Figs. 5 and 6 indicates that Hc/τ is on the order of 5×10^{-6}. Since c is about 10^{-2}, the turbidity of 1% polymer solutions amounts to about 2×10^{-3}, which indicates that a column about 5 meters long of such solutions will scatter enough light to reduce the initial intensity of the primary beam to $1/e$ of its value. Table 2 summarizes the turbidities of a few characteristic systems and indicates the distances l that are necessary to reduce the initial intensity of a light beam to $1/e$ of its original value.

Table 2   Order of Magnitude of Light Scattering in a Few Typical Systems

System	Order of Magnitude of Turbidity (cm^{-1})	Order of Magnitude of l (cm)
Ideal crystal	Virtually zero	Infinite
Ideal gas or vapor	10^{-7}	10^7
Pure liquids	10^{-5}	10^5
Solutions of ordinary small molecules in 1% concentration	10^{-4}	10^4
Polymer solutions in 1% concentration	10^{-3}	10^3
Rubber latex in 1% concentration	10^{-1}	10
Milk	10	10^{-1}

Table 2 shows that the turbidity of polymer solutions in the range of concentration in which measurements are usually made (between 0.1 and 1.0%) is about 100 times larger than the turbidity of pure liquids and indicates that the concentration fluctuations in polymer solutions are 100 times more efficient in creating refracting inhomogeneities than the density fluctuations of the solvent. This is fortunate because it permits neglecting the contribution of the solvent in most cases, as given by Eq. 16, at the side of the contribution of the solute as expressed in Eq. 17. The *total* scattering of a solution is represented by the sum of these two contributions:

$$\tau = \frac{32\pi^3}{3} \frac{\mu^2 kT}{\lambda^4} \left[\frac{1}{K} \left(\frac{\partial\mu}{\partial p} \right)^2 + \frac{1}{c} \frac{\partial c}{\partial \pi} (\gamma c)^2 \right] \tag{26}$$

where, for polymer solutions, the first term in the brackets is usually small and has to be subtracted from the total effect.

In considering the two working equations, Eqs. 24 and 25, which serve to determine M_2 and B, it becomes apparent that there are a few significant differences between them.

1. It is worthwhile to note that at very low concentrations the osmotic pressure π is inversely proportional to M_2, whereas the turbidity τ is directly

proportional to it. The quantities π and τ are those that must be measured experimentally and should not be too small if the measurement is to remain accurate. This indicates that the osmotic method is particularly suitable for measuring small molecular weights (between 20,000 and 200,000), whereas the turbidity method is more appropriate for larger molecular weights (between 100,000 and 1,000,000). Hence the two methods complement each other in a very satisfactory manner.

2. The different relation of M_2 with π and τ, respectively, also has a very important consequence as soon as one investigates not sharp fractions, which have a comparatively uniform molecular weight, but unfractionated samples with a broad distribution of molecular weights. Let us assume that we measure the osmotic pressure of a solution containing a given *amount* (say 1 gram) of solute. If we subdivide the dissolved material into smaller and smaller units, the osmotic pressure increases because it is proportional to the *number* of independent solute particles in the system. In a mixture of molecules with different molecular weights one measures, therefore, the *number-average* molecular weight. On the other hand, if we measure the turbidity of a solution containing a given *amount* of solute, it is immediately apparent from Eq. 23 that τ increases as the weight of the independent solute particles increases. If the turbidity method is applied to a mixture, therefore, the *weight-average* molecular weight of the various species of dissolved molecules is determined.

These two averages are identical only if one operates with sharp fractions; whenever the distribution curve extends over a wider range of molecular weights, its weight average becomes larger than its number average; the ratio $\overline{M}_w/\overline{M}_n$ can even be taken as a first approximate measure of the heterogeneity of a given sample. In case of a *statistical distribution function*, as it occurs in synthetic polymerization and polycondensation products and also in degraded natural polymers, the ratio $\overline{M}_w/\overline{M}_n$ is equal to 2 and assumes even larger values if the spread of molecular weights becomes larger than indicated by random building-up or random degradation of long-chain molecules. The combination of the osmotic-pressure and the light-scattering methods can therefore be used to obtain a first idea about the molecular homogeneity of a given polymer sample.

d. Scattering from Solutions Containing Particles Comparable in Magnitude to the Wavelength of Light

Equation 12 was derived under the assumption that the solute molecules are small in comparison with the wavelength of the incident light, so that each molecule can be considered to be a *point source* of secondary radiation, which has then the character of a simple *spherical wave*; the dependency of the intensity of the scattered light on the angle of observation, θ, is symmetrically distributed about the 90° plane.

If, however, any linear dimension of the molecule approaches the magnitude of the wavelength, the radiations from all of its component dipoles vary in phase, and the molecule cannot any longer be considered as a simple point source. Since the radiation from the component parts of any one molecule is coherent, the resulting scattered intensity will be proportional to the square of the *vector sum* of the amplitude of the scattered rays. This will result in an interference pattern that is characteristic of the size and shape of the particle.

This phenomenon can be readily illustrated by a simple example. The sphere in Fig. 7 represents a scattering particle with dimensions that are of the

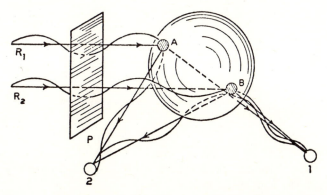

Fig. 7 Lateral scattering of light by a particle whose dimensions are comparable to the wavelength of light.

order of magnitude of the wavelength. Let us consider two scattering volume elements at A and B, and two incident rays, R_1 and R_2, which are in phase at plane P. In reaching a forward observation point, O_1, ray R_1, which travels the *shorter* distance to reach element A, travels the longer distance, AO_1, from A to O_1; whereas ray R_2, which travels the *longer* distance to element B, travels the *shorter* distance to the observation point. Thus the differences in path lengths are partially compensated for, and the rays do not become very much out of phase in the forward direction, O_1.

For a backward observation point, O_2, however, the situation is reversed; the ray (R_2) that travels the *longer* distance from the plane to the element also travels the *longer* distance to the observation point. If, as in this illustration, the molecular dimensions are something less than one wavelength, the path difference will be such as to produce a phase lag that will result in destructive interference of the two rays in the backward direction, O_2.

Therefore it can be seen that, as a result of this interference effect, there will be an all-around decrease in the scattered intensity. Since destructive interference is more probable between rays scattered in the backward direction, the radiation envelope will no longer be symmetrical.

A typical interference pattern of this type is shown in Fig. 8.

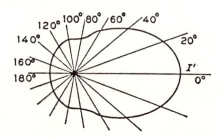

Fig. 8   The radii of the curve are proportional to the intensity of the scattered light in the direction of the radius. The envelope shows considerable dissymmetry in favor of forward scattering, as explained in Fig. 7.

If polymer molecules consist of comparatively stiff chains or are strongly solvated, their shape can be approached by a rigid-rod model of length L that leads to the scattering function

$$I\left(\theta, \frac{L}{\lambda}\right) = \frac{I^0}{\chi} \int_0^{2x} \frac{\sin \chi}{\chi} \, d\chi - \left(\frac{\sin \chi}{\chi}\right)^2 \tag{27}$$

The factor $(1 + \cos^2 \theta)$ has been omitted from the right side because it is symmetrical about $90°$; the quantity χ is defined by

$$\chi = 2\pi \frac{L}{\lambda} \sin \frac{\theta}{2}$$

and I^0 represents the intensity scattered by particles that are small in comparison with λ.

Expression 27 represents a function that decreases as the angle θ increases; the rate of this decrease depends on the parameter L/λ. Hence by measuring $I(\theta, L/\lambda)$ at several values of θ, one can first verify that the dissolved particles approach the shape of rigid rods and then determine the average length L of the particles in terms of λ.

If polymer molecules consist of flexible chains and are not noticeably solvated their shape can be approached by *randomly kinked coils* of roughly spheroidal shape. For such particles the intensity distribution is

$$I\left(\theta, \frac{h}{\lambda}\right) = \frac{2I^0}{v}\left[e^{-v} - (1 - v)\right]$$

$$v = \frac{2}{3}\left(\frac{h}{\lambda}\right)^2\left(2\pi \sin\frac{\theta}{2}\right)^2 \tag{28}$$

Equation 28 also represents a function that decreases with increasing observation angle θ; the rate of decrease depends on the parameter h/λ, where h is the root-mean-square separation of the ends of the chain and represents the largest distance that occurs in the most probable configuration of such molecules. By measuring $I(\theta, h/\lambda)$ at various angles, one can again confirm the existence of randomly coiled particles in solution and determine their largest average extension. A precise definition of h is given in Chapter 3.

Practical measurements of light scattering from polymer solutions are made at various finite concentrations c and at a variety of scattering angles θ. To obtain the important molecular quantities $\overline{M}_w$, h, and B (interaction parameter) it is necessary to extrapolate to zero concentration and zero scattering angle. A systematic graphical method for achieving this was devised by Zimm (6).

The fascinating and important subject of light scattering is treated extensively in reference 7.

Light-scattering measurements can be carried out with practically any solvent over a wide range of temperatures and give significant results over a wide range of molecular weights (i.e., from 5000 to 5,000,000). The only important requirement is that the solution be free from any contaminations, such as dust, or colloidal components other than the solute. In some instances, such as for emulsion polymers, it is difficult to fulfill this requirement, but whenever it can be fulfilled the light-scattering methods is one of the most powerful tools for the study of dilute polymer solutions.

4. ULTRACENTRIFUGATION

Ultracentrifugal methods provide another powerful tool for the investigation of solutions and suspensions. Svedberg was the first to use high-speed centrifuges for the analysis of colloidal systems and has brought the method to a high state of perfection. The original apparatus has been simplified by various workers.

Under suitable conditions ultracentrifugation provides a means of measuring molecular weight or particle size in solution, and, if the material is not uniform in particle size, the degree of nonuniformity can be determined.

Information regarding the aggregation, state of solvation, and size and shape of the particles can also be inferred from ultracentrifugal data. The technique also permits the fractionation by purely mechanical means of mixed macromolecules from the fluids in which they are found. This procedure has been particularly useful in connection with biological fluids, as in the fractionation of various proteins from human and animal blood.

The evaluation of the molecular properties is made possible by measuring (a) the sedimentation equilibrium, (b) the sedimentation rate, and (c) the coefficient of diffusion.

Two types of ultracentrifugation instrument are generally useful: one for sedimentation-equilibrium measurements in which a centrifugal acceleration not larger than 15,000 g is sufficient, and the other for sedimentation rates in which a 10 to 20 times greater centrifugal acceleration is required.

The rate of sedimentation is determined by observing the rate of movement of a boundary that separates the solution from the pure solvent. In the majority of cases the color of the solution, its absorption in the ultraviolet, or its refractive index is the means by which the moving boundary is followed.

The centrifugation can be continued until the concentration as a function of the distance from the axis of rotation no longer changes, so that a balance is set up between the centrifuging forces and the thermal diffusion.

Observations with the ultracentrifuge permit the calculation of the true coefficient of diffusion. If the particles are sufficiently small, the sedimentation boundary will spread because of diffusion, and from this spreading the coefficient of diffusion can be calculated.

a. Sedimentation Equilibrium

The analysis of sedimentation equilibrium is based on the following thermodynamic equation, which applies when the concentration of each independently variable component of the solution becomes an unchanging function of the distance from the axis of rotation:

$$RTd \ln a_i - M_i(1 - v_{\chi i}\rho_\chi)\omega_\chi^2 \, d\chi = 0 \tag{29}$$

where a_i is the activity of component i with molecular weight M_i and partial specific volume $v_{\chi i}$, and ρ_χ is the density of the solution at a distance χ from the axis of rotation of the rotor revolving with an angular velocity ω.

In order to make possible calculations of the molecular weight from sedimentation-equilibrium data it is necessary to work in sufficiently dilute solutions so that the activity is proportional to the concentration. Under these conditions Eq. 29 can be integrated, giving

$$M = \frac{2RT \ln (c/c_0)}{(1 - v\rho)\omega^2(\chi^2 - \chi_0^2)} \tag{30}$$

where c is the concentration at the distance χ from the center of rotation and c_0 is the concentration at a reference point in the solution χ_0 centimeters from the rotor center. The concentration can be determined by optical methods, such as light absorption. If the molecular weight of the substance in solution or suspension is uniform, a plot of $\ln c$ versus χ^2 should give a straight line, from whose slope the molecular weight of the dissolved substance can be determined by Eq. 30.

If the material in solution or suspension is polydisperse with regard to molecular weight, the $\ln c$ versus χ^2 line is curved in such a way that the slope is greater at the smaller values of χ.

The other important application of the ultracentrifuge is the sedimentation-velocity method; it is discussed in Chapter 5.

5. MOLECULAR WEIGHT DISTRIBUTION

In previous sections we discussed the fundamental methods by which $\overline{M}_n$ and $\overline{M}_w$ are obtained. The ratio $\overline{M}_w/\overline{M}_n$, called the heterogeneity index (HI), is a measure of the polydispersity of a polymer sample. The greater the value of HI, the greater the polydispersity. For a truly monodisperse sample the HI is obviously 1.0. A widely encountered molecular weight distribution is the so-called random distribution, which arises from random condensation, random degradation, or equilibrium polymerization. For this distribution, discussed more thoroughly in Chapter 16, the HI is approximately 2.0.

Commercial polymers such as plastics must have a very carefully controlled weight distribution if they are to be fabricated rapidly and reproducibly by a certain type of machine into a specified object. The desired breadth of the molecular weight distribution varies depending on whether the fabrication technique is extrusion, injection molding, or blow molding (see Chapter 14). Depending on the synthetic route by which the polymer is made, the range of HI values that are available for a given polymer is to some extent pre-determined. For example, commercial polystyrenes made by free radical polymerization have an HI range from 1.5 to 3.0. On the other hand, high density polyethylene made by the Ziegler catalyst or by surface catalysis may have HI values as large as 30.

Very often it is necessary to have a more detailed assessment of the molecular weight distribution than is given by the averages. For example, it might be very useful to know what weight percent of polymer is in the very high molecular weight range or in the very low molecular weight range, because these "tails" of the distribution often have pronounced effect on the fabrication and on the final properties of polymers, even when comparing two polymers of the same $\overline{M}_w$ and $\overline{M}_n$.

The most detailed description of molecular weight distribution is a plot of $W(M)$ versus M where $W(M)$ is the weight fraction of polymer having a molecular weight M. Sometimes $X(M)$ versus M is plotted where $X(M)$ is the mole fraction of polymer having a molecular weight M. The $X(M)$ is clearly proportional to $W(M)/M$.

The obvious way to obtain $W(M)$ versus M is by a fractionation technique which isolates polymer fractions of narrow molecular weight distribution. Fractional precipitation is one of the oldest and once was the most widely used of fractionation methods. It is based on the addition of a precipitant to a solution of a polymer, for example, methanol to a solution of polystyrene in benzene. The high molecular weight portion precipitate is first removed and then successive precipitations follow. The molecular weight of each precipitated fraction is determined by intrinsic viscosity or some absolute technique. Refractionations are generally required.

To obtain good fractionations the polymer solution should be dilute and the precipitant must be added slowly with as violent agitation as possible. The amount of precipitant used at each fractionation episode controls the size of the fractions obtained.

The simplest technique for molecular weight evaluation of the individual fractions is the measurement of intrinsic viscosity (Chapter 5). For a given polymer-solvent system intrinsic viscosity measurements are calibrated by means of one of the absolute measurements discussed in this chapter—osmometry, light scattering, or ultracentrifugation. Fractional precipitation by addition of nonsolvent, by cooling, and by other related techniques are discussed in reference 8.

Gel permeation chromatography has now been established as a very widely used, accurate, and convenient method for obtaining molecular weight distributions. The apparatus consists of a column of beads of porous glass or porous crosslinked polystyrene. The micropores are of roughly the same size as are the polymer molecules which will be fractionated. Typically three columns each 4 ft long are connected in series and the inner diameter of the columns is approximately 1 cm. In a typical case the solvent tetrahydrofuran flows through the column at the rate of 1 cm³/min. At a given instant a dilute solution of polystyrene in tetrahydrofuran is injected over a period of 1 min, while the flow of solvent continues. As the solution percolates through the column the solute molecules permeate into the pores of the beads. The smaller molecules will permeate the beads more deeply than the larger molecules. The flow of pure solvent that follows the injection of polymer solution elutes the polymer molecules from the packing. The larger molecules elute first since they are nearer the surface of the beads. The volume (or time at constant flow) required to elute a portion of the polymer is a function of the molecular weight of that portion. A very convenient feature of GPC is that the eluant is continuously and automatically monitored by measurement of the differential refractive index. Hence as polymer elutes, the magnitude of the refractive index change at a given elution volume is proportional to the polymer content at that elution volume. Thus the data are automatically depicted in the form of a weight distribution plot.

In the case of polystyrene the column is calibrated by injection of mono-disperse polystyrene, prepared by anionic polymerizations. These samples have accurately determined molecular weights and are eluted in a very narrow range of eluted volume.

The penetration of polymer molecules into the beads is inversely related to the "hydrodynamic volume" of the polymer molecules, so that hydrodynamic volume is inversely related to elution volume. In Chapter 5 it is shown that the hydrodynamic volume is proportional to $[\eta]M$ where $[\eta]$ is the intrinsic volume of the polymer molecules. It is, therefore, not surprising that for a given column a plot of $[\eta]M$ versus elution volume is nearly the same for a wide variety of polymers (9).

References 10 through 12 provide more thorough discussion of GPC.

There are other methods for determination of molecular weight distributions. For volatile polymers of low molecular weight gas chromatography is very effective, and adsorption chromatography is useful for a few low molecular weight polymers with polar terminal groups.

REFERENCES

1. R. M. Fuoss and D. J. Mead, *J. Phys. Chem.* **47**, 59 (1943).

2. M. L. Huggins, *J. Chem. Phys.* **9**, 440 (1941); P. J. Flory, *J. Chem. Phys* **9**, 660 (1941).

3. J. W. Strutt (Lord Rayleigh), *Phil. Mag.* **41**, 107, 447 (1871).

4. A. Einstein, *Ann. Physik*, **33**, 1275 (1910).

5. P. Debye, *J. Appl. Phys.* **15**, 338 (1944).

6. B. H. Zimm, *J. Chem Phys.* **16**, 1099 (1948).

7. M. Kerker, *The Scattering of Light and Other Electromagnetic Radiation*, Academic Press, New York, 1969.

8. M. J. R. Cantow, *Polymer Fractionation*, Academic Press, New York, 1967.

9. Z. Grubsic, P. Rempp, and H. Benoit, *Polymer Lett.* **5**, 753–759 (1967).

10. J. F. Johnson and R. S. Porter, in *Progress in Polymer Science*, Vol. II, A. D. Jenkins, Ed., Pergamon Press, Oxford and New York, 1970, pp. 201–256.

11. J. Cazes, *J. Chem. Ed.* **47**, Part I, No. 7, A461 (1970); *ibid.*, Part II, No. 8, A505 (1970).

12. R. F. Boyer, *The Influence of G.P.C. on Polymer Science and Technology*, Waters Associates, Framingham, Mass., 1969, p. 48.

Conformations of Polymer Molecules

3

A. Peterlin

As a consequence of special conditions of polymerization or polycondensation the polymer molecule possesses a certain *configuration*, that is, a sequence and orientation of chain groups of every single monomer. As a rule the configuration remains with the molecule and cannot be changed easily by the usual processes the molecule is subjected to in technical handling. In the case of the very simple vinyl and vinylidene homopolymers with different substituents on the two chain carbons, as, for instance, CH_2CHX or CH_2CXY with X and Y for any substituent like Cl, OH, CH_3, and so on, one has two orientations of the monomer: head–tail (ht) or tail–head (th). These can be arranged into any combination of positional sequences: isotactic, ·ht·ht·ht· or ·th·th·th·; syndiotactic, ·ht·th·ht·th·; and atactic, ·ht·ht·th· or ·ht·ht·th· (Fig. 1). If the chain is positionally isotactic—that is, if it contains only either ht or th monomer orientation (both chains are identical) but has different substituents X and Y—one can still have two different sterical

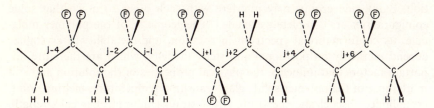

Fig. 1 Three isotactic and two consecutive syndiotactic positional pairs in a molecule of polyvinylidene fluoride: ht·ht·ht·th·ht·ht·.

arrangements of the substituent in subsequent vinyl monomers (Fig. 2); that is, left–left or right–right and left–right or right–left, yielding three different steric triads: lll or rrr (isotactic), lrl or rlr (syndiotactic), and llr, lrr, rrl or rll (heterotactic or atactic). In many respects the heterotactic sequences act as a different " mer," so that a homopolymer molecule with such grouping resembles a copolymer.

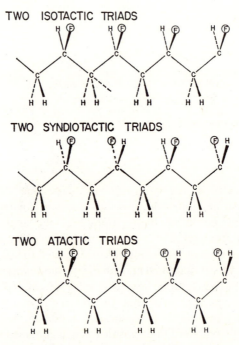

TWO   ISOTACTIC   TRIADS

TWO   SYNDIOTACTIC   TRIADS

TWO   ATACTIC   TRIADS

Fig. 2 Polyvinylfluoride chains with isotactic, syndiotactic, and atactic steric triads.

The position of side groups influences the energy of interaction and hence the probability of occurrence of different configurations during polymerization. Bulky side groups may interfere with each other so much that some configurations are completely excluded. The geometry of the polymer molecule, its conformational spectrum in solution, and the ability to crystallize depend markedly on positional and steric tacticity, which are therefore important factors that influence the physical properties of the solution and still more those of the polymer solid: glass-transition temperature, melting point, crystallinity, viscoelastic spectrum, stress–strain curve, mechanical strength, and processability. In what follows the analysis is concentrated on the *conformations* a macromolecule can assume as a consequence of its configuration.

If one examines the orientation of two consecutive chain groups connected by a single chemical bond, one finds as a rule a certain freedom depending on the character of the chemical bond and that of the chain groups, particularly of the substituents attached to the chain atoms. There is first a more or less fixed valency angle α between two consecutive chain bonds and a certain degree of preference for the orientation of three consecutive chain bonds. The valency angle and the orientation are to some extent affected by the environment—that is, by the chemical and geometrical character of the chain and the embedding medium (e.g., solvent in solution or molecules of the same kind in melts and crystals).

The carbon–carbon chemical bond can be a single, a double, or a triple valency bond (Fig. 3). A triple bond demands the remaining valency bond

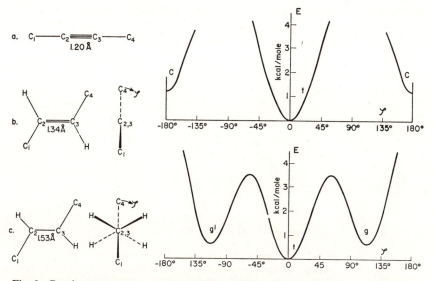

Fig. 3 Bond arrangements and potential energy curves $E(\varphi)$ in a triple-, double-, and single-bonded carbon-carbon pair of a chain molecule with identical substituents on each carbon atom (symmetrical chain). The angle φ is zero if the four consecutive carbon atoms are in the same plane in the most extended (*trans*) position.

of the two carbons in the straightforward continuation of the triple bond (Fig. 3a). A double bond permits two conformations, *cis* and *trans*, with a certain degree of rotational freedom, as indicated by the potential energy curve (Fig. 3b). The probability of the orientation of an angle α is given by $\exp(-E(\phi)/kT)$. A single valency bond has a potential energy curve with three minima (Fig. 3c). The deepest, corresponding to the most stable conformation, is the *trans* position. The two *gauche* positions (g and g') have generally a higher energy than the *trans* position (*t*) and differ from each other if any carbon atom has two different substituents. Hence they are identical in the case of $-CH_2-CH_2-$, $-CH_2-CF_2-$, but not in the case of $-CH_2-CHF-$. In the case of polyethylene the *gauche* conformation has about 500 ± 100 cal/mole more energy than the *trans* conformation (1).

To some extent the thermodynamic equilibrium distribution of angular orientation can be derived from infrared absorption, and hence one can calculate the energy difference between the preferred orientations and the height of intervening energy barriers. On the other side, the energy curve $E(\phi)$ can be derived from interaction among the atoms and atomic groups attached to the chain atoms. In the simplest case one takes into consideration only the four subsequent chain atoms and the atoms directly bound to the two central chain atoms (the triad approximation corresponding to *cis*, *trans*, and *gauche* preferential orientations). The pentad approximation considering the atoms and groups up to two valency bonds from the central two chain atoms needs a more fully specified terminology. One immediately sees that in a polymer the energy curve $E(\phi)$ is strongly affected by substituents or side groups attached to the chain atoms. With asymmetrical monomers $-CH_2-CXY-$ the g and g' minima differ in energy content (Fig. 4).

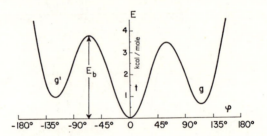

Fig. 4 The potential energy curve $E(\varphi)$ for a single-bonded carbon–carbon pair of a polymer chain molecule with different substituents on one carbon (asymmetrical chain).

In thermodynamic equilibrium the distribution probability of orientation of consecutive valency bonds, $\exp(-E/kT)$, yields a wide variety of chain shapes. The actual number of possible conformations of a single macromolecule with m chain bonds and z positions of a single bond is z^m. This is such an enormously high number that the average polymer molecule with $m \simeq 1000$ may never go through all conformations. Nevertheless the average shape obtained on the supposition that all conformations occur with the thermodynamic probability forms the basis of the theoretical treatment of mechanical, thermodynamical, optical, electrical, and magnetic properties of polymers.

There are two important cases in which the conformational analysis is usually applied. One is the *polymer crystal*, with a very regular conformation identical for all molecules with equivalent lattice position. The other is the *polymer melt* or *solution*, with a more or less random chain conformation governed by the thermodynamic equilibrium and space requirement of the chain and by the interaction with the solvent, which is the polymer itself in the case of melts and a low-molecular-weight liquid in the case of solutions.

In polymer crystals the chain conformations are very simple: zigzag, that is, *trans* conformations only (polyethylene, polytetrafluoroethylene, polyamides); helices (polypropylene, polyoxymethylene, polystyrene, polypeptides, polynucleotides); or very simply kinked chains (polyethylene terephthalate). By considering all the atoms of a single macromolecule, that is, neglecting completely the interaction with adjacent chains and taking into account the periodicity requirements of the crystal lattice, one can calculate the energy functions for different helical conformations of many polymers. As a rule the minima obtained correspond rather well to the conformations observed in polymer crystals and to the eventual phase transformations with temperature (2–4).

This chapter is devoted mainly to the random conformation of macromolecules in melts and solutions. The average shape determines a great many properties of melts and solutions, particularly their hydrodynamic properties (viscosity, translational diffusion, and transport of macromolecules in gravitational fields and, if the molecule has a net charge, in electric fields) and rheo-optical properties (streaming birefringence). A minor effect is found in electrical and magnetic properties (rotational diffusion). The influence of the isolated macromolecule's shape on most of these properties was indeed calculated. Hence from the excess effects of dilute solutions through concentration extrapolated to infinite dilution (intrinsic effects) it has been possible to derive the average shape of the macromolecule and to compare it with the very direct data obtained from light-scattering measurements.

No such theoretical treatment exists for concentrated solutions and melts, and hence there is practically no reliable information about chain conformations in them. Important cases of melts are rubbers and polymer glasses, two kinds of amorphous polymer. The former contain more or less permanent crosslinks, which are responsible for shape stability under deformation. The chain section between two consecutive crosslinks of the macromolecule obeys very nearly the same rules as a similar section of the completely free molecule. In glasses, in the temperature range below the glass transition, the mobility of polymer chains is so drastically reduced that the conformations correspond not to the actual temperature T but to the higher glass-transition temperature T_g. The thermodynamic equilibrium distribution is frozen in at T_g and does not adjust to that of the lower temperatures T of the glassy state, $T < T_g$. Exactly speaking there is a very slow relaxation, or approach to equilibrium, but it is so slow that it can be very nearly always neglected if the temperature is sufficiently below T_g.

In that which follows most of the emphasis is given to the conformational analysis of the random-coil model of linear unbranched molecules.

1. RANDOM-COIL MODEL OF LINEAR MACROMOLECULES

The chain contains $N + 1$ chain atoms and N links connecting two subsequent chain atoms. The link b_j connects the $(j - 1)$th with the jth atom, j running from 1 to N. In the absence of an outside force field for a random chain of freely jointed links each of length b, the orientation of every link is completely at random and independent of that of the rest of the chain. The average square of the length of the coil is hence the sum of b^2

$$h^2 = \langle r^2 \rangle = Nb^2 \tag{1}$$

with the root-mean-square (rms) end-to-end distance $h = N^{1/2}b$ proportional to the square root of the number of links.

In the case of vinylic polymers the degree of polymerization P is connected with N by the relationship $N + 1 = 2P + f$ if there are end groups with altogether f chain atoms. The vector connecting the jth atom with the kth atom reads

$$\mathbf{r}_{jk} = \mathbf{b}_{j+1} + \cdots + \mathbf{b}_k = \sum_{p=1}^{m} b_{j+p} \qquad m = k - j \tag{2}$$

with the average square value

$$h_{jk}^2 = \langle r_{jk}^2 \rangle = \sum_{p=1}^{m} \sum_{q=1}^{m} \langle \mathbf{b}_{j+p}\mathbf{b}_{j+q} \rangle \tag{3}$$

$$= \langle \mathbf{b}_{j+1}\mathbf{b}_{j+1} \rangle + \langle \mathbf{b}_{j+1}\mathbf{b}_{j+2} \rangle + \cdots + \langle \mathbf{b}_{j+1}\mathbf{b}_k \rangle$$
$$+ \langle \mathbf{b}_{j+2}\mathbf{b}_{j+1} \rangle + \langle \mathbf{b}_{j+2}\mathbf{b}_{j+2} \rangle + \cdots + \langle \mathbf{b}_{j+2}\mathbf{b}_k \rangle + \cdots$$
$$+ \langle \mathbf{b}_k\mathbf{b}_{j+1} \rangle + \langle \mathbf{b}_k\mathbf{b}_{j+2} \rangle + \cdots + \langle \mathbf{b}_k\mathbf{b}_k \rangle$$

The double sum was first written by Eyring (5) in the form of a matrix that permits a very easy interpretation and in some cases also a very easy evaluation of the rms distance h_{jk}. The diagonal elements are the square of the lengths of all the links of the section.

A few simple cases permit a straightforward evaluation. Let us first assume that all links have the same length b. If the valency angle is the same for all chain atoms and there is completely free rotation about the chemical bond [i.e., $U(\phi)$ is independent of ϕ], one has the very simple relationship (Fig. 5)

$$\langle \mathbf{b}_j\mathbf{b}_{j+p} \rangle = b^2 \cos^p(180° - \alpha) = (-1)^p b^2 \cos^p \alpha \tag{4}$$

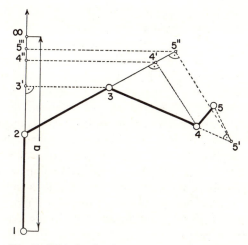

Fig. 5 With fixed valency angle α and free rotation around the valency bond the average projections of bond $j+1, j+2, j+3, \ldots$ on bond j equal the sections $23' = b \cos (180 - \alpha)$, $3'4'' = b \cos^2 (180 - \alpha)$, $4''5'' = b \cos^3 (180 - \alpha)$, respectively. The sum of all projections is $1\infty = b/(1 + \cos \alpha) = a$.

which yields

$$h_{jk}^2 = h_m^2 = mb^2 \left[\frac{1 - \cos \alpha}{1 + \cos \alpha} + \frac{2 \cos \alpha}{m} \frac{1 + \cos^m \alpha}{(1 + \cos \alpha)^2} \right] \qquad m = |k - j| > 0 \quad (5)$$

For very large m Eq. 5 reduces to

$$h_{m, \infty}^2 = mb^2 \frac{1 - \cos \alpha}{1 + \cos \alpha} = mb_{\text{eff}}^2 \qquad m \gg 1 \tag{6}$$

with the effective link length

$$b_{\text{eff}} = b \left(\frac{1 - \cos \alpha}{1 + \cos \alpha} \right)^{1/2} = b \tan \left(\frac{\alpha}{2} \right) \geq b \tag{7}$$

which is always larger than or equal to the true length b because the valency angle is never smaller than $90°$ and hence $\cos \alpha \leq 0$. The ratio $b_{\text{eff}}/b \geq 1$ is a very good measure of chain extension. It is $2^{1/2} = 1.4$ for the tetrahedral C—C bond angle ($\cos \alpha = -\frac{1}{3}$).

If $\alpha = 90°$, $b = b_{\text{eff}}$ and hence

$$h_m^2 = mb^2 \qquad \alpha = 90° \tag{8}$$

independently of m, because according to Eq. 5 all the nondiagonal elements of the matrix in Eq. 2 are zero. Exactly the same solution (Eq. 8) is obtained in the case that the valency angle may assume with equal probability any value between 0 and 180°. This is the case of completely unrestricted random walk with a single step length b (Eq. 1), which plays a very important role in the hydrodynamic theories of polymer solutions [*necklace model*, first introduced by Kuhn (6)]. In the case of the random-walk chain and the chain with a 90° valency angle the average square end-to-end distance is exactly proportional to the number of links. The 90° valency angle corresponds to the cubic lattice, which is often applied for the study of macromolecular conformations, particularly for the consideration of the excluded-volume effect.

The exact expression for the average square end-to-end distance (Eq. 5) relatively slowly approaches the limiting value (Eq. 6), as was first systematically investigated by Sadron and Benoit (7–9). In Fig. 6 the values $h/N^{1/2}b$

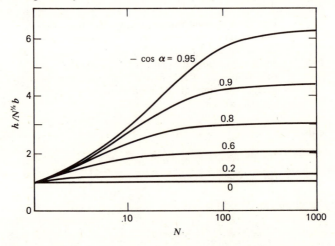

Fig. 6 The ratio of mean-square end-to-end length of a chain with fixed valency angle α and a chain with free joints as a function of the number N of links for different values of cos α (10).

are plotted versus N for cos α values between 0 and -1. A similar dependence is found with chains exhibiting a hindered rotation or a limited number of permitted angular orientations of subsequent bonds. In real chains with strongly interdependent bond rotations the dependence on m of the ratio between the exact and asymptotic average square length is more complicated and evades simple generalization beyond the certainty that it approaches unity with increasing m. This general property of polymer chains leads to two basic models of the randomly coiled macromolecule: (a) the *random-walk necklace model* (6) and (b) the *wormlike chain model* (11, 12).

As a rule the experiments (light scattering, intrinsic viscosity) yield not the mean-square end-to-end distance but the so-called gyration radius R. The latter is defined as

$$R^2 = (N+1)^{-1} \sum_{j=0}^{N} \langle r_j^{*2} \rangle = \tfrac{1}{2}(N+1)^{-2} \sum_{j}^{N} \sum_{k=0}^{N} \langle r_{jk}^2 \rangle$$

$$= (N+1)^{-2} \sum_{m=1}^{N} (N-m+1) h_m^{2} \tag{9}$$

where r_j^* is the distance between the jth chain atom and the center of mass of the molecule. The definition is straightforward only if all chain atoms are identical and the substituents on them are identical and so small that one can in comparison with r neglect their displacement away from the chain. Such is indeed the situation with polyethylene even at small degrees of polymerization and with homopolymers in general at sufficiently large P. In the remaining cases—low P, heteropolymers, large side chains—one has to introduce appropriate refinements in the definition of the gyration radius.

By introducing h_m^2 from Eq. 6, we obtain

$$R^2 = Nb^2 \left[\frac{(N+2)(1-\cos\alpha)}{6(N+1)(1+\cos\alpha)} + \frac{\cos\alpha}{(N+1)(1+\cos\alpha)^2} \right.$$

$$\left. + \frac{2\cos^2\alpha}{(N+1)^2(1+\cos\alpha)^3} + \frac{2\cos^3\alpha(1+\cos^N\alpha)}{N(N+1)^2(1+\cos\alpha)^4} \right] \tag{10}$$

with the asymptotic value

$$R^2 = \frac{Nb^2}{6} \frac{1-\cos\alpha}{1+\cos\alpha} = \frac{h^2}{6} \tag{11}$$

Here R and h both refer to a chain with N links. Again the asymptotic value is approached relatively slowly, and more so the more extended the chain (i.e., the closer α to 180°, $\cos\alpha$ to -1). Of particular importance is the ratio $6R^2/h^2$ with the limiting value 1. In Fig. 7 it is plotted versus N for different valency angles. Again Eq. 11 is the exact solution for $\alpha = 90°$, $\cos\alpha = 0$ and for the random-walk chain. The situation becomes less simple with consideration of hindered rotation in practically the same manner as in the case of the average end-to-end distance.

An important property of the chain is the persistence length a introduced by Kratky and Porod as the average projection of the infinitely long chain on the direction of the first bond, that is, on the initial tangent of the chain (Fig. 5). One obtains the general expression

$$a = \frac{1}{b}(\langle \mathbf{b}_1\mathbf{b}_1 \rangle + \langle \mathbf{b}_1\mathbf{b}_2 \rangle + \cdots) = \frac{1}{b} \sum_{j=1}^{\infty} \langle \mathbf{b}_1\mathbf{b}_j \rangle \tag{12}$$

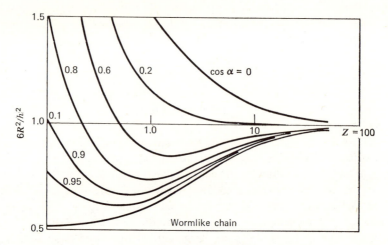

Fig. 7 The ratio of mean-square gyration radius and end-to-end length for chains with fixed valency angle α and for a wormlike chain as a function of the number Z of equivalent segments (10).

which in the case of a constant bond length, valency angle, and free rotation around the chemical bond reduces to

$$a = b \sum_{j=1}^{\infty} (-\cos \alpha)^j = \frac{b}{1 + \cos \alpha} = \frac{b}{2} \cos^2 \frac{\alpha}{2} = \frac{b_{\text{eff}}}{\sin \alpha} \qquad (13)$$

a value that rapidly increases as α approaches $180°$. The persistence length can be used as a measure of chain extension in very nearly the same way as the ratio b_{eff}/b.

a. Random-Walk Necklace Model (6)

The limiting values for the average square end-to-end distance (Eq. 6) and gyration radius (Eq. 11) are strictly valid for a model consisting of N freely jointed links of equal length b_{eff}. The dimensions of the model are the same as those obtained in the case of the three-dimensional random-walk problem in a cubic lattice or in free space with N steps if the step length is constant. The coil dimensions so calculated, however, become very unrealistic with shorter chain sections and fail completely with adjacent chain atoms, yielding $b_{\text{eff}} > b$ instead of $b_{\text{eff}} < b$ if the valency angle is not $90°$.

One automatically excludes from consideration such short sections if one assumes (13) a model with Z freely jointed links (segments) of length A, with the additional condition that the extended length L of the molecule equal that of the model. One hence has two equations

$$h^2 = Nb_{\text{eff}}^2 = ZA^2 = 6R^2 \tag{14a}$$

$$L = Nb = ZA \tag{14b}$$

The extended length is here calculated as if the valency angle were not α but $180°$. The characteristic length and number of segments are obtainable from h (or R) and L as follows:

$$Z = \frac{L^2}{h^2} = \frac{L^2}{6R^2} \tag{15a}$$

$$A = \frac{h^2}{L} = \frac{6R^2}{L} \tag{15b}$$

The number of chain atoms in a segment is $\zeta = N/Z = 6R^2/Lb = h^2/Lb$. If the monomer contains n chain atoms, the number of monomers per segment is ζ/n. In vinylic polymers $n = 2$. If the monomer contains chain atoms with different bond lengths, as, for instance, $-CH_2-CH_2-O-$, one can replace b by the sum of all the bonds in order to obtain the number of monomers in the segment. Table 1 shows the data for A and ζ/n for different polymers as derived from the experimentally observed h or R values in θ-solvents in the limit of very high N or P.

The model is evidently very good at large N (i.e., large Z) and does not yet fail too much even at $Z = 1$. Of course it cannot be applied to subsegmental sections (i.e., to $m < \zeta$). The larger A or ζ, the more extended the chain. A comparison with Eq. 8 shows the complete identity of such an estimate with that obtained from the ratio

$$\frac{A}{b} = \zeta = \left(\frac{b_{\text{eff}}}{b}\right)^2 = \frac{1 - \cos\alpha}{1 + \cos\alpha} \tag{16}$$

The physical foundation of Kuhn's model is based on the fact that in a real molecular chain the orientation of any bond becomes completely independent from the orientation of another bond if the bonds are separated by a sufficient number of chain atoms. The minimum number determines ζ and hence the length of the segment. Condition 14b, however, is very arbitrary and yields too large a value for Z and hence too small a value for ζ. On the other hand, it permits the straightforward calculation of Z and A if M and h are known (Table 1). But one must not forget that Z and A are not true molecular parameters, although they rather well characterize the molecule. It is particularly important to remember that Eqs. 14a and 6 as derived from Eq. 5 are valid only in the limit of very large N or Z. Hence one is faced with significant

Table 1 Parameters A, a, ζ, and ζ/n of the Random Walk Model Derived from the Limiting Experimental Values $h/M^{1/2}$ of Polymers in θ Solvents[a]

Polymer	M_{mon}	$h/M^{1/2}$ (Å)	n	b (Å)	$A = h^2/L$ (Å)	$a = A/2$ (Å)	$\zeta = A/b$	ζ/n [b]
Polyethylene	28	0.950 ± 0.040	2	1.53	8.1	4.1	5.4	2.7
Polyethylene oxide	44	0.750 ± 0.030	3	4.39/3	5.6	2.7	3.75	1.25
Polystyrene	104	0.670 ± 0.015	2	1.53	15.3	7.7	10.2	5.1
Polymethyl methacrylate	100	0.650 ± 0.060	2	1.53	13.4	6.7	8.8	4.4
Cellulose	162	0.900 ± 0.150	1	5.1	25.7	12.9	5	5
Methyl cellulose	186	1.500 ± 0.060	1	5.1	81.0	40.5	16	16

[a] Unperturbed dimensions after Kurata and Stockmayer (14).
[b] Number of monomers in a segment.

deviations from the limiting values of Z and A if one calculates them from the data of shorter molecules or applies them for shorter intramolecular distances.

Such a random-walk necklace model turned out to be very well suited for the calculation of physical properties of the macromolecular coil. At the freely rotating joints one concentrates the hydrodynamic resistance; that is, one places there beads with a hydrodynamic radius a_h yielding a resistance coefficient $6\pi a_h = \zeta\Lambda$, where Λ is the resistance coefficient of a chain atom or chain group (e.g., CH_2). The beads may be also the seat of optical, electrical, and magnetic polarizability. If one wishes to study the effect of anisotropy of polarizability or of electric or magnetic dipole orientation, it is preferable to locate it in the segment, that is, on the link connecting two consecutive beads. The independent orientation of the segments immediately shows that the coil orientation in applied electric or magnetic fields is independent of molecular weight: the dipole moment and intrinsic birefringence in such a field are simply those of the segment.

With the random-walk necklace model the probability for the occurrence of an end-to-end vector $\mathbf{r}$ is easily calculated (6). In first approximation it turns out to be a Gaussian distribution function

$$\phi(r) = \left(\frac{\mu}{\pi}\right)^{3/2} \exp(-\mu r^2) \tag{17}$$

$$\mu = 3/2h^2$$

The probability function for the occurrence of a length r is obtained by multiplying $\phi(r)$ by the corresponding volume element factor $4\pi r^2$.

$$w(r) = 4\pi r^2 \phi(r) \tag{18}$$

The distribution function $\phi(r)$ is closely connected with the entropy S of the ensemble of all conformations with the same end-to-end vector $\mathbf{r}$:

$$S(\mathbf{r}) = k \ln \phi(r) = \tfrac{3}{2}k \ln \frac{\mu}{\pi} - k\mu r^2 \tag{19}$$

This value is independent of the state of the ends: they may be free (molecule in solution or melt) or fixed (molecular chain between two crosslinks of a network). For the chain loop on the surface of the crystal one has to consider the fact that the halfspace occupied by the crystal is not available for chain conformation. This fact drastically reduces the number of permitted conformations and hence the entropy.

Corrective terms are needed if the end-to-end distance approaches the extended chain length (15). The distribution function can be imagined as the result of the diffusion of the free ends of the molecule under the influence of an entropic restoring force F_S

$$F_S = -2\mu kT\mathbf{r} \qquad (20)$$

The force is a consequence of the fact that the free ends are parts of the same molecule and hence not only cannot diffuse farther away from each other than the extended chain length but are confined to relatively small distances favored by the probability distribution of the random walk with Z steps. Hence the force is not created by any elastic forces within the molecule but results from Brownian motion, which imposes a never ending change of location for every segment.

Analytically the force F_S is Hookean in the whole range of validity of Eq. 15. The elastic modulus of a network containing ν such chains per unit volume turns out to be

$$E_S = \nu kT = \frac{cN_A kT}{M} \sim 10^7 \text{ dynes/cm}^2 \qquad (21)$$

where c is the concentration, N_A is Avogadro's number, and M is the molecular weight of the free chain between two crosslinks. The estimated value 10^7 is obtained with $c = 1$ and $M = 2500$. Such a network model with entropic forces is the basis of the theory of rubber elasticity. If, however, the end-to-end distance r approaches the extended chain length, the force F_S must become infinite in order to prevent r from getting larger than L.

The distribution function ϕ from Eq. 17 and the ensuing restoring force F_S from Eq. 20 make possible a straightforward analytical treatment of the deformation and orientation of the macromolecular coil in a flow field in terms of the location of the free ends. The mathematical problem is always reduced to the consideration of one free end, the other one or the center of mass being fixed at the origin of the coordinate system. Such a one-point dynamics is that of the elastic dumbbell model and yields correct general expressions as long as only one relaxation time is involved, but the numerical coefficients need correction (16, 17). A more realistic approach needs the consideration of all segments or beads of the necklace. The treatment of the model with constant link length is rather unhandy and was mainly used for the calculation of the translational motion [sedimentation and diffusion (18, 21)] and occasionally for viscosity (16, 18, 19, 20, 22–27) and streaming birefringence (16, 28).

Bueche (29) introduced ideally elastic links with an average square length b_0^2 instead of links with constant length. The length-distribution function (Eq. 17) and the elastic restoring force (Eq. 20) apply to every link with $\mu_0 = \frac{3}{2}b_0^2$. In order to have such an elasticity one has to assume that every link is composed of a sufficiently large number of freely jointed sublinks, so that in the whole range of deformation occurring in an applied hydrodynamic field the link never gets fully extended. With such a model the motion in laminar flow can be analytically formulated by a system of linear partial differential equations, and the solution can be obtained for the case of free-draining (30) partially or completely impermeable coils (31) and the whole transition between them.

b. Wormlike Chain (*11, 12*)

Instead of a finite bond length with a discontinuity of bond orientation by an angle $(180° - \alpha)$ at every chain atom Porod and Kratky (11, 12) subdivided the chain so much farther that the change in orientation becomes continuous. During the limiting process that reduces the bond length b and the angle $(180° - \alpha)$ the length $L = Nb$ of the molecule and the persistence length a

$$a = \frac{b}{1 + \cos \alpha} = \lim \frac{2b}{\pi^2} \left(1 - \frac{\alpha}{180°}\right)^2 \tag{22}$$

remain constant. Under these conditions the mean-square end-to-end distance and gyration radius turn out to be

$$h^2 = 2a^2(x - 1 + e^{-x}) \tag{23a}$$

$$R^2 = a^2 \left[\frac{2}{x^2}(e^{-x} + x - 1) - 1 + \frac{x}{3}\right] \tag{23b}$$

$$x = \frac{L}{a}$$

One sees very well from Fig. 7 that the wormlike chain is the limiting case of the random-coil model with α approaching 180°. The other limit is the random-walk necklace model with constant values 1 for the ratio h^2/mb^2 and $h^2/6R^2$ independent of m or L.

For large x one obtains

$$h^2 = 2aL\left(1 - \frac{1}{x}\right) = 2aL - 2a^2 \to 2aL$$

$$R^2 = \frac{aL}{3}\left(1 - \frac{3}{x} + \frac{6}{x^2} - \frac{6}{x^3}\right) \to \frac{h^2}{6} \qquad x \to \infty \tag{24}$$

that is, the same limiting values as for the random-walk necklace model if $A = 2a$. For small x we have

$$h^2 = L^2 \left(1 - \frac{x}{3} + \frac{x^2}{12} \cdots \right) \qquad x \to 0$$

$$R^2 = \frac{L^2}{12} \left(1 - \frac{x}{5} + \frac{x^2}{30} \cdots \right) = \frac{h^2}{12} \left(1 - \frac{x}{5} + \cdots \right) \tag{25}$$

yielding at $x = 0$ the values of a rigid rod. As one sees from Eqs. 24 and 25, the persistence-length model describes pretty well the whole transition from a fully extended rod to the random coil—that is, from the monomer to the polymer. It is therefore particularly suited for the treatment of oligomers, where the random-walk necklace model deviates too much from reality or even fails completely. It has indeed been very often applied for the calculation of translational resistance (32), viscosity (33, 34), light scattering, and small-angle X-ray scattering (11, 12, 35–41) of low-molecular-weight polymers (i.e., of molecules with short chains that do not yet form sufficiently well-developed coils). A good example are oligomers of most polymers, which have a highly curved chain and hence a relatively small persistence length a or segment length A or effective bond length b_{eff} and relatively-high-molecular-weight members of more extended polymers with large A, as, for instance, cellulose, cellulose derivatives, and particularly the double- or triple-stranded helices of polynucleotides (DNA, RNA).

In the treatment of oligomers one has to keep in mind that the wormlike chain model in the form presented here considers only the atoms on the chain and completely neglects the lateral dimensions of the chain. This does not cause troubles as long as the characteristic length L and a are sufficiently large in comparison with the thickness of the chain. Such is the situation with the medium-molecular-weight cellulose and polynucleotides. With oligomers, however, the lateral dimensions are of the same order of magnitude as the length L and therefore have to be explicitly considered in the calculation of average dimensions and of the effects depending on them.

A direct experimental determination of the persistence length is based on small-angle X-ray scattering of dilute polymer solutions (11, 12). If one plots the product of scattering intensity I and the square of scattering angle as function of scattering angle, one obtains curves (Fig. 8) with an inflection at the point δ^* separating the central horizontal section from the linear region with the asymptote passing through the origin. The position δ^* corresponding to $t^* = ks^*a = 2\pi a\delta^*/\lambda = 2$ hence yields $a = \lambda/\pi\delta^*$. Theoretical curves calculated by the Monte Carlo method (35–37) for a model with $\cos \alpha = -0.9$ and $a = b/(1 + \cos \alpha) = 10b$ also show that the molecule must be pretty long, with a contour length at least 20 times the persistence length

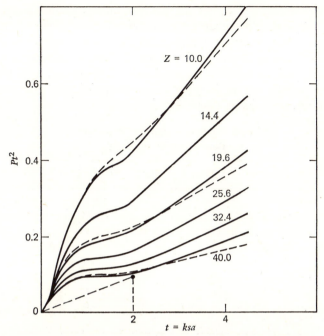

Fig. 8 The product $t^2P(t)$ with $t = ksa$ and $P(t) = I(t)/I(0)$ calculated for different numbers of chain elements and valency angle $\cos\alpha = -0.9$ (solid curves) and -0.8 (broken curves) (38).

if one wants to obtain a well-developed inflection point. The finite width of the actual macromolecular chain and a locally enhanced conformational order (helical structure of short chain sections) may eventually completely obscure the inflection at δ^* and produce additional minima in the $I\delta^*$ versus δ plots. This was demonstrated convincingly by the very extensive and precise measurements of Kirste, Kratky, and Wunderlich (41–43) on polymethyl methacrylate solutions in acetone.

2. HINDERED ROTATION ABOUT THE CHEMICAL BOND

In real polymer chains the rotation about the chemical bond is no longer completely free. The sterical hindrance and the energetic interaction of the atomic groups introduce a bond rotational potential U, which with simple molecules like CH_3CH_3, CH_2ClCH_3, or CH_2ClCH_2Cl is a function of the angle ϕ between the planes through the C—C bond and a characteristic atom of the first and the second group, respectively. In the case of CH_3CH_3 the potential energy has a threefold symmetry. The energy minima occur when the hydrogens of the respective groups are in the staggered, and the maxima at the eclipsed, conformations. A similar consideration applies to the case of one or more hydrogens substituted by other atoms. Such substitution, as a rule, destroys some symmetry as far as position and height of energy maxima and minima are concerned. Heights of rotational barriers (i.e., energy differences between maxima and minima) are derived from microwave frequency, thermodynamic equilibrium distribution, and infrared or Raman intensity studies (Table 2).

Table 2   Barrier Height for Representative Bond-Rotation Potentials[a]

Compound and Bond	Barrier Height (kcal/mole)
CH_3-CH_3	2.9
CH_3-CH_2F	3.31
CH_3-CH_2Cl	3.69
CH_3-CH_2Br	3.57
CH_3-CHF_2	3.18
CH_3-CHO	1.17
$CH_3-CH=CH_2$	1.98
CH_3-OCH_3	2.72
$CH_3CH_2-CH_2CH_3$	3.5

[a] From the Tabulations of Herschback (44).

In first approximation the situation is of the same type in a polymer chain with the angle ϕ_j between the planes defined by the consecutive bond pairs b_{j-1}, b_j, and b_j, b_{j+1}, respectively (Figs. 3 and 4). If the energy function is symmetrical, [i.e., $E(\phi) = E(-\phi)$], then the average value of $\cos \phi$

$$\eta = \langle \cos \phi \rangle = \frac{\int_0^{2\pi} \cos \phi \, \exp[-E(\phi)/kT] \, d\phi}{\int_0^{2\pi} \exp[-E(\phi)/kT] \, d\phi} \tag{26}$$

is a good measure of hindrance of bond rotation. The value $\eta = 1$ means complete immobilization at the *trans* conformation and $\eta = 0$ is completely free rotation. For such a chain one obtains the mean-square end-to-end distance (45)

$$h^2 = Nb^2 \left(\frac{1 - \cos \alpha}{1 + \cos \alpha} \frac{1 + \eta}{1 - \eta} + \frac{\eta \cos \alpha - \lambda_1}{\lambda_1 - \lambda_2} P_1 - \frac{\eta \cos \alpha - \lambda_2}{\lambda_1 - \lambda_2} P_2 \right) \tag{27}$$

and gyration radius (46)

$$R^2 = \frac{Nb^2}{b} \left(\frac{N+2}{N+1} \frac{1 - \cos \alpha}{1 + \cos \alpha} \frac{1 + \eta}{1 - \eta} + \frac{\eta \cos \alpha - \lambda_1}{\lambda_1 - \lambda_2} Q_1 - \frac{\eta \cos \alpha - \lambda_2}{\lambda_1 - \lambda_2} Q_2 \right) \tag{28}$$

with

$$P_{1,2} = \frac{2\lambda_{1,2}}{N} \frac{1 - \lambda_{1,2}^{N}}{(1 - \lambda_{1,2})^2} \qquad (29)$$

$$Q_{1,2} = \frac{\lambda_{1,2}}{N + 1(1 - \lambda_{1,2})^2} - \frac{2\lambda_{1,2}^{2}}{(N + 1)^2(1 - \lambda_{1,2})^3} + \frac{2\lambda_{1,2}^{3}(1 - \lambda_{1,2}^{N})}{N(N + 1)^2(1 - \lambda_{1,2})^4}$$

$$\lambda_{1,2} = \frac{-(1 - \eta)\cos\alpha \pm (1 - \eta)^2 \cos^2\alpha + 4\eta}{2}$$

For a sufficiently large number of skeletal atoms N the limiting values read (47)

$$h^2 = Nb^2 \frac{1 - \cos\alpha}{1 + \cos\alpha} \frac{1 + \eta}{1 - \eta} = Nb^2_{\text{eff}}$$

$$R^2 = \frac{h^2}{6} \qquad (30)$$

$$b^2_{\text{eff}} = b^2 \frac{1 - \cos\alpha}{1 + \cos\alpha} \frac{1 + \eta}{1 - \eta}$$

Limited rotation around the chemical bond increases the chain extension b_{eff}/b in a very similar manner as the valency angle if both are measured by η and $-\cos\alpha$, respectively. The dependence is identical in the limit of very large N. Therefore the replacement of such a chain by the random-walk necklace or wormlike model has the same advantages and limitations as mentioned above in connection with the chain exhibiting free rotation about the chemical bond.

If the chain contains atoms or groups of different kinds, as, for instance, $-CH_2-CH_2-O-$ or $-CH_2-CHF-$, then bond lengths, bond angles, and rotational potential energies will differ correspondingly. In the case of periodically repeating units, which is the rule with man-made homopolymers and some natural products (cellulose, caoutchouc), one may first calculate the average square dimensions for the smallest repeating unit (i.e., for the monomer) and then construct the whole molecule from such elements.

3. NEXT-NEIGHBOR INTERACTION IN REAL MACROMOLECULAR CHAINS

In the derivations of Eqs. 26 through 30 one assumed that the bound rotations in the polymer chain are mutually independent and symmetrical about the planar (*trans*) conformation. The latter condition is fulfilled in polymers with no structural asymmetry, as, for instance, $-CH_2-CH_2-$,

$-CF_2-CH_2-$, $-CH_2-O-$. The former, however, is more an exception than the rule. In real chain molecules the bond vectors are subject to strong mutual correlations. The orientation of a given bond j is influenced by the orientations of its next neighbors in the chain. The correlation is reflected in the dependence of the energy function on ϕ_j, ϕ_{j-1}, ϕ_{j+1}, ϕ_{j-2}, $\ldots$. The extension to second and even higher neighbors is rarely needed. In calculating the conformational energy and hence the corresponding statistical weight one benefits from the fact that the potential energy as a function of ϕ has two, mostly three, deep minima separated by relatively high barriers. Consequently the bond orientation is practically restricted to the immediate neighborhood of the minima. One does not commit a significant error by choosing the positions of the minima as the only available conformations. The continuous distribution of orientations is hence replaced by a discrete spectrum—for example, *trans* (t), *gauche+* (g), and *gauche−* (g'), each with a proper energy value E and corresponding statistical weight $\exp(-E/kT)$.

Since we are interested only in the ratios of the weight functions, we designate by E the energy difference between the chosen and the *trans* conformation, which is usually the position of lowest energy. The consideration of first-neighbor interaction hence leads to a statistical weight matrix

$$
U = \begin{array}{c} \\ (t) \\ (g) \\ (g') \end{array}
\begin{array}{ccc}
(t) & (g) & (g') \\
\left| \begin{array}{ccc}
1 & u_{12} & u_{13} \\
u_{21} & u_{22} & u_{23} \\
u_{31} & u_{32} & u_{33}
\end{array} \right|
\end{array}
\tag{31}
$$

States of the preceding bond are shown to the left of each row, those for the bond considered are shown above the columns. The interaction with the subsequent bond is taken into account at the consideration of the next bond. The bulkiness of the main chain usually demands that u_{23} and/or u_{32} be very small; that is, the conformational pairs gg' and/or $g'g$ are very nearly excluded (Fig. 9). The virtual suppression of *gauche* rotations of opposite sign for pairs of bonds that are first neighbors is a feature manifested in most chain molecules. It supersedes all other characteristics of bond-rotational potentials in its effects on the configurations of chain molecules. Bulky side groups can have a similar effect and in addition favor specific conformations.

The following matrix appreciably simplifies in the case of structural symmetry ($u_{jk} = u_{kj}$, $u_{12} = u_{13}$, $u_{22} = u_{33}$, $u_{23} = u_{32} \sim 0$):

$$
U = \left| \begin{array}{ccc}
1 & \sigma^{1/2} & \sigma^{1/2} \\
\sigma^{1/2} & \sigma & 0 \\
\sigma^{1/2} & 0 & \sigma
\end{array} \right|
\tag{32}
$$

Here $\sigma = \exp(-E_{gg}/kT)$ and the statistical weight of gg' and $g'g$ is approximated by zero, which means complete exclusion of this sequence.

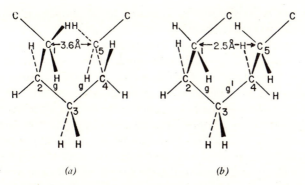

Fig. 9 Conformations generated by rotations about the chain C—C bonds (a) gg, (b) gg', yielding a C_1—C_5 distance of 3.6 and 2.5 Å, respectively. The overcrowding in the gg' pair practically excludes this combination.

The statistical weight matrices make possible a mathematically correct and relatively simple calculation of the partition function, of the probability of occurrence of single-bond conformations and of bond pairs, and of different moments, as, for instance, of mean-square end-to-end distance, gyration radius, optical polarizability tensor, dipole moment even for such complicated asymmetrical molecules as vinyl chains, polypeptides, copolymers, and proteins (4).

If one considers three bond sequences only, for polyethylene $h^2/nb^2 = 2/3 + 4/3 \exp(E/RT)$, where E is the energy difference between trans and gauche.

4. EXCLUDED-VOLUME EFFECT (LONG-RANGE INTERACTIONS)

The consideration of bond angle and hindered rotation results in an extension of the chain as manifested in the increase of the ratio of effective and true bond length. But the limiting expressions for the average square end-to-end distance and gyration radius always turn out to be proportional to the number of chain elements, that is, to the degree of polymerization or to the molecular weight. Actual measurements of gyration radius by light scattering or by intrinsic viscosity, however, show that in most cases the average coil dimensions increase faster:

$$R^2 = KM^{1+\varepsilon} \qquad \varepsilon \geq 0 \qquad (33)$$

The constancy of ε usually extends over a limited range of molecular weight. As far as one can conclude from intrinsic viscosity, it tends to go to zero with low but not vanishing M when the molecule is still a good coil (48).

At lower M one is faced with the transition to the monomer, which is best described by the wormlike chain.

The above treated cases with $\varepsilon = 0$ correspond to the situation in ideal or θ-solvents, characterized by the disappearance of the second virial coefficient $A_2 = 0$ in the concentration dependence of osmotic pressure and light-scattering intensity. In good solvents, however, ε is positive and can reach values up to 0.34. The θ-solvent is a poor solvent; it is indeed a precipitant for a polymer with $M = \infty$. With finite M one can even decrease a little the solvent power below that of the θ-solution before precipitation starts, either by lowering the temperature or by adding a small amount of a mild precipitant.

Very often (49, 50) one characterizes the solvent effect by the ratio $\alpha = R/R_\theta$ between the observed rms gyration radius in that solvent (R) and the value corresponding to the unperturbed dimensions in a θ-solvent (R_θ). As a rule α increases with molecular weight. Moreover, R_θ depends to some extent on temperature. If one measures R_θ in different θ-solvents, one generally obtains a little different values as a consequence of the fact that the temperature and the solvent affect not only the second virial coefficient but also the rotational energy $E(\phi)$ and hence the length of the statistical segment A or the effective bond length b_{eff}. This effect was not considered in the data of Table 1. The temperature dependence of the unperturbed gyration radius can be best judged from the experimentally observed values of $h_\theta^{-2} \, dh_\theta^2/dT = d \ln h_\theta^2/dT$ listed in Table 3. The data were obtained either from the change of intrinsic viscosity $d[\eta]/dT$ or from the change of entropic restoring force dF_S/dT of the pure or swollen melt (rubber). Thus R_θ is not a constant for a polymer, and consequently α is not a sufficiently constant parameter for a general characterization of the polymer–solvent system.

One hence has enough evidence that the quality of the solvent (i.e., the solvent–polymer interaction) influences the average dimensions of the single macromolecule, and this in a manner that goes beyond that what can be described by the valency angle and limited rotation. The immediate interaction of adjacent skeletal atoms and atomic groups and its modification by the solvent cannot explain a nonvanishing ε. In contrast to these *short-range interactions* one has to consider effects that for the whole chain are increasing with the chain length because they have to explain deviations from the random conformation that are becoming larger with M.

Such an effect is the volume requirement of the chain. Every segment occupies a certain volume that is therefore no more available to any other segment of the chain. In the random-walk treatment one does not pay any attention to this problem and simply assumes that any position of the new segment (next step) is equally probable and permitted. If, however, the volume requirement of all the segments is considered, one easily concludes that the conformations leading to a larger coil dimension are favored in comparison

Table 3 Temperature Coefficients of Unperturbed Chain Dimensions[a]

Polymer	Method	Temperature range (°C)	$10^3 d \ln h_\theta^2/dT$ (per degree Celsius)
Polymethylene	dF_s/dT	140–190	−1.0 (±0.1)
	dF_s/dT, swollen with n-$C_{32}H_{66}$	120–170	−1.15 (±0.1)
	dF_s/dT, swollen with DEHA[b]	130–180	−1.0 (±0.2)
	$d[\eta]/dT$ in n-$C_{16}H_{34}$ and n-$C_{28}H_{58}$	110–170	−1.2 (±0.2)
Polystyrene, atactic	dF_s/dT	120–170	0.37
	$[\eta]$ in Cl—$(CH_2)_m$—H ($m = 10, 11, 12$)	6.6–58.6	0.44
Polyisobutylene	dF_s/dT	20–95	−0.08 (±0.06)
	dF_s/dT, swollen with n-$C_{16}H_{34}$	20–60	−0.09 (±0.07)
	$d[\eta]/dT$ in n-$C_{16}H_{34}$	30–130	−0.22 (±0.10)
Poly(n-butene-1):			
Atactic	dF_s/dT	140–200	0.50 (±0.04)
Isotactic	dF_s/dT	140–200	0.09 (±0.07)
Poly(n-pentene-1):			
Atactic	dF_s/dT	40–140	0.53 (±0.05)
Isotactic	dF_s/dT	80–140	0.34 (×0.04)
	$d[\eta]/dT$ in n-$C_{16}H_{34}$	35–90	0.52 (±0.2)
Polydimethylsiloxane	dF_s/dT	30–100	0.78 (±0.06)
	$d[\eta]/dT$ in silicone fluid	30–105	0.71 (±0.13)
Polyoxyethylene	dF_s/dT	30–90	0.23 (±0.02)
Natural rubber	dF_s/dT	−20–+25	0.41 (±0.04)

[a] Data from Flory (4).

[b] DEHA = di-(2-ethylhexyl) azelate.

with those leading to a more compact structure. A very rough and inefficient procedure to prove this consequence would be to construct by the random-walk procedure a series of conformations for the whole molecule and discard all those that contain one or more overlapping contacts. Since according to the random-walk model the average volume per molecule is increasing as $M^{3/2}$ ($\sim h^3$) and the number of contacts per molecule as M^2 (each segment can get in contact with any other segment), the increase of contact density per molecule is proportional to $M^{1/2}$. Hence the resulting modification of molecular dimensions increases with M. According to this rough estimate the effect would be matched by coil expansion if $\varepsilon = \frac{1}{3}$, which indeed agrees well with the maximum experimental value reported (0.34). One also sees that the volume effect is based on *long-range interaction*, that is, on contacts between distant sections of the chain in contrast with the short-range interaction between adjacent chain groups that influences the segment length A or the ratio b_{eff}/b.

As a consequence of the long-range-interaction effects the expansion of the macromolecule in a good solvent must be rather nonuniform: practically none for the single segment and maximum for the free ends. A good description for such a case is (51).

$$h_m{}^2 = b^2 m^{1+\varepsilon} \tag{34}$$

which yields the limiting value

$$R^2 = \frac{h^2}{(2 + \varepsilon)(3 + \varepsilon)} \tag{35}$$

instead of $R^2 = h^2/6$ for a uniformly expanded coil.

The long-range interaction considers not only the volume requirements of chain elements but also the forces between any two of them—exactly speaking, the excess of these forces over the interaction forces between a chain element and the solvent. The interaction energy is schematically portrayed in Fig. 10: increased attraction between chain elements up to the closest approach at r_0 and strong repulsion below r_0 as a consequence of the impenetrability of the elements. The attraction may just so much increase the probability of close approach that it compensates the effect of excluded volume. This is the situation in a θ-solvent, where the molecular conformations are distributed as if there were no limitation by the volume requirement of chain elements. With decreasing attraction the compensation is not more complete, so that the excluded volume causes an expansion of the average shape with a positive ε. The full effect of excluded volume occurs if there is no excess force field; that is, with a constant E independent of r. The expansion, of course, gets still larger, if the excess forces between elements of the chain are repulsive; that is, E increases with decreasing r.

The excluded-volume effect on molecular dimensions, taking into account the volume requirement and the excess interaction energy, was very thoroughly

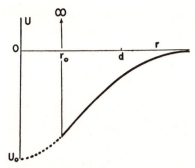

Fig. 10 Interaction energy $U(r) = -U_0$ $\exp(-3r^2/2d^2)$ for $r > r_0$ between two monomers according to Bueche (29) and Yamakawa and Kurata (52). The closest possible approach is $r_{min} = r_0$.

treated in the last 20 years by analytical procedures and by the Monte Carlo method. The connection of the effect with the second virial coefficient yields the possibility of experimentally checking the calculated values of coil expansion as a function of the solvent–solute interaction. The results are rather satisfactory, although a perfect fit between theory and experiment has not yet been achieved (53).

REFERENCES

1. S. Mizushima and H. Okazaki, *J. Am. Chem. Soc.* **71**, 3411 (1949).
2. T. M. Birstein and O. B. Ptitsyn, *Conformations of Macromolecules*, Interscience, New York, 1966.
3. P. Corradini, Proc. R. A. Welch Found. Conf. Chem. Res. X. Polymers, Houston, Texas, 1967, p. 91.
4. P. J. Flory, *Statistical Mechanics of Chain Molecules*, Wiley, New York, 1968.
5. H. Eyring, *Phys. Rev.* **39**, 746 (1932).
6. W. Kuhn, *Kolloid-Z.* **68**, 2 (1934).
7. C. Sadron, *Mem. Lab. Central Serv. Chem.*, 1943, p. 92.
8. C. Sadron, *J. Chim. Phys.* **43**, 145 (1946).
9. H. Benoit and C. Sadron, *J. Polymer Sci.* **4**, 473 (1949).
10. A. Peterlin, *Ann. N.Y. Acad. Sci.* **89**, 578 (1961).
11. G. Porod, *Monatsh. Chem.* **80**, 251 (1949).
12. O. Kratky and G. Porod, *Rec. Trav. Chim.* **68**, 1106 (1949).
13. W. Kuhn, *Kolloid-Z.* **76**, 258 (1936); *ibid.*, **87**, 3 (1939).
14. M. Kurata and W. H. Stockmayer, *Adv. Polymer Sci.* **3**, 196 (1964).
15. W. Kuhn and H. Grün, *Kolloid-Z.* **101**, 248 (1942).
16. J. J. Hermans, *Physica* **10**, 477 (1943).

17. W. Kuhn and H. Kuhn, *Helv. Chim. Acta* **26**, 1394 (1943).
18. J. G. Kirkwood and J. Riseman, *J. Chem. Phys.* **16**, 565 (1948).
19. A. Peterlin, *Les Grosses Molécules en Solution, Hommage Nat. P. Langevin and J. Perrin*, Collège de France, 1948, p. 70.
20. A. Peterlin, *J. Polymer Sci.* **5**, 473 (1950).
21. A. Peterlin, *J. Chim. Phys.* **48**, 13 (1951).
22. M. L. Huggins, *J. Phys. Chem.* **42**, 911 (1938); *ibid.*, **43**, 439 (1939).
23. M. L. Huggins, *J. Appl. Phys.* **10**, 700 (1939).
24. H. A. Kramers, *Physica* **11**, 1 (1944).
25. H. A. Kramers, *J. Chem. Phys.* **14**, 415 (1946).
26. P. Debye, *J. Chem. Phys.* **14**, 636 (1946).
27. A. Peterlin, *J. Chim. Phys.* **47**, 669 (1950).
28. J. G. Kirkwood and J. Riseman, *J. Chem. Phys.* **17**, 442 (1949).
29. F. Bueche, *J. Chem. Phys.* **21**, 205 (1953).
30. P. E. Rouse, *J. Chem. Phys.* **21**, 1272 (1953).
31. B. H. Zimm, *J. Chem. Phys.* **24**, 269 (1956).
32. J. E. Hearst and W. H. Stockmayer, *J. Chem. Phys.* **37**, 1425 (1962).
33. A. Peterlin, *Z. Naturf.* **10a**, 412 (1955).
34. J. E. Hearst, E. Beals, and R. A. Harris, *J. Chem. Phys.* **48**, 5371 (1968).
35. A. Peterlin, *Nature* **171**, 259 (1953).
36. A. Peterlin, *Makromol. Chem.* **9**, 244 (1953).
37. A. Peterlin, *J. Polymer Sci.* **10**, 425 (1950).
38. A. Peterlin, *J. Polymer Sci.* **47**, 403 (1960).
39. S. Heine, O. Kratky, G. Porod, and P. J. Schmitz, *Makromol. Chem.* **44–46**, 682 (1961).
40. S. Heine, O. Kratky, and J. Roppert, *Makromol. Chem.* **56**, 150 (1962).
41. R. G. Kirste, *Z. Phys. Chem.* N.F. **42**, 351, 358 (1964).
42. R. G. Kirste and O. Kratky, *Z. Phys. Chem.* N.F. **31**, 363 (1962).
43. W. Wunderlich and R. G. Kirste, *Ber. Bunsenges. Phys. Chem.* **68**, 646 (1964).
44. D. R. Herschbach, *Intern. Symp. Mol. Struct. Spectroscopy, Tokyo 1962*, Butterworths, London, 1963.
45. H. Benoit, *J. Chim. Phys.* **44**, 18 (1947).
46. H. Benoit and P. M. Doty, *J. Phys. Chem.* **57**, 958 (1953).
47. S. Oka, *Proc. Phys. Math. Soc. Japan* **24**, 657 (1942).
48. U. Bianchi and A. Peterlin, *J. Polymer Sci.* A-2, **6**, 1759 (1968).
49. P. J. Flory, *J. Chem. Phys.* **17**, 303 (1949).
50. P. J. Flory and T. G. Fox, Jr., *J. Amer. Chem. Soc.* **73**, 1904 (1951).
51. A. Peterlin, *J. Chem. Phys.* **23**, 2464 (1955).
52. H. Yamakawa and M. Kurata, *J. Phys. Soc. Japan* **13**, 78 (1958).
53. G. C. Berry and E. F. Casassa, *Macromol. Rev.* **4**, 1 (1969).

Thermodynamics of Polymer Solutions

4

A. V. Tobolsky and I. V. Yannas

1. THERMODYNAMIC FUNCTIONS

The properties of solutions are thermodynamically described by partial molar quantities of the following kind:

$$\overline{V}_1 = \left(\frac{\partial V}{\partial n_1}\right)_{T,\, p,\, n_2} ; \qquad \overline{S}_1 = \left(\frac{\partial S}{\partial n_1}\right)_{T,\, p,\, n_2}$$

$$\overline{H}_1 = \left(\frac{\partial H}{\partial n_1}\right)_{T,\, p,\, n_2} ; \qquad \overline{F}_1 = \left(\frac{\partial F}{\partial n_1}\right)_{T,\, p,\, n_2}$$

$$(1)$$

In the above equations V, H, S, and F represent volume, enthalpy, entropy, and Gibbs free energy, respectively. A binary solution is assumed with n_1 moles of solvent and n_2 moles of solute. Similar definitions apply for $\overline{V}_2$, $\overline{H}_2$, and so on.

In a sense the partial molar quantities $\overline{V}_1$, etc., are analogs of the molar quantities V_1, etc., for pure solvent. For ideal solutions it is a fact that $\overline{V}_1$ is equal to V_1, and $\overline{H}_1$ to H_1. On the other hand, for a stable solution $\overline{S}_1$ is always larger than S_1, and $\overline{F}_1$ is always smaller than F_1. This results from the fact that the entropy of a solution is always larger than the entropy of the unmixed components; this excess entropy is called the entropy of mixing.

Similarly the free energy of a stable solution is always less than the free energy of the unmixed components; this difference is called the free energy of mixing. For nonideal solutions there is also a nonzero enthalpy (heat) of mixing. Quite obviously,

$$\Delta F_m = \Delta H_m - T \Delta S_m \tag{2}$$

At equilibrium under constant pressure and temperature the partial molar free energy of a given component must be the same in all phases. For this reason the partial molar free energy is frequently referred to as the chemical potential μ and the molar free energy as μ°. A very important and useful quantity is the free energy of dilution $\overline{\Delta F_1}$, defined below along with related quantities:

$$\overline{\Delta F_1} = \overline{F_1} - F_1 = \left(\frac{\partial \Delta F_m}{\partial n_1}\right)_{T,p,n_2} = \mu_1 - \mu_1^\circ = RT \ln a_1 \tag{3}$$

$$\overline{\Delta S_1} = \overline{S_1} - S_1$$

$$\overline{\Delta H_1} = \overline{H_1} - H_1$$

Equation 3 also defines a_1, the thermodynamic activity of the solvent.

The colligative properties of solutions are all related to the free energy of dilution as follows:

$$\frac{p_1}{p_1^\circ} = e^{\overline{\Delta F_1}/RT} \tag{4}$$

$$\pi = -\frac{\overline{\Delta F_1}}{\overline{V_1}} \tag{5}$$

$$\Delta T = \frac{T_t \overline{\Delta F_1}}{L_t} \tag{6}$$

In Eq. 4 p_1 is the vapor pressure of solvent in the solution and p_1° is the vapor pressure of pure solvent. In Eq. 5 π is the osmotic pressure of the solution. In Eq. 6 ΔT is the boiling-point elevation or the freezing-point depression, L_t is the heat of condensation or of fusion, and T_t is the boiling point or the freezing point.

We now confront the basic content of this chapter by presenting the equations for $\overline{\Delta F_1}$ for ideal solutions and for solutions of flexible macromolecules. For ideal solutions

$$\Delta F_m = RT(n_1 \ln x_1 + n_2 \ln x_2)$$

$$\overline{\Delta F_1} = \mu_1 - \mu_1^\circ = RT \ln x_1 \tag{7}$$

$$\overline{\Delta F_2} = \mu_2 - \mu_2^\circ = RT \ln x_2$$

where x_1 and x_2 are the mole fractions of solvent and solute in solution, respectively. Comparing Eqs. 7 with Eq. 3 we see that the activities of an ideal solution are equal to its mole fractions.

For flexible linear macromolecules the expressions for ΔF_m, $\overline{\Delta F_1}$, and $\overline{\Delta F_2}$ are given by the very well-known Flory–Huggins equations (1):

$$\Delta F_m = RT(n_1 \ln v_1 + n_2 \ln v_2 + \chi n_1 v_2)$$

$$\overline{\Delta F_1} = RT\left[\ln(1 - v_2) + \left(1 - \frac{1}{r}\right)v_2 + \chi v_2^2\right] \tag{8}$$

$$\overline{\Delta F_2} = RT[\ln v_2 - (r - 1)(1 - v_2) + \chi r(1 - v_2)^2]$$

where v_2 is the volume fraction of the polymer, r is the ratio of molar volumes of polymer and solvent, and χ is an interaction parameter that generally varies from -1.0 to slightly over 0.5. Experimental results on polymer solutions can often be approximated by Eqs. 8. However, χ usually increases with polymer concentration, and in many instances the change is large; the original and simplest theory suggests that χ should be constant. Furthermore, for polymer solutions that are nearly athermal the experimental value of χ is typically in the range of 0.3 to 0.5 as v_2 approaches zero, whereas the "theoretical" value of χ for athermal polymer solutions should be close to zero. In spite of these limitations Eqs. 8 represent the most convenient framework presently available for discussion of polymer solutions. Note that volume fractions are used rather than the mole fractions employed in Eqs. 7.

The expressions for the colligative properties can now be obtained by inserting the value of $\overline{\Delta F_1}$ into Eqs. 4, 5, and 6. For reasonably dilute solutions it is very useful to express the colligative properties in terms of the weight concentration c of solute in grams per cubic centimeter. When the logarithms in Eqs. 8 are expanded in a power series and inserted into Eqs. 4, 5, and 6, the following equations result (note that $c = v_2 d_2$, where d_2 is the polymer density):

$$\ln \frac{p_1}{p_1^\circ} = -\frac{V_1}{M_2} c + \frac{(\chi - \frac{1}{2})c^2}{d_2^2} \tag{9}$$

$$\pi = \frac{RT}{M_2} c + \frac{RT}{V_1 d_2^2}(\tfrac{1}{2} - \chi)c^2 \tag{10}$$

$$\Delta T = \frac{RT_t^2}{L_t}\left[-\frac{V_1}{M_2} c + \frac{(\chi - \frac{1}{2})c^2}{d_2^2}\right] \tag{11}$$

where M_2 is the molecular weight of solute (polymer). It is noteworthy that the equations for ideal solutions are in all cases obtained by considering only the terms that are first power in c in the above equations. Data on colligative properties provide a means of obtaining the molecular weight M_2 and the interaction parameter χ. For osmotic pressure a plot of π/c versus c should give a straight line whose intercept is RT/M_2 and whose slope is $RT(\frac{1}{2} - \chi)/V_1 d_2{}^2$.

2. IDEAL SOLUTIONS

In this and the following sections we outline some very simple statistical thermodynamic derivations for solution properties.

The thermodynamic criterion for a spontaneous process occurring at constant temperature and pressure is that the Gibbs free energy F should decrease during the process.

Consider n_1 moles of gas 1, separated by a partition from n_2 moles of gas 2, all at constant temperature and pressure. If the partition is removed, the gases interdiffuse. This is a spontaneous irreversible process occurring at constant temperature and pressure, and also at constant energy and volume, since no heat is absorbed and no work is produced in the environment. The entropy change of each gas can be considered as the entropy change of expansion.

The entropy change ΔS_m and the change in Gibbs free energy ΔF_m are given by the following formulas:

$$\Delta S_m = -R(n_1 \ln x_1 + n_2 \ln x_2) \tag{12}$$

$$\Delta F_m = RT(n_1 \ln x_1 + n_2 \ln x_2) \tag{13}$$

where x_1 and x_2 are the mole fractions of components 1 and 2, respectively. Equation 12 can be derived from the fact that ΔS of expansion from V_A to V_B is $nR \ln(V_B/V_A)$.

The positive ΔS_m and negative ΔF_m persist even if the gases being mixed are as similar as two isotopes. However, if the gases are identical, the process of mixing does not give rise to a new state and ΔS_m and ΔF_m are zero. This apparent discontinuity is known as the Gibbs paradox and has to do with the interchangeability of identical particles.

If N_1 particles of solid lattice 1 are mixed with N_2 particles of solid lattice 2, the number of distinguishable lattice configurations Ω is

$$\Omega = \frac{(N_1 + N_2)!}{N_1! N_2!} \tag{14}$$

The entropy change on mixing is, according to Boltzmann's law,

$$\Delta S_m = k \ln \Omega = -R(n_1 \ln x_1 + n_2 \ln x_2) \tag{15}$$

where k is Boltzmann's constant and R is the gas constant. Stirling's law for factorials is used in developing Eq. 15.

If there is no enthalpy (heat) of mixing, as is true for ideal solutions, then

$$\Delta F_m = -T\,\Delta S_m = RT(n_1 \ln x_1 + n_2 \ln x_2) \tag{16}$$

It will be noted that Eqs. 12 through 16, which are the laws of ideal mixing, are the same for ideal gases and solids. These laws are assumed to be valid for the mixing of ideal liquids.

By differentiating the above equations, we obtain:

$$\overline{\Delta S_1} = -R \ln x_1; \qquad \overline{\Delta F_1} = RT \ln x_1 \tag{17}$$

By definition for ideal solutions $\overline{\Delta H_1} = 0$ and $\overline{\Delta V_1} = 0$.

3. NONIDEAL ENTROPY OF MIXING

If an ideal gas has an excluded volume b per mole, the equation of state is

$$p = \frac{nRT}{V - nb} \tag{18}$$

The entropy of expansion from state A to state B is

$$\Delta S = nR \ln \frac{V_B - nb}{V_A - nb} \tag{19}$$

When n_1 moles of gas 1 and n_2 moles of gas 2 both of the above type are mixed at constant temperature and pressure, the entropy change is

$$\Delta S_m = n_1 R \ln \frac{V - n_1 b_1 - n_2 b_2}{n_1(V_1 - b_1)} + n_2 R \ln \frac{V - n_1 b_1 - n_2 b_2}{n_2(V_2 - b_2)} \tag{20}$$

where V_1 and V_2 are the molar volumes at the fixed temperature and pressure, and V is the final volume equal to $n_1 V_1 + n_2 V_2$.

For the mixing of two liquids we can use Eq. 20 as an approximate expression with the additional proviso that $b_1/V_1 = b_2/V_2$.

Under these conditions we obtain (2)

$$\Delta S_m = -R(n_1 \ln v_1 + n_2 \ln v_2) \tag{21}$$

$$\overline{\Delta S_1} = -R \ln v_1 - Rv_2\left(1 - \frac{1}{r}\right) \tag{22}$$

$$\overline{\Delta F_1} = RT \ln v_1 + RTv_2\left(1 - \frac{1}{r}\right) \tag{23}$$

where v_1 and v_2 are the volume fractions and r is the ratio of molar volumes (i.e., $r = V_2/V_1$). Note that if r is unity, volume fractions are equal to mole fractions and the above equations are identical with those for ideal solutions, as indeed they should be. Any nonideal behavior implicit in Eqs. 22 and 23 results from unequal molar volumes of the two components.

The entropy of mixing for flexible macromolecules with a solvent can be obtained by a lattice-model generalization of Eq. 14. The number of configurations Ω can be shown to be given by

$$\Omega = \frac{(N_1 + rN_2)!}{N_1!N_2!} \left(\frac{z - 1}{N_1 + rN_2} \right)^{(r-1)N_2} \tag{24}$$

where r is the ratio of molar volumes of polymer and solvent or, nearly equivalently, the degree of polymerization of the polymer. Starting with S (configurational) $= k \ln \Omega$, Eq. 24 leads directly to Eqs. 21 and 22. Note that in order to obtain ΔS_m one must also consider S (configurational) at $N_1 = 0$.

4. HEAT OF MIXING

The simplest calculation of the heat of mixing treats a lattice of coordination number z containing N_1 solvent and N_2 solute molecules of nearly equal size. We consider that each (1,1) pair contributes w_{11} to the lattice energy and similarly define w_{12} and w_{22}. (All these quantities are in general negative since energy zero refers to the separated molecules, and the pair energies are on balance more largely due to attractive, rather than repulsive, forces.)

If the number of (1,1) pairs, etc., is N_{11}, N_{12}, and N_{22}, the energy change on mixing a pure lattice of N_1 molecules with a pure lattice of N_2 molecules is easily shown to be (2, 3)

$$\Delta E_m = N_{12} w$$
$$w = w_{12} - \tfrac{1}{2}(w_{11} + w_{22}) \tag{25}$$

It is readily shown that the most probable and expected value of N_{12} is equal to $zN_1N_2/(N_1 + N_2)$. The expected value of ΔE_m is therefore

$$\Delta E_m = \frac{zwN_1N_2}{N_1 + N_2} = (N_1 + N_2)zwx_1x_2 \tag{26}$$

where x_1 and x_2 are mole fractions.

Hildebrand (2) and Scatchard (4) have presented a more general formula for the heat of mixing for molecules of unequal volume, which ultimately goes back to the van der Waals equation of state for gases. The history and the development of the theory are fully presented by Hildebrand and Scott (2).

The result is

$$\Delta H_m = (n_1 V_1 + n_2 V_2)(\delta_2 - \delta_1)^2 v_1 v_2 = n_1 V_1 v_2 (\delta_2 - \delta_1)^2 \qquad (27)$$

where n_1 and n_2 are the moles of solvent and solute and v_1 and v_2 are the volume fractions of solvent and solute in the solution. The quantities δ_1 and δ_2 are the solubility parameters of solvent and solute. These interesting quantities, which are the core of the Hildebrand theory for the heat of mixing, are defined as follows:

$$\delta_1 = \left(\frac{E_1'}{V_1}\right)^{1/2} \qquad (28)$$

$$\delta_2 = \left(\frac{E_2'}{V_2}\right)^{1/2} \qquad (29)$$

where E_1' is the molar energy of vaporization of solvent and similarly for the solute. The molar energy of vaporization of the solvent is equal to $H_1' - RT$, where H_1' is the molar heat of vaporization of the solvent, and similarly for the solute. The heats of vaporization are readily obtained by direct measurement or from the variation of vapor pressure with temperature (Clapeyron–Clausius equation). The solubility parameters for many simple liquids have been calculated; a partial tabulation is given in Table 1.

As can be seen from Eq. 27, the Hildebrand theory predicts only endothermic (positive) heats of mixing. The heat of mixing is zero (athermal) when δ_1 and δ_2 are equal; the solution process becomes more endothermic as the difference between δ_1 and δ_2 becomes greater. Increasing endothermicity means increasing opposition to the process of solution. Consider the formula

$$\Delta F_m = \Delta H_m - T \Delta S_m \qquad (2)$$

The process of solution is increasingly favored as the value of ΔF_m becomes increasingly negative. The value of ΔS_m is always positive, which favors the solution process. However, the calculations of the preceding section show that ΔS_m for polymer solutions is smaller than ΔS_m for a solution in which the polymer segments are separated from each other. A positive value of ΔH_m opposes the process of solution because it makes ΔF_m less negative. Although exothermic heats of solution are observed (negative ΔH_m), they arise from specific solvent–solute interactions and are not predicted by the Hildebrand theory.

In general liquids of low molecular weight are miscible even if $|\delta_1 - \delta_2|$ is reasonably large. However, polymers will generally not dissolve appreciably in a solvent if $|\delta_1 - \delta_2|$ is as great as 2.0.

This brings us to the interesting question of how one determines the

Table 1   Solubility Parameters of Various Liquids[a]

Solvent	δ_d	δ_p	δ_h	δ
Water	6.00	15.3	16.7	23.50
Methanol	7.42	6.0	10.9	14.28
Ethanol	7.73	4.3	9.5	12.92
n-Butanol	7.81	2.8	7.7	11.30
Ethylene glycol	8.25	5.4	12.7	16.30
Dioxane	9.30	0.9	3.6	10.00
Carbon disulfide	9.97	0.0	0.0	9.97
Dimethyl sulfoxide	9.00	8.0	5.0	12.93
γ-Butyrolactone	9.26	8.1	3.6	12.78
Acetone	7.58	5.1	3.4	9.77
Methyl ethyl ketone	7.77	4.4	2.5	9.27
Tetrahydrofuran	8.22	2.8	3.9	9.52
Ethyl acetate	7.44	2.6	4.5	9.10
Acetonitrile	7.50	8.8	3.0	11.90
Nitroethane	7.80	7.6	2.2	11.09
Aniline	9.53	2.5	5.0	11.04
Dimethylformamide	8.52	6.7	5.5	12.14
Pyridine	9.25	4.3	2.9	10.61
Carbon tetrachloride	8.65	0	0	8.65
Chloroform	8.65	1.5	2.8	9.21
Trichloroethylene	8.78	1.5	2.6	9.28
Benzene	8.95	0.5	1.0	9.15
Toluene	8.82	0.7	1.0	8.91
Tetralin	9.35	1.0	1.4	9.50
Hexane	7.24	0	0	7.24
Cyclohexane	8.18	0	0	8.18

[a] Data from Hansen (5, 6).

solubility parameter of a polymer. The most widely used method is to prepare a slightly crosslinked sample of the polymer that will swell, but not dissolve, in any solvent because of its three-dimensional nature. The extent of swell is measured in a large number of solvents of varying δ. The solubility parameter of the polymer is presumed to be equal to the solubility parameter of the solvent that causes the largest equilibrium swelling. Solubility parameters for various polymers obtained in this manner are shown in Table 2.

The partial molar heat of dilution $\overline{\Delta H_1}$ is given by

$$\overline{\Delta H_1} = \overline{H}_1 - H_1 = \left(\frac{\partial \Delta H_m}{\partial n_1}\right)_{T, p, n_2} = (\delta_1 - \delta_2)^2 V_1 v_2^2 \qquad (30)$$

Table 2   Solubility Parameters of Various Polymers

Polymer	δ_d	δ_p	δ_h	δ
Teflon	—	—	—	6.2
Silicone rubber	—	—	—	7.3
Rubber	8.15	0	0	8.15
Polyisobutylene	7.7	0	0	7.7
Polystyrene	8.95	0.5	1.6	9.11
Polybutadiene	8.3	0	0.5	8.32
Polyethylene	8.1	0	0	8.1
Polyvinyl chloride	8.16	3.5	3.5	8.88
Polyvinyl acetate	7.72	4.8	2.5	9.43
Polymethyl methacrylate	7.69	4.0	3.3	9.28
Polypropylene oxide	7.02	2.5	1.0	7.52
Butadiene–styrene (75 : 25)	—	—	—	8.1
Buna-N [butadiene–acrylonitrile (75 : 25)]	—	—	—	8.9
Ethyl cellulose	—	—	—	8.3
Polymethacrylonitrile	—	—	—	10.7
Polyethylene terephthalate	—	—	—	10.7
Cellulose diacetate	—	—	—	10.9
Epoxy resin	—	—	—	10.9
Polyvinylidene chloride	—	—	—	12.2
Nylon 66	—	—	—	13.6
Polyacrylonitrile	—	—	—	15.4

5. FREE ENERGY OF DILUTION

The basic equation for the free energy of dilution for flexible macromolecules has already been given as

$$\overline{\Delta F_1} = RT\left[\ln(1 - v_2) + \left(1 - \frac{1}{r}\right)v_2 + \chi v_2{}^2\right] \tag{8}$$

We have already derived the equations for $\overline{\Delta S_1}$ and $\overline{\Delta H_1}$:

$$\overline{\Delta S_1} = -R\left[\ln(1 - v_2) + \left(1 - \frac{1}{r}\right)v_2\right] \tag{22}$$

$$\overline{\Delta H_1} = (\delta_1 - \delta_2)^2 V_1 v_2{}^2 \tag{30}$$

From a direct comparison of the above three equations it would appear that

$$\chi = \frac{(\delta_1 - \delta_2)^2 V_1}{RT} \tag{31}$$

It must be pointed out, however, that none of the three equations given above is exact and that there appear to be compensating errors in $\overline{\Delta S_1}$ and $\overline{\Delta H_1}$. Certain authors prefer the following equation for χ:

$$\chi' = \frac{1}{z} + \frac{(\delta_1 - \delta_2)^2 V_1}{RT} \tag{32}$$

where z is the coordination number of the lattice.

Consider now the expansion of Eq. 8 in a Taylor series:

$$\overline{\Delta F_1} = -\frac{RTv_2}{r} + RT(\chi - \tfrac{1}{2})v_2^2 - \frac{RTv_2^3}{3} - \cdots \tag{33}$$

Inasmuch as χ is dependent on temperature, there is a temperature θ for most poor solvents at which $\chi = 0.5$ and the second term on the right-hand side vanishes. The θ-temperature is called the "ideal" temperature. At the temperature $T = \theta$ [Flory's θ-point (7)] the chemical potential due to solvent–segment interactions vanishes, and in a sense deviations from ideality also vanish; for example, a plot of π/c versus c at the θ-temperature will give a line of zero slope, at least for low values of c. It is believed that the free energy of interaction of polymer segments within a volume element of solution is also zero at the θ-temperature and that the polymer chain assumes its *unperturbed* dimensions at this condition. Certainly the θ-temperature represents a well-defined state of the polymer–solvent interaction. For this reason θ-temperatures for a variety of polymer–solvent systems have been tabulated in many books.

The root-mean-square (rms) end-to-end dimensions of a polymer at its θ-temperature are related to this same quantity at temperature T by the following equation:

$$\frac{h}{h_\theta} = \alpha \tag{34}$$

$$\alpha^5 - \alpha^3 = 2C\left(1 - \frac{\theta}{T}\right)M^{1/2} \tag{35}$$

where C is approximately unity.

6. SOLUBILITY AND SWELLING

For a polymer to dissolve in a liquid it is a thermodynamic requirement that $\overline{\Delta F_1}$ be negative, and examination of Eq. 33 shows that this requirement is met only if the value of χ is less than approximately 0.5. A more careful examination shows that there is a critical value χ_c above which the polymer precipitates (i.e., where there is a formation of a second phase), given by

$$\chi_c = \frac{1}{2} + \frac{1}{r^{1/2}} \tag{36}$$

At χ_c the volume fractions of solvent and polymer in the liquid (solution) phase are

$$(v_1)_c = \frac{r^{1/2}}{1 + r^{1/2}} \tag{37}$$

$$(v_2)_c = \frac{1}{1 + r^{1/2}} \tag{38}$$

If r, and hence the molecular weight of the polymer, is high, the liquid in equilibrium with the swollen polymer will be practically pure liquid. The polymer phase is highly swollen with solvent.

Equations 36, 37, and 38 are derived from the condition that at equilibrium $\mu_1' = \mu_1''$ and $\mu_2' = \mu_2''$, where the primes and double primes refer to the two phases in equilibrium.

If the value of χ is equal to or greater than the critical value, two phases coexist. The condition for equilibrium between a swollen insoluble polymer and the liquid that causes swelling is that $\overline{\Delta F_1}$ equal zero. Equating the right-hand side of Eq. 8 to zero and neglecting v_2/r, we obtain

$$v_2 + \chi v_2^2 = -\ln(1 - v_2) \tag{39}$$

where v_2 is the volume fraction of polymer in the swollen polymer phase. Equation 39 has a positive solution if χ is larger than χ_c (or in fact larger than 0.5).

If it is crosslinked, the polymer can only swell and not dissolve, no matter how good a solvent the liquid is for the noncrosslinked (linear) polymer. An additional term due to the elastic deformation during swelling must be added to Eq. 8, namely, $RTv_2^{1/3}/M_c$, where M_c is the molecular weight of the portion of the chain between crosslinks. If we neglect v_2/r, we obtain

$$\overline{\Delta F_1} = RT \left[\ln(1 - v_2) + v_2 + \chi v_2^2 + \frac{v_2^{1/3}}{M_c} \right] \tag{40}$$

At equilibrium swelling $\overline{\Delta F_1} = 0$ and therefore

$$\ln(1 - v_2) + v_2 + \chi v_2^2 + \frac{v_2^{1/3}}{M_c} = 0 \tag{41}$$

If χ is known from experiments with the linear polymer, M_c can be determined from Eq. 41.

Crystalline polymers can absorb limited amounts of solvent, thereby changing their crystalline melting temperature from T_m° to T_m. The relevant equation is

$$\frac{1}{T_m} - \frac{1}{T_m^\circ} = \frac{R}{\Delta H_u} \frac{V_u}{V_1} (v_1 - \chi v_1^{2}) \tag{42}$$

where ΔH_u is the heat of fusion per mole of repeating unit of polymer, V_u is the molar volume of the repeating unit of polymer, V_1 is the molar volume of solvent, v_1 is the volume fraction of absorbed solvent, and χ is the interaction parameter.

Equation 42 is closely related to the classical melting-point depression and solubility expression for solutions of simple molecules:

$$\frac{1}{T_m} - \frac{1}{T_m^\circ} = - \frac{R}{\Delta H_2} \ln x_2 \tag{43}$$

where ΔH_2 is the heat of fusion of the major component 2 and x_2 is the mole fraction. If the solution contains only a small amount of component 1, $\ln x_2 \approx -x_1 \approx (V_2/V_1)v_1$; hence

$$\frac{1}{T_m} - \frac{1}{T_m^\circ} = \frac{R}{\Delta H_2} \frac{V_2}{V_1} v_1 \tag{44}$$

For an ideal solution $\chi = 0$, and therefore Eqs. 42 and 43 reduce to the same form for small proportions of the noncrystallizing component.

7. COILS, RODS, SPHERES, ELLIPSOIDS

In this section we treat the properties of solutions of polymers other than flexible linear macromolecules. Frequently solution properties are expressed in terms of a so-called virial expansion of the osmotic pressure, which is valid for any experimental results or for any theoretical model:

$$\frac{\pi}{RTc} = - \frac{\overline{\Delta F}_1}{RTV_1 c} = \frac{1}{M_2} + A_2 c + A_3 c^2 + \cdots \tag{45}$$

where A_2 is called the second virial coefficient. The second virial coefficient for ideal solutions and for solutions of flexible linear macromolecules can be immediately obtained from Eq. 10.

$$A = {}_2 0 \qquad \text{ideal solutions} \tag{46}$$

$$A_2 = \frac{\frac{1}{2} - \chi}{V_1 d_2^{2}} \qquad \text{flexible chains} \tag{47}$$

The second virial coefficients for solutes that can be approximated as large rigid spheres, rigid rods, and prolate or oblate ellipsoids are taken from a review article by Isihara and Guth (8). In all cases the heat of mixing is neglected and A_2 arises from a nonideal entropy of mixing.

$$A_2 = \frac{4}{V_2 d_2{}^2} \qquad \text{rigid spheres} \tag{48}$$

$$A_2 = \frac{1}{V_2 d_2{}^2}\frac{L}{D} \qquad \text{rigid rods} \tag{49}$$

$$A_2 = \frac{4f}{V_2 d_2{}^2} \qquad \text{prolate or oblate ellipsoids} \tag{50}$$

$$f = 1 + \frac{1}{15}\varepsilon^4 + \cdots$$

$$\varepsilon = \left(\frac{1 - b^2}{a^2}\right)^{1/2}$$

In Eq. 48 V_2 is the molar volume of the solute. In Eq. 49 L is the length of the rod and D is the diameter. In Eq. 50 b is the minor axis of the ellipse and a is the major axis. Equation 50 is valid when the eccentricity is much smaller than unity.

A consequence of the thermodynamics of solutions of rigid rods is that a phase separation should occur in reasonably concentrated solutions. An approximate evaluation of the critical volume fraction v_c at which phase separation occurs gives

$$v_c = \frac{4D}{L} \qquad \text{rigid rods} \tag{51}$$

This phenomenon of phase separation in concentrated solutions has been observed in solutions of tobacco mosaic virus and in solutions of polypeptides. The phase more concentrated in the rigid-rod solute molecules tends to show an ordering of these molecules, as demonstrated by birefringence phenomena; the dilute phase shows no ordering of the solute.

A more detailed study of the statistical thermodynamics of solutions of rigid rods leads to the following equation (9, 10):

$$\overline{\Delta F_1} = RT\left[\ln(1 - v_2) + \frac{y - 1}{r}v_2 + \frac{2}{y} + \chi v_2{}^2\right] \tag{52}$$

where v_2 is the volume fraction of solute, r is the ratio of molar volumes of solute and solvent, χ is the interaction parameter, and y is a parameter that varies between unity and r. In very dilute solutions where the rods are disoriented with respect to one another y equals r. In very concentrated solutions where the rods align with one another y equals unity. The equation that gives the value of y as a function of v_2 and r is

$$v_2 = \frac{r}{r-y}\left[1 - \exp\left(\frac{-2}{y}\right)\right] \tag{53}$$

Finally an equation has been proposed for $\overline{\Delta F_1}$ for solutions of rodlike molecules containing flexible side groups. A molecule of this type is the synthetic polypeptide polybenzyl glutamate. The equation for $\overline{\Delta F_1}$ for solutions of this molecule in helicogenic solvents is (11).

$$\overline{\Delta F_1} = RT[\ln(1 - v_2') + v_2' + \chi(v_2')^2] \tag{54}$$

where v_2' is the volume fraction in solution of the flexible side groups.

More recent developments in the theory of polymer solutions introduce approximate equations of state for liquids (12).

8. THE GENERALIZED SOLUBILITY PARAMETER

Recently Hansen (5, 6) introduced a three-dimensional solubility parameter that appears to give very good agreement with experimental data on solubility of polymers. It is assumed that the energy of vaporization of a liquid can be divided into contributions from van der Waals dispersion forces E_d, dipole–dipole forces E_p, and hydrogen-bonding forces E_h:

$$E = E_d + E_p + E_h \tag{55}$$

Dividing by the molar volume of the solvent gives

$$\frac{E}{V} = \frac{E_d}{V} + \frac{E_p}{V} + \frac{E_h}{V}$$

$$\delta^2 = \delta_d^2 + \delta_p^2 + \delta_h^2 \tag{56}$$

$$\delta_d = \left(\frac{E_d}{V}\right)^{1/2}; \quad \delta_p = \left(\frac{E_p}{V}\right)^{1/2}; \quad \delta_h = \left(\frac{E_h}{V}\right)^{1/2}$$

The methods for evaluation of E_d, E_p, and E_h for common solvents are presented by Hansen (5, 6). The final result is that one can obtain for any liquid a solubility parameter composed of three parameters, δ_d, δ_p, and δ_h. These can be represented as coordinates in a three-dimensional space. The values of these parameters for many common liquids are shown in Table 1.

The solubility of a polymer can also be represented by three parameters δ_d, δ_p, and δ_h. In this three-dimensional δ-space the polymer can be represented by a point; so too can the numerous liquids shown in Table 1.

For each polymer there is a characteristic radius originating from its point in δ-space which encloses all liquids that are solvents for the polymer. In general the order of magnitude of this radius is two δ-units; however, relatively large deviations from this value can sometimes occur.

As a fair approximation, one can ignore the δ_p-coordinate and consider the problem in a two-dimensional map of δ_d and δ_h, with each polymer having a characteristic circle with a well-defined center and radius.

Values of δ_d, δ_p, δ_h, and δ for a number of polymers are listed in Table 2. For certain polymers listed only δ is known.

9. THE HELIX–COIL TRANSITION

Synthetic polypeptides that are helical rigid rods in many solvents, will, at sufficiently high temperatures, convert into the random-coil configuration. This phenomenon of denaturation seems to have the characteristics of a diffuse melting. Zimm and Bragg (13) have presented a simple statistical thermodynamic treatment of this phenomenon, and others have extended this treatment and also considered the problem of chain dimensions during denaturation (14).

Consider a synthetic polypeptide to consist of random-coil sequences where each segment is denoted by r and helical sequences in which each segment is denoted by h. A particular conformation of the macromolecule can be written as

$$rrkhhhkrrrrkhkrrrkhhhhh \cdots \qquad (57)$$

Note that the first and last helical segment of each sequence is marked k to emphasize that these segments are boundary segments.

The segments r, h, and k can be assigned partition functions f_r, f_h, and f_k. The partition function of the chain as a whole must be a sum of terms over all conformations of type 57, which is a form of the Ising problem. The partition function of a chain of n segments is the nth power of the maximum root of the matrix

$$M = \begin{vmatrix} f_r^{1/2}f_r^{1/2} & f_r^{1/2}f_k^{1/2} \\ f_h^{1/2}f_k^{1/2} & f_h^{1/2}f_h^{1/2} \end{vmatrix} \qquad (58)$$

From the partition function one can obtain at any temperature x_r, x_h, and $\bar{r}$ and $\bar{h}$, which are mole fraction of r-segments, mole fraction of h-segments, average sequence length of r-segments, and average sequence length of h-segments, respectively. These results are expressed as functions of ΔH_f, ΔS_f, ΔF_k, and T, where ΔH_f and ΔS_f represent the enthalpy and entropy for the conversion of an h-segment to an r-segment and ΔF_k represents the free energy of formation for the conversion of an r-segment to a k-segment.

The mole fraction of crystalline segments plotted against temperature starts at $x_h = 1$, gives a value of $x_h = 0.5$ at the transition temperature $T_f = \Delta H_f / \Delta S_f$, and then approaches zero at high temperatures. The width of the transition region depends on the value of ΔF_k: if ΔF_k is very large, the transition is very sharp, becoming infinitely sharp as ΔF_k approaches infinity; if ΔF_k is small, the transition is diffuse.

The mean-square end-to-end distance of the chain in the region of transition is given by the formula (14)

$$R^2 = \left[\frac{x_h}{3} (2\bar{h} + 1) + 1 - x_h \right] n l_0^2 \tag{59}$$

where n is the number of chain segments and l_0 is the length of each segment.

REFERENCES

1. P. J. Flory, *Principles of Polymer Chemistry*, Cornell University Press, Ithaca, N.Y., 1953, Chapter XII; M. L. Huggins, *J. Chem. Phys.*, **9**, 440 (1941); P. J. Flory, *ibid.* **9**, 660 (1941).

2. J. H. Hildebrand and R. L. Scott, *Solubility of Non Electrolytes*, Reinhold, New York, 1950.

3. J. H. Hildebrand and R. L. Scott, *Regular Solutions*, Prentice-Hall, 1950.

4. G. Scatchard, *Chem. Revs.*, **8**, 321 (1931).

5. C. M. Hansen, *Ind. Eng. Chem., Prod. Res. Develop.* **8**, 2 (1969).

6. C. M. Hansen, *J. Paint Technol.* **39**, 104 (1967).

7. P. J. Flory, reference 1, page 523.

8. A. Isihara and E. Guth, *Advances in Polymer Science*, **5**, 233–260 (1967).

9. P. J. Flory, *Proc. Roy. Soc. (London)* A**234**, 60, 73 (1956).

10. P. J. Flory, *J. Polymer Sci.*, **49**, 105 (1961).

11. P. J. Flory and W. J. Leonard, Jr., *J. Am. Chem. Soc.* **87**, 2102 (1965).

12. B. E. Eichinger and P. J. Flory, *Trans. Faraday Soc.* **64**, 2035 (1968).

13. B. H. Zimm and J. K. Bragg, *J. Chem. Phys.* **28**, 1246 (1959).

14. A. V. Tobolsky and V. D. Gupta, *Indian J. Phys.* **37**, 625 (1963).

Hydrodynamics of Polymer Solutions and Melts

5

A. V. Tobolsky

1. VISCOSITY OF SIMPLE LIQUIDS

A liquid is a condensed phase that under normal experimental conditions responds to a shear stress by flowing. At sufficiently low shear stresses both liquids and gases obey Newton's law:

$$f = \eta \frac{ds}{dt} \tag{1}$$

where f is the shearing stress, ds/dt is the rate of shear, and η, the viscosity coefficient obtained from Eq. 1, is expressed in poises (P). Viscosity has the dimensions of m/lt, where m is mass, l is length, and t is time.

Viscometers are instruments used for measuring viscosities. They can be of several designs, depending on the range of viscosity to be measured. The following types are very common: capillary flow, rotating concentric cylinders, falling (or rising) ball, and parallel-plate plastometer.

The range of liquid viscosities is enormous. At room temperature the viscosities of water, castor oil, and pitch are approximately 10^{-2}, 10^{1}, and 10^{10} P, respectively. By comparison, the viscosity of air at 23°C is 1.84×10^{-4} P.

The viscosity of gases at pressures less than 100 atm is satisfactorily derived from the principles of the kinetic theory of gases, using the concept of transfer of momentum resulting from molecular motion between neighboring planes moving with different velocities. The result for the fluidity (reciprocal viscosity) of a gas is

$$\frac{1}{\eta} = \left(\frac{9\pi^3}{8mkT}\right)^{1/2} d^2 \tag{2}$$

where m is the molecular mass and d is the molecular diameter.

The interpretation of the viscosity of simple liquids whose molecules are essentially spherical has proved to be far more intractable. It has, however, been clear that the fluidity (reciprocal viscosity) increases with increasing free volume. The free volume per molecule, v_f, is defined in the following way:

$$v_f = v - v_0 \tag{3}$$

where v is the average molecular volume and v_0 is the van der Waals volume of the molecule.

In order for a molecule in a liquid to be able to move from one plane to another (and thereby transfer its momentum) it must find in its immediate vicinity a hole of size Bv or larger, B being close to unity. According to Cohen and Turnbull (1), the probability of finding such a hole is $\exp(-Bv/v_f)$. By using this factor to modify Eq. 2, we arrive at the following equation for the fluidity of a liquid:

$$\frac{1}{\eta} = \left(\frac{9\pi^3}{8mkT}\right)^{1/2} d^2 \exp\left(-\frac{Bv}{v_f}\right) \tag{4}$$

This is very similar to the equation of Cohen and Turnbull (1) and also to an empirical equation proposed by Doolittle (2).

The preexponential term in Eq. 4 is nearly constant, and the major effect of temperature and pressure on fluidity is through the exponential term. It is clear that fluidity is essentially a function of specific volume. The effect of temperature and pressure on fluidity is therefore through their effect on specific volume; this was noted as an experimental fact by Batchinski (3) over 50 years ago.

2. TEMPERATURE DEPENDENCE OF LIQUID VISCOSITY

The temperature dependence of the viscosity of liquids is best expressed by relating viscosity at temperature T to the viscosity at the glass-transition temperature T_g. This is especially pertinent to the viscosity of amorphous polymeric liquids. The quantity T_g can be obtained from volume-versus-temperature measurements.

Equation 4 can be expressed as follows:

$$\ln \eta(T) = \ln A + \frac{Bv(T)}{v_f(T)} \tag{5}$$

$$\ln \eta(T_g) = \ln A + \frac{Bv(T)}{v_f(T_g)} \tag{6}$$

Subtracting Eq. 6 from Eq. 5 and setting B equal to unity, we obtain

$$\ln \left[\frac{\eta(T)}{\eta(T_g)}\right] = \frac{1}{f} - \frac{1}{f_g} \tag{7}$$

where f is the fractional free volume at temperature T and f_g is the fractional free volume at T_g.

The dependence of free volume on temperature is taken to be the difference between the thermal expansion coefficients a_r and a_g above and below T_g. Therefore

$$f = f_g + (a_r - a_g)(T - T_g) = f_g + a_2(T - T_g)$$
$$a_2 = a_r - a_g \tag{8}$$

If Eq. 8 is substituted into Eq. 7, we obtain

$$\log \left[\frac{\eta(T)}{\eta(T_g)}\right] = -\frac{1}{2.303 f_g} \frac{T - T_g}{(f_g/a_2) + (T - T_g)} \tag{9}$$

Empirical data on numerous organic and inorganic polymers have established the approximate validity of the equation

$$\log \left[\frac{\eta(T)}{\eta(T_g)}\right] = \frac{-17.44(T - T_g)}{51.6 + (T - T_g)} \tag{10}$$

Equating the constants of Eqs. 10 and 9, we obtain

$$f_g = 0.025; \quad a_2 = a_r - a_g = 4.8 \times 10^{-4} \ \text{deg}^{-1} \tag{11}$$

The development shown in Eqs. 5 through 11 is due to Williams, Landel, and Ferry (4). Equation 10, the Williams–Landel–Ferry equation, is a most valuable working tool, but it must be regarded as a successful semiempirical equation rather than a rigorous result.

Although it was developed from data based on polymers, Eq. 10 probably also holds fairly well for simple liquids, provided that these could be successfully supercooled through the melting temperature T_m all the way to T_g.

For simple liquids the viscosity at T_g is 10^{13} P.

3. VISCOSITY OF POLYMER SOLUTIONS

We consider dilute solutions where the solute is composed of compact spherical particles. Einstein (5) showed that the viscosity η of the solution (suspension) relative to that of the solvent, η_0, is a linear function of the volume fraction ϕ of the solute and is independent of the particle size:

$$\frac{\eta}{\eta_0} = \eta_r = 1 + 2.5\phi \tag{12}$$

$$\eta_r - 1 = \eta_{sp} = 2.5\phi$$

where η_r is the relative viscosity and η_{sp} is the specific viscosity. These definitions of the η_r and η_{sp} are used for all solutions, no matter what the shape of the particles.

Guth and Simha (6) extended Eq. 12 to somewhat higher concentrations:

$$\eta_r = 1 + 2.5\phi + 14.1\phi^2 \tag{13}$$

It has proved very useful to define an intrinsic viscosity for all polymer solutions by means of the following equation:

$$[\eta] = \left(\frac{\eta_{sp}}{c}\right)_{c \to 0} \tag{14}$$

where c is usually expressed as grams of solute per 100 cc of solution. The intrinsic viscosity of an Einstein solution for compact spherical particles is

$$[\eta] = 0.025v \tag{15}$$

where v is the specific volume of the solute.

For nonspherical particles subject to strong Brownian movement Simha (7) obtained the following limiting relations for prolate and oblate ellipsoids in terms of the axial ratio p (defined so that p is larger than unity in both cases):

Prolate ellipsoids; rods:

$$\frac{[\eta]}{0.01v} = \frac{p^2}{15(\ln 2p - \frac{3}{2})} + \frac{p^2}{5(\ln 2p - \frac{1}{2})} + \frac{14}{15} \tag{16}$$

Oblate ellipsoids; disks:

$$\frac{[\eta]}{0.01v} = \frac{16}{15}\frac{p}{\tan^{-1} p} \tag{17}$$

where v is the specific volume of the solute.

For random-coil molecules of high molecular weight one may imagine that the solvent is trapped within the swollen coil; that is, the flow lines of the surrounding solvent do not penetrate the swollen coil. Thus for random coils one can write the intrinsic viscosity as follows:

$$[\eta] = 0.025 \, \frac{\Omega}{\overline{M}} \tag{18}$$

where Ω is the "hydrodynamic volume" of the coil, including the trapped solvent, and $\overline{M}$ is the molecular mass of the polymer molecules.

The intrinsic viscosities of compact spherical particles, such as globular proteins or spherical viruses, are on the order of 0.025 since v is on the order of unity. The intrinsic viscosities of natural rubber, high-molecular-weight polystyrene, and the like are about 100 times greater because of the greater space-filling power of random-coil molecules and also their inclusion of trapped solvent. The intrinsic viscosity of a rodlike molecule, obeying Eq. 16, is even greater than that of a random coil of the same chain length (molecular weight). In fact one of the ways by which it was discovered that synthetic polypeptides exist as rigid rods in dilute solution was by demonstrating the applicability of Eq. 16 to a polypeptide of known molecular weight (8).

4. RELATIONS BETWEEN INTRINSIC VISCOSITY AND MOLECULAR WEIGHT

The first empirical relation between intrinsic viscosities and molecular weights was given by Staudinger (9):

$$[\eta] = KM \tag{19}$$

where K is a constant for a homologous series of polymers of a given molecular structure.

This empirical relationship was modified by Mark and Houwink (10) as follows:

$$[\eta] = KM^a \tag{20}$$

where a is an exponent that varies between 0.5 and 2.0.

This relation when calibrated against absolute molecular weight measurements such as given in Chapter 2 provides the most rapid and convenient way of measuring molecular weights. For polydisperse samples it gives a viscosity average molecular weight as given in Chapter 2, Eq. 5.

The theoretical situation is as follows: for random-coil molecules in very poor solvents a is 0.5; for high-molecular-weight random coils in good solvents a is about 0.8; for low-molecular-weight random coils a is 1.0; for rigid rods a is 2.0 from equation (22) or 1.7 from equation (16).

The theories for low-molecular-weight random coils or for rigid rods use a free-draining model. It is assumed that the lines of flow in the solvent are unperturbed. The shear gradient produces a rotation of the molecules that dissipates energy and causes the enhanced viscosity. Huggins (11) showed that for random coils

$$[\eta] = \frac{\pi}{600} \left(\frac{bR_0^2}{m} \right) \tag{21}$$

where b is the radius of the chain segment, m is its mass, and R_0^2 is the mean-square end-to-end distance of the polymer chain. Equation 21 is valid whether or not the random coil undergoes internal rotational motion. Also for rigid rods Eq. 21 applies if R_0 is replaced by R, the length of the rod. (Note that R_0^2 is equivalent to h^2 of chapter 3.)

$$[\eta] = \frac{\pi}{600} \left(\frac{bR^2}{m} \right) \tag{22}$$

Comparing Eq. 18 with Eqs. 21 and 22, we observe that the former assumes hampered flow of solvent and molecular entrapment whereas the latter equations do not. Hampered flow and entrapment of solvent are important for random coils of high molecular weight. Considering Eq. 18, we observe that for random coils

$$\Omega \sim \tfrac{4}{3}\pi R_0^3 \sim M^{3/2} \tag{23}$$

$$[\eta] \sim M^{1/2}$$

Kirkwood and Riseman (12) devised for random coils a theory that encompasses Eqs. 21 and 18, that is, the free-draining model and the hampered-flow model. Their final result is

$$[\eta] = \frac{\pi}{600} \frac{bR_0^2}{m} F(x)$$

$$x = \left(\frac{6}{\pi} \right)^{1/2} N^{1/2} \tag{24}$$

$$F(x) = \frac{6}{\pi^2} \sum_{k=1}^{\infty} \left[k^2 \left(1 + \frac{x}{\sqrt{k}} \right) \right]^{-1}$$

The values of $F(x)$ have been tabulated by Kirkwood and Riseman (12).

Equation 24 yields Eq. 21 for small values of N. For large values of N the intrinsic viscosity goes to

$$[\eta] = 0.0060 \frac{R_0^3}{\bar{M}} = 3.6 \times 10^{21} \frac{R_0^3}{M} \tag{25}$$

where $\bar{M}$ is mass and M is molecular weight.

It can be postulated that

$$[\eta] = \frac{\Phi R_0{}^3}{M} \tag{26}$$

where Φ can be determined from simultaneous measurements of $R_0{}^2$ (light scattering), $[\eta]$, and M (osmometry or light scattering) for several polymers and solvents. The best experimental value for Φ is 2×10^{21} (13), which is close to the theoretical value of Eq. 25. Conversely Eq. 26 can be used to estimate $R_0{}^2$ for random-coil molecules from $[\eta]$ and M.

In θ-solvents $R_0{}^2$ varies as M, and therefore from Eq. 26 $[\eta]$ varies as $M^{0.5}$. For good solvents $R_0{}^2$ varies as $M^{1+\varepsilon}$, where ε has theoretical values as large as 0.33. This would have $[\eta]$ vary as $M^{0.8}$ in good solvents.

For fairly dilute solutions Huggins (11) proposed that the concentration dependence of specific viscosity is given by

$$\frac{\eta_{sp}}{c} = [\eta] + k'[\eta]^2 c \tag{27}$$

For polymers of quite low molecular weight Lyons and Tobolsky (14) have found that the following formula fits over the entire concentration range, including undiluted polymer:

$$\frac{\eta_{sp}}{c} = [\eta] \exp\left(\frac{k'[\eta]c}{1 - bc}\right) \tag{28}$$

The viscosity of polyelectrolyte solutions (e.g., polysodium acrylate or poly-N-methyl-4-vinylpyridonium bromide) presents an interesting extension of the principles of dilute-solution viscosity. At low concentrations of polymer in water the charges carried on the polymer backbone repel each other and cause the polymer chain to extend. This results in very high values of η_{sp}/c at low values of c. At higher values of c the effect of the counterions is such as to return the polyelectrolyte chain dimensions to normal, and the quantity η_{sp}/c decreases with increasing c. Fuoss and Strauss (15) showed that the following equation fits the data for polyelectrolytes quite well:

$$\frac{\eta_{sp}}{c} = \frac{A}{1 + Bc^{1/2}} + D \tag{29}$$

The quantity $A + D$ is the limit approached at zero concentration and is the analog of the intrinsic viscosity of ordinary polymers. The quantity $Bc^{1/2}$ depends on electrostatic interaction between polyions and counterions and causes η_{sp}/c to decrease with increasing c.

The anomalous effects displayed in Eq. 29 can be eliminated by adding simple electrolytes, such as sodium chloride, to the polyelectrolyte solutions, whereupon the behavior of η_{sp}/c versus c becomes normal, obeying Eq. 27.

5. VISCOSITY OF POLYMER MELTS

It has been found that the viscosity of polymers can be expressed as a product of separable functions of molecular weight and temperature. The temperature dependence of viscosity has been discussed in Section 2. It has been found empirically that for low molecular weights η varies as $M_w^{1.0}$ up to a so-called entanglement value of molecular weight. The entanglement value varies from polymer to polymer but is on the order of 15,000. Above the entanglement value η varies with $M_w^{3.4}$ (16, 17, 18).

An approximate formula for the melt viscosity of high polymers can therefore be expressed as follows:

$$\log_{10} \eta = 3.4 \log M_w - \frac{17.44(T - T_g)}{51.6 + (T - T_g)} + C \tag{30}$$

The constant C varies from one polymer to the other. The first term on the right-hand side can be expressed in terms of R_0^2 rather than M_w:

$$\log_{10} \eta = 3.4 \log R_0^2 - \frac{17.44}{51.6 + (T - T_g)} + D \tag{31}$$

The variation of D from polymer to polymer is smaller than that of C.

6. DIFFUSION IN SIMPLE LIQUIDS

If a soluble substance is placed in a liquid in such a way that it initially occupies only a part of the volume, it will eventually disperse itself uniformly throughout the entire space occupied by the liquid. This phenomenon, which acts independently of gravitation and depends only on random thermal collisions, is called free diffusion. The fundamental definition of the diffusion constant comes from Fick's law:

$$Q = -D \frac{dc}{dx} \tag{32}$$

where Q is the diffusion current (i.e., the net amount of material that passes a plane of unit area in unit time), dc/dx is the concentration gradient, and D is the diffusion constant. Fick's law holds only for dilute solutions. The coefficient of selfdiffusion in a liquid can be measured by using a minute amount of radioactive isotope of the liquid as the dissolved solute.

The diffusion constant is measured in square centimeters per second. In an idealized experimental setup a solution containing the particles whose diffusion is under investigation is placed in the lower part of a tube and separated from the pure solvent (in the upper part of the tube) by a partition. The partition must then be removed rapidly without causing turbulence. The concentration changes at various points along the tube are noted as a function of time, usually by some optical method. From these data the diffusion constant is calculated.

Einstein (19) showed that the diffusion constant for dissolved particles could be calculated by molecular dynamics and the theory of Brownian motion. He obtained the equation

$$D = \frac{kT}{F} \tag{33}$$

where k is Boltzmann's constant, T is the absolute temperature, and F is the frictional force acting on the dissolved particle if it moves in the liquid with unit velocity. For a spherical particle (larger than the solvent molecules) the friction coefficient is given by Stokes' law:

$$F = 6\pi\eta_0 r \tag{34}$$

where η_0 is the viscosity of the solvent and r is the radius of the spherical particle. By combining Eqs. 33 and 34, we obtain for spherical particles

$$D = \frac{kT}{6\pi\eta_0 r} \tag{35}$$

By inserting the absolute values of the Boltzmann constant and assuming $T = 300°K$ and $\eta_0 = 0.01$ P, we obtain

$$Dr \approx 2 \times 10^{-13} \, \text{cm/sec}$$

which enables us to estimate the diffusion constant for spherical solutes of differing radius.

The molecular theory of diffusion relates to the theory of Brownian motion. If particles with the dimension of 1 micron or less are placed in a liquid, an irregular random-flight motion known as Brownian movement occurs. Einstein postulated that the motion of the suspended particles was acquired by the transfer of momentum during collisions with the molecules of the liquid. The average kinetic energy of translation of a suspended particle in the liquid is given by the equipartition principle:

$$E = \frac{3kT}{2} \tag{36}$$

If at time zero the particle starts out in a plane $x = 0$, the probability dW that at time t the particle will be in a plane between x and dx is

$$dW = \frac{1}{\sqrt{\pi Dt}} \exp\left(\frac{-x^2}{4Dt}\right) dx \tag{37}$$

The mean-square displacement $\overline{x^2}$ in the x-direction in a time interval t is given by the equation due to Einstein:

$$\overline{x^2} = 2Dt = \frac{2kT}{F}t \tag{38}$$

For spherical particles

$$\overline{x^2} = \frac{kT}{3\pi\eta_0 r}t \tag{39}$$

Microscopic observations of suspensions of colloidal particles and the employment of Eq. 39 enabled Perrin (20) to determine Boltzmann's constant and hence Avogadro's number.

From measurements of the diffusion constant the friction factor F can be determined by Eq. 33. Different shapes of molecules give different formulas for F. The value of F for a sphere has already been given by Eq. 34. The theoretical value of F for random coils given by Kirkwood and Riseman (12) is

$$F = \frac{Nf}{(1 + Nf/6\pi\eta_0 R'_0)}$$
$$R'_0 = 0.27R_0 \tag{40}$$
$$f = 6\pi b\eta_0$$

where f is the frictional factor of the chain segment, b is the radius of the chain segment, η_0 is the solvent viscosity, N is the number of chain segments, and R_0 is the rms end-to-end distance of the polymer coil. For small values of N, F equals Nf; for large values of N, F equals $6\pi\eta_0 R'_0$. The quantity R'_0 (equal to $0.27R_0$) can be interpreted as the effective frictional radius of the swollen polymer coil.

Perrin and Herzog (21) obtained the following values for F for ellipsoids of revolution in a solvent of viscosity η_0 :

Prolate ellipsoid (rod); semiaxes a, b, b:

$$F = \frac{6\pi\eta_0(ab^2)^{1/3}[1 - (b^2/a^2)]^{1/2}}{(b/a)^{2/3}\ln\left\{\dfrac{1 + [1 - (b^2/a^2)]^{1/2}}{b/a}\right\}} \tag{41}$$

Oblate ellipsoid (disk); semiaxes a, a, b:

$$F = \frac{6\pi\eta_0(a^2b)^{1/3}[(a^2/b^2) - 1]^{1/2}}{(a/b)^{2/3} \tan^{-1}[(a^2/b^2) - 1]^{1/2}} \qquad (42)$$

Values of the frictional constant F obtained from diffusion or sedimentation velocity (next section) are a very important way of determining the particle shape. The molecular weight must be known from other methods, and then the experimental value of F should be interpreted in terms of Eqs. 35, 40, 41, or 42.

7. SEDIMENTATION VELOCITY

In a high-speed centrifuge a dissolved polymer of uniform molecular weight will sediment with a more or less sharp boundary, depending on the amount of thermal diffusion. The rate at which the mean position of the boundary moves depends on the rate of centrifugation and can be determined experimentally. Svedberg (22) defined the "sedimentation constant" s by the equation

$$s = \frac{dx/dt}{\omega^2 x} \qquad (43)$$

where dx/dt is the instantaneous rate of sedimentation at point x and ω is the angular velocity of centrifugation. The quantity s has the dimensions of time. One Svedberg unit is defined as 10^{-13} sec.

The fundamental equation for the calculation of molecular weights from sedimentation-rate data is obtained by setting the centrifugal acceleration times the buoyancy-corrected mass of the particle, $\overline{M}(1 - vd)$, equal to its velocity times the frictional coefficient F.

$$\overline{M}(1 - vd)\omega^2 x = F\frac{dx}{dt}$$

$$\overline{M} = \frac{Fs}{1 - vd} = \frac{skT}{(1 - vd)D} \qquad (44)$$

where $\overline{M}$ is the mass of the particle, v is its specific volume, d is the density of the solution, and D is the diffusion constant of the particle. For monodisperse systems D can be obtained from ultracentrifugal data from the rate of spread of the boundary layer.

If the particle mass $\overline{M}$ is known from osmometry or light scattering, Eq. 44 presents another way of determining F or D.

A more detailed discussion of the hydrodynamics and the thermodynamics of polymer solutions can be found in references 23 and 24.

REFERENCES

1. M. H. Cohen and D. Turnbull, *J. Chem. Phys.* **31**, 1164 (1959).
2. A. K. Doolittle, *J. Appl. Phys.* **23**, 236 (1952).
3. A. J. Batchinski, *Z. Physik Chem.* **84**, 643 (1913).
4. M. L. Williams, R. F. Landel, and J. D. Ferry, *J. Am. Chem. Soc.* **10**, 375 (1955).
5. A. Einstein, *Ann. Physik* **34**, 591 (1911).
6. E. Guth and R. Simha, *Kolloid Z.* **74**, 266 (1936).
7. R. Simha, *J. Phys. Chem.* **44**, 25 (1940).
8. P. Doty and J. T. Yang, *J. Am. Chem. Soc.* **78**, 2650 (1956).
9. H. Staudinger and R. Nodzu, *Ber.* **63**, 721 (1930).
10. H. Mark, "Der feste Körper," Hirzel, Leipzig (1938); R. Houwink, *J. Prakt. Chem.* **157**, 5 (1940).
11. M. L. Huggins, *J. Phys. Chem.* **42**, 911 (1938).
12. J. G. Kirkwood and J. Risemann, *J. Chem. Phys.* **16**, 565 (1948).
13. P. J. Flory *Principles of Polymer Chemistry*, Cornell University Press, 1953, p. 616.
14. P. F. Lyons and A. V. Tobolsky, *Polymer Eng. and Sci.* **10**, 1 (1970).
15. R. M. Fuoss and U. P. Strauss, *J. Polymer Sci.* **3**, 602 (1948).
16. R. D. Andrews and A. V. Tobolsky, in *Physical Chemistry of High Polymeric Systems* by H. Mark and A. V. Tobolsky, Interscience, New York, 1950, p. 343.
17. T. G. Fox and P. J. Flory, *J. Appl. Phys.* **21**, 581 (1950); *J. Phys. Chem.* **55**, 221 (1951).
18. T. G. Fox and V. R. Allen, *J. Chem. Phys.* **41**, 344 (1964).
19. A. Einstein, *Ann. Physik* **77**, 549 (1905).
20. J. Perrin, *Kolloid-Beihefte* **1**, 221 (1910).
21. E. Perrin, *J. Phys. Radium* **7**, 1 (1936); R. O. Herzog, R. Illig, and H. Kudar, *Z. Phys. Chem.* **A167**, 329 (1933).
22. T. Svedberg and K. O. Pederson, *The Ultracentrifuge*, Clarendon Press, London, 1940.
23. C. Tanford, *Physical Chemistry of Macromolecules*, Wiley, New York, 1961 Chapter 6.
24. H. Morawetz, *Macromolecules in Solution*, Interscience, New York, 1965, Chapter 6.

Phase Transitions and Vitrification 6

A. Bondi and A. V. Tobolsky

Visual and tactile impressions convince us that solids, especially molecular solids, differ widely in "solidity." Naively, we shall define as a solid any substance that is able to support a shear stress. Later on we shall have to qualify this simple definition somewhat. The first part of this chapter deals with crystalline solids, defined as substances whose X-ray diffraction patterns reveal the presence of long-range order and which—if they melt below their thermal decomposition temperature—melt at a fixed temperature and pressure, independently of the time scale of observation. The second part of this chapter deals with glassy solids, defined as substances whose X-ray diffraction patterns show no evidence of long-range order (i.e., with order at most to the third row of neighbors) and which become liquids (or rubbers) at temperatures and pressures whose exact magnitudes depend on the time scale of the observation.

The technological importance of temperatures that mark discontinuities in the mechanical properties of materials can hardly be exaggerated, as these temperatures will often determine the utility of a material for a given application. The less obvious solid-to-solid transitions are sometimes more important in this respect than "inspection" properties, such as the melting point T_m or the vitrification (glass-transition) temperature T_0.

In low-molecular-weight substances one can observe four condensed states of matter: crystalline, mesomorphic (liquid crystals), liquid, and glassy. For high polymers the regularity of chain structure determines whether the polymer can form a crystalline phase at low temperatures; isotactic polypropylene is crystalline, whereas atactic polypropylene is amorphous at room temperature. The glassy state is somewhat of a rarity among pure low-molecular-weight organic compounds and often can be achieved only through special subcooling. The glassy state is very common for organic polymers. Finally, polymers rarely achieve true thermodynamic equilibrium between crystalline and amorphous regions.

1. THERMAL MOTIONS AND PHASE TRANSITIONS IN MOLECULAR SOLIDS

The thermal oscillations of molecules on the lattice of molecular crystals acquire a significant amplitude and anharmonicity at far lower temperature than do the more familiar ionic and metallic solids because of the comparatively weak cohesive forces within molecular solids. Hence thermal expansion and density fluctuations provide opportunities not only for large-amplitude torsional oscillations of comparatively large molecules but even for their rotational jumps into new equilibrium positions.

Highly symmetrical molecules (such as benzene) or rotatable molecule parts (such as methyl groups) execute such rotational jumps at very low temperatures (Table 1) with sufficient frequency and in sufficient numbers to be easily detected by narrowing of the nuclear-magnetic-resonance (NMR) signals (1, 2) and by selective energy absorption in oscillatory deformation measurements (3). Yet neither the frequency nor the number of jumps in a unit volume is high enough to affect the thermodynamic properties of the crystal (1). However, mechanical transport properties, such as creep, may reflect any concentration of mobile elements that is large enough to cause readily observable mechanical energy absorption peaks. Systematic research on this has yet to be conducted.

Asymmetrical or only slightly symmetrical molecules must overcome higher barriers to external rotation and require greater lattice expansion and thus higher temperatures to jump into new rotational equilibrium positions. The concentration of jumping molecules is then so large that the thermodynamic properties of the solid—namely, volume and energy (heat capacity)—rise discontinuously at the temperature at which such jumps become just possible. The molecular and crystal structure conditions leading to the two basically different types of motion have been analyzed by Darmon and Brot (1), who proposed the criteria for their incidence enumerated in Table 1. Needless to say, the mechanical properties, especially creep, change drastically

Table 1   Relation of Molecular Motions in Crystals to
Thermodynamic Phase Transitions

Substance[b]	T_m (°K)	T_t (°K)	T_t/T_m	Molecular Motions[a]		
				From (°K)	To (°K)	T_n/T_m
Group 1:						
Benzene	278	—	—	100	130	0.41
Adamantane	541	209	0.39	140	170	0.29
Quinuclidine	431	198	0.46	140		0.33
Triethylenediamine	433	351	0.81	190		0.44
Cyclohexane	280	186	0.66	155	180	0.60
Group 2:						
Cyclopentene	138	87	0.63	87	110	0.63
Furan	188	150	0.80	150		0.80
Cyclohexene	170	139	0.82	139		0.82

[a] Expressed as NMR line-width narrowing, beginning at one temperature (T_n) and ending at the other; the absolute level of these temperatures depends on the frequency employed. Most data are from Darmon and Brot (1).
[b] Division into symmetry groups proposed by Darmon and Brot (1). Group 1: the molecule possesses the symmetry of the allowed orientations; group 2: the molecule possesses a lower symmetry than the allowed orientations.

at most such first-order transition points, the best known of which is, of course, the melting point.

First-order transitions exhibit a discontinuity in volume and enthalpy at T_t. Second-order transitions are defined as those in which the coefficients of expansion and specific heat exhibit a discontinuity at T_t. In addition, one observes heat capacity (or thermal expansion) versus temperature curves of a few "basic" shapes, most of which have yet to be described by clear-cut theoretical treatments. The transition temperature of all of these true thermodynamic phase transitions is independent of heating, cooling, or deformation rate, provided these rates are lower than the rate of nucleation and of growth of the new phase. The incidence temperature for the nonthermodynamic motions mentioned earlier and for the slow motions characteristic of vitrification, on the other hand, depends strongly on the frequency of deformation. Hence these motions are called rotational processes rather than transitions, the word "transition" being reserved for true thermodynamic phase changes. The common name "glass transition" for vitrification is thus misleading but is probably too deeply embedded in technical jargon for effective replacement. Glass transition should be interpreted as a narrow temperature region whose location depends on the cooling rate and on the time scale of the property measurement under consideration and within which most physical properties of the vitrifying system undergo a drastic change.

2. VOLUME EXPANSION AT TRANSITION POINTS

Typical examples of the magnitude of volume expansion at phase transitions are shown in Table 2. Although there is as yet no basis for precise predictions of this expansion, qualitative generalizations can be made and rationalized (4):

1. The sum of the expansion discontinuities up to and including the melting dilation increases with the relative melting point T_m/T_b, where T_b is the atmospheric boiling point. Very polar substances with an inordinately expanded solid lattice, such as ice, are the only exceptions to this generalization.

2. Solids composed of nearly spherically symmetrical rigid molecules often expand strongly at a single solid-to-solid transition point and comparatively little at T_m.

3. Solids composed of nonpolar or slightly polar elongated flexible-chain molecules expand about as much at a single solid-to-solid phase transition as at the melting point, which is generally only a few degrees higher.

4. Solids composed of very polar symmetrical, more or less flexible, elongated molecules may pass through a whole series of solid-to-solid phase transitions before finally reaching the melting point.

In all of the enumerated categories large expansions at the solid-to-solid phase transition (T_t) are tantamount to marked softening at $T > T_t$, so that solids at $T_t < T < T_m$ are generally called plastic crystals (5). The molecules of class 4 crystals are held in a state of solidlike order in two dimensions by strong dipole sheets, at $T_t^{(2)} < T_t^{(1)} < T < T_m$. Hence when T is greater than the first or the second solid-to-"solid" transition point, the material is mechanically nearly liquidlike and is called liquid crystal.

The yield stress of plastic crystals at $T_t < T < T_m$ is between one-fourteenth and one-half their yield stress at $T < T_t$. Their yield stress at $T < T_t$ is of the same magnitude as the yield stress of ordinary "hard" crystals near $T \lesssim T_m$ (5). Typical examples for plastic crystals are carbon tetrabromide at 226 to 250°K or d-camphor at 245 to 453°K (5). Liquid crystals generally exhibit zero or negligible yield stress at $T_t^{(1)} < T < T_m$, and their viscosity in that temperature range is nearly as low as that of ordinary liquids composed of similar compounds (5). The liquid crystals of metallic soaps are often composed of association polymers that are extremely viscous.

The molecular origin of plastic crystals is the very high degree of symmetry (often approaching spherical symmetry) of their molecules. Although their low-temperature crystal form can be monoclinic, rhombic, etc., their high-temperature modification has a very symmetrical and open cubic or hexagonal

Table 2   Relation of Generalized Volume Change at Melting and Transition Points to Molecular Structure

Substance	$\dfrac{\Delta V_t}{V_W}$	$\dfrac{\Delta V_m}{V_W}$	$\dfrac{\Delta V_t + \Delta V_m}{V_W}$	$\dfrac{T_m}{T_b}$
Simple substances:				
Ar		0.203	0.203	0.96
N_2				0.82
O_2	0.083	0.071	0.154	0.61
F_2	0.079	0.066	0.145	0.58
CCl_4	0.194	0.079	0.273	0.71
Alkanes:				
Ethane		0.187	0.187	0.41
n-Hexane		0.175	0.175	0.52
n-Octane		0.217	0.217	0.54
n-Dodecane		0.239	0.239	0.54
n-Octadecane		0.246	0.246	0.51
Hexamethylethane	0.023	0.26	0.28	0.99
Aromatics:				
1,2-Dimethylbenzene				0.59
1,4-Dimethylbenzene		0.314	0.314	0.70
1,2-Terphenyl		0.141	0.141	0.55
1,4-Terphenyl		0.280	0.280	0.76
Naphthalene		0.255	0.255	0.72
Polar substances:				
Water		−0.167	−0.167	0.73
Methanol		0.078	0.078	0.52
Phenol		0.089	0.089	0.69
Trimethylacetic acid	0.170	0.113	0.283	0.71
Succinonitrile	0.128	0.045	0.173	0.61

Nomenclature:

ΔV_t = volume change at first-order solid-to-solid transition.

ΔV_m = volume change at melting point.

V_W = van der Waals volume, the volume occupied by the molecules as calculated from bond distances and from nonbonded distances observed by X-ray (or neutron) diffraction crystallography (given in Chapter 14 of reference 4).

T_m = melting point (degrees Kelvin).

T_b = atmospheric boiling point (degrees Kelvin).

crystal structure. The great molecular mobility on this open lattice accounts for the softness of the plastic crystal. Liquid crystals, by contrast, are composed of far less symmetrical, polar, elongated molecules, which are arranged (at $T_t < T < T_m$) in orderly array only in two dimensions, normal to the long axes of the molecules, and are completely or nearly completely disordered in the remaining space coordinate. This disorder accounts for the absence of a finite yield stress and for the low viscosity of the majority of liquid crystals. Liquid crystals are more fully discussed in Chapter 7.

3. ENTROPY AND ENTHALPY OF FUSION

The absolute magnitude of the melting temperature is given by the relation $T_m = \Delta H_m / \Delta S_m$, where ΔH_m and ΔS_m are the enthalpy and entropy of fusion, respectively. The enthalpy change ΔH_m is essentially the change in potential energy of the system due to the expansion ΔV_m and is just as difficult to relate to first principles. By contrast, the entropy change ΔS_m is readily related to a simple model of the liquid state and can be conveniently related to molecular structure. Since the melting point of a compound is generally known or easily determined, the correlated, estimated entropy change is easily converted to the enthalpy change by $\Delta H_m = T_m \Delta S_m$. The reverse program, an a priori guess of T_m from molecular structure, is basically hopeless, although trends can be discerned in each given class of compounds.

First-order solid-to-solid phase transitions that are evident from the volume changes mentioned earlier are accompanied by enthalpy and entropy changes, whose magnitude generally parallels that of the volume change. Since the sum of all these entropy changes characterizes the transition from the "hard" solid to the liquid state, it is not surprising to find that only this sum of entropy changes can be correlated with molecular structure (4). The incidence and magnitude of the solid-to-solid phase transitions cannot at present be predicted from molecular structure information. Hence we first examine the correlations of the sum of entropies.

a. Monomeric Rigid and Flexible Molecules

The entropy of fusion (plus transition) of rigid molecules is illustrated in Table 3. For more complex rigid molecules this quantity is dominated by their moments of inertia and their symmetry, as is well illustrated in reference 4, which also considers the effect of rotating side groups.

Molecules composed of several mobile segments gain extra degrees of freedom on transition from a crystalline to a liquid environment because the segments remain in fixed positions relative to each other when located in

Table 3   Total Entropy of Fusion and Transition of Various Simple Compounds[a]

Compound	$\Sigma \Delta S_{m,t}/R$	Compound	$\Sigma \Delta S_{m,t}/R$
Diatomic molecules:		Nonlinear polyatomic	
H_2	1.04	molecules—	
N_2	2.12	*continued*:	
NO	2.53	SO_2	4.51
CO	2.72	NH_3	3.48
O_2	3.47	PH_3	3.61
F_2	3.41	CH_4	1.61
Cl_2	4.47	SiH_4	2.13
Br_2	4.86	GeH_4	1.99
I_2	4.92		
Diatomic molecules		Rigid single rings:	
(associating):		Furan	4.08
HF	2.90	Thiophene	5.01
HCl	2.97	Benzene	4.25
HBr	2.62	Pyridine	4.25
HI	2.56	β-Sulfolene	4.86
		Sulfolane	4.55
Linear polyatomic			
molecules:		Rigid condensed	
FCN	4.00	double rings:	
HCN	3.89	Indene	4.52
$(CN)_2$	3.98	Indane	4.67
N_2O	4.32	Naphthalene	6.48
CO_2	4.66	1,4-Dihydronaphthalene	4.96
COS	4.24	Tetralin	6.33
CS_2	3.28	Azulene	6.18
Nonlinear polyatomic		Rigid condensed rings:	
molecules:		Acenaphthene	6.85
H_2O	2.65	Phenanthrene	6.02
H_2S	3.30	Anthracene	7.09
H_2Se	4.55	Chrysene	7.50

[a] Data from Bondi (4).

a crystal lattice but rotate (into new equilibrium positions) in a liquid environment. Hence the entropy of fusion (plus transition) of substances composed of flexible molecules rises linearly with the number of chain segments per molecule, as shown by the homologous-series data in Table 4. The differences in the entropy contribution per methylene group illustrate the effect of crystal structure on the entropy change on fusion.

Table 4   Relations between the Total Entropy of Fusion of Long-Chain Compounds and N_c, the number of Carbon Atoms per Molecule[a] (4)

Homologous Series[b]	$\Sigma \, \Delta S_{m,\,t}/R$
n-Paraffins (N_c = even)	$0.80 + 1.33N_c$
n-Paraffins (N_c = odd)	$1.10 + 1.18N_c$
2-Methyl-n-alkanes	$-1.24 + 1.2N_c + 20/N_c^2$
2,2-Dimethyl-n-alkanes	$-6.26 + 1.33N_c + 83/N_c^2$
n-Alkylcyclopentane	$-5.1 + 1.30N_c + 131/N_c^2$
n-Alkylcyclohexane	$-6.3 + 1.45N_c + 56/N_c^2$
n-Alkylbenzene[c]	$-5.6 + 1.18N_c + 100/N_c^2$
n-Alkane thiols (N_c = even)	$3.3 + 1.2N_c$
n-Alkane thiols (N_c = odd)	$3.90 + 1.33N_c$
n-Alkyl bromide (N_c = odd)	$2.4 + 1.38N_c + 0.54/N_c^2$
n-Alkanoic acid (N_c = even)	$-2.56 + 1.33N_c + 36/N_c^2$
n-Alkanoic acid (N_c = odd, >5)	$-2.7 + 1.25N_c$
Na-n-Alkanoates (N_c = even)	$-6.6 + 1.35N_c$

[a] Generally *not valid* for the *first two* members of the series.
[b] The available data for 1-alkenes and 1-alkanols are too irregular (and probably unreliable) for representation.
[c] The datum for ethyl benzene is appreciably higher than predicted by this value.

The effect of permanent electric dipoles (functional groups) on the entropy of fusion (plus transition) has been studied most extensively on aliphatic compounds. The effect of such groups is recognized particularly well as the difference $\delta = \Delta S_{m,\,t}$ (polar compounds) $- \Delta S_{m,\,t}$ (homomorph), where the homomorph is a hydrocarbon of the same shape as the polar molecule (e.g., ethane for methanol, toluene for thiophenol). Many of these group increments δ have been collected in reference 4. The magnitude of δ is the difference between the reduction in ΔS_m, when external rotation in the liquid state is hindered by dipole–dipole interaction, and that when internal rotation is increased by virtue of the lowered barrier to hindered internal rotation.

b. High Polymers

The experimental determination of the enthalpy or the entropy of fusion of high polymers is more difficult than that of solids composed of simple molecules. Not only is the degree of crystallinity of the original sample generally just a crudely determined number, but it often changes as a result of and during the calorimetric or solution-melting-point measurement. One of the basic reasons for this complication is of course the difficulty of achieving "true" crystallization equilibrium (see Chapter 8).

Considering these inherent difficulties, one might recoil from quoting any experimental data. In several instances there is, however, sufficient agreement with the data for low-molecular-weight compounds to encourage presentation of the available data in Table 5. All data are, of course, per repeating unit. The first datum in Table 5, $\Delta S_m (CH_2)$, is virtually identical with the value for odd-numbered n-paraffins in Table 4. The item $\Delta S (X) \equiv [\Delta S_m - n_m \Delta S_m (CH_2)]$, where n_m is the number of methylene groups per repeating unit. In view of the large experimental uncertainties in ΔS_m, not too much should be made of small differences in these data. Practically all functional groups shown reduce the entropy contribution of conformational changes—accompanying the phase change—below that of the methylene group (per unit volume).

In the case of hydrocarbons and slightly polar groups this drop in ΔS_m is due to rigidity of the backbone structure, whereas with strongly polar groups this reduction is due to strong association in the melt phase. Conversion of

Table 5   Entropy of Fusion per Repeating Unit for Polymers (4)

Polymer	$\Delta S_m/R$	Functional Group (X)	$\Delta S(X)/R$
Polyethylene	1.18	—	—
Polypropylene (isotactic)	2.1	$>C-CH_3$	0.9
Polybutene-1 (isotactic)	2.1	$>C-C_2H_5$	0.9
Polystyrene (isotatic)	2.1	$>C-C_6H_5$	0.9
Polyvinyl fluoride	1.6–1.9	$>C-F$	0.4–0.7
Polyvinyl chloride:			
Isotactic	0.86	$>C-Cl$	−0.3
Syndiotactic	2.49	$>C-Cl$	+1.3
Polyacrylonitrile	1.07	$>C-C=N$	−0.1
Polyvinyl alcohol	1.65	$>C-OH$	0.47

Table 5—*Continued*

Polymer	$\Delta S_m/R$	Functional Group (X)	$\Delta S(X)/R$
Polybutadiene; 1,4-*cis*	3.87	$\overset{\displaystyle H}{\underset{}{}}\ \overset{\displaystyle H}{\underset{}{}}$ C=C	1.5
Polyisoprene: 1,4-*cis*	1.74	H CH$_3$ C=C	−0.6
1,4-*trans*	4.40	H$_3$C C=C	2.0
Polychloroprene; 1,4-*trans*	2.86	Cl C=C	0.5
Polyoxymethylene	1.8	O	0.6
Polyethylene oxide	2.9	O	0.5
Alkanedioic glycol esters Terephthalic glycol esters	— —	$-O-\overset{\displaystyle O}{\overset{\|}{C}}-$ $-O-\overset{O}{\overset{\|}{C}}-C_6H_4-\overset{O}{\overset{\|}{C}}-O$	-1.7 ± 0.7 $+2.6 \pm 0.3$
Poly-*p*-xylene	5.60	$-C_6H_4-$	$+3.2$
Nylon 6	5.60	$-\overset{H}{\overset{\|}{N}}-\overset{O}{\overset{\|}{C}}-$	−0.25
Nylon 66	10.3	$-\overset{H}{\overset{\|}{N}}-\overset{O}{\overset{\|}{C}}-$	−0.75
Nylon 610	13.6	$-\overset{H}{\overset{\|}{N}}-\overset{O}{\overset{\|}{C}}-$	−1.4
Sebacoyl piperazine	6.9	$-\overset{}{\underset{O}{\overset{\|}{C}}}-N\overset{C-C}{\underset{C-C}{\diagup\diagdown}}N-\overset{}{\underset{O}{\overset{\|}{C}}}-$	−2.5
Polytetrafluoroethylene	0.6	CF$_2$	(0.6)
Polymonochlorotrifluoroethylene	1.25	$\overset{Cl}{\underset{F}{C}}$	0.65[a]

[a] Relative to CF$_2$.

Table 6   Values of δ for Various Functional Groups

Functional Group	δ (Units of R)
$-O-$	-0.6
$-F$	-0.2 to -0.5
$-Cl$:	
Isotactic	-1.2
Syndiotactic	$+0.4$
$\begin{matrix} O \\ \parallel \\ -O-C- \end{matrix}$	-3.8
$\begin{matrix} H \quad O \\ \mid \quad \parallel \\ -N-C- \end{matrix}$	-2.4 to -3.5

ΔS_m (X) to δ by comparison with the appropriate (hydrocarbon) homomorph repeating unit yields the series of increments shown in Table 6.

Many of the increments shown in Table 6 are reassuringly similar to those obtained from data on low-molecular-weight compounds. The difference between isotactic and syndiotactic polyvinyl chloride is surprisingly similar to that between the corresponding classes of n-alkyl halides (4). However, the ester-group decrement could not have been guessed from data on low-molecular-weight esters, whose decrements rarely exceed $1.5R$ and are usually close to $0.5R$. The same applies to the amide group. These groups appear to interfere more strongly with the random conformation of the polymer coil than with the external rotation of monomeric molecules in their melts.

4. ENTROPY AND ENTHALPY CHANGES AT SOLID-TO-SOLID TRANSITIONS

The incidence and location of first-order solid-to-solid phase transitions can often be guessed by inspection of the magnitude of T_m and ΔS_m. If the melting point of a compound is significantly higher than it "should be," and ΔS_m is substantially smaller than it "should be," in comparison with other compounds of similar molecular structure, one has reason to suspect the incidence of a solid-to-solid phase transition (far) below the melting point. Comparison of the observed ΔS_m with the estimated $\Sigma \Delta S_{m,t}$ shows quite well whether ΔS_m is too small. Moreover, in the case of globular molecules ΔS_m is generally about $1.2 \pm 0.4R$, so that one can guess the entropy of transition (ΔS_t) by difference. The location of T_t is usually near the melting temperature of the "normal" crystals of compounds with similar—isomeric—molecular structures.

The situation is far more unpredictable when one deals with polar elongated molecules that form "liquid crystals," often with many solid-to-"solid" transitions. Many of these compounds form association polymers of rather high molecular weight. The soaps are the best known examples.

Few first-order solid-to-solid transitions have been established for polymer crystals, possibly because of the sluggishness of equilibration, were it to occur. Since the different solid phases of a polymorphic crystalline polymer can differ substantially in physical properties, especially specific volume, such slow approach to the phase stable at use temperature can cause serious problems in practical applications as well as in research and in quality control. The best known solid-to-solid phase transitions of crystalline polymers to date are polytetrafluoroethylene, 19, 30°C; *trans*-1,4-polybutadiene, 44, 76°C; isotactic polybutene-1, 100°C; and isotactic polypropylene, somewhere between 30 and 70°C, probably 60°C.

5. THERMODYNAMIC RELATIONS FOR TRANSITION POINTS

The rise of the melting or of the transition temperature with application of hydrostatic pressure P depends on the ratio of the respective changes in entropy and volume:

$$\frac{dT_m}{dP} = \frac{\Delta V_m}{\Delta S_m} \quad \text{and} \quad \frac{dT_t}{dP} = \frac{\Delta V_t}{\Delta S_t}$$

Although exactly valid only at $P \to 0$, these relations are adequate up to pressures of about 500 atm. Somewhat more complicated expressions are required at higher pressures.

Simple thermodynamic relations also apply for mixtures. When the components are completely miscible in the liquid state and completely immiscible in the solid state, one obtains the well-known phase diagram with a eutectic point.

Mixed crystals with T_m between $T_m^{(1)}$ and $T_m^{(2)}$ are formed when both components are of very similar molecular geometry and cohesive energy density. The location of solid-to-solid phase-transition temperatures in mixed molecular crystals, at least one component of which exhibits such transition, generally follows a eutectic-shaped curve with a smoothed minimum.

Molecular compound formation or the fit of molecules of the two mixture components into a new compromise lattice leads to a maximum in the curve for T_m versus mole fraction.

6. VITRIFICATION (GLASS TRANSITION)

Most liquids can be made to solidify in the glassy state if they are cooled through the crystallization-temperature range more rapidly than the time required for crystal nuclei to form. This is easy to do if the symmetry of the molecules is of a low order or if the rotational isomerization from the liquid state to that required for crystallization is quite slow at $T \leq T_m$ (6). In those cases the liquid-phase rotamer distribution persists in the glassy state. The most common cause of easy vitrification, however, is a high melt viscosity at or below the melting point.

In the glass-transition-temperature (T_g) range (it is not a precisely defined temperature point) the viscosity of the melt increases by several orders of magnitude within 10°K. The viscosity becomes so high that even the volume change with temperature experiences a significant delay (7, 8). Hence a glass can be considered as fully described only if the cooling rate that prevailed during its preparation is specified. A rapidly cooled liquid becomes glassy at a higher temperature and is likely to exhibit a lower density than one that has been cooled slowly from the melt state (Fig. 1). As the density determines nearly all other physical properties of a glass, the cooling rate can have a profound effect on all of the physical properties of a glass. The finite, and rather low, thermal conductivity of molecular glasses also means that the cooling rate, and therefore the physical properties of a glass, may depend strongly on the geometry that prevails during the cooling process. Worse yet, the properties may differ substantially from the ouside to the inside of any massive piece of glass cooled faster than 10^{-5}°C/sec.

An example of the effect of the cooling rate on physical properties was shown for mechanical properties (10), where we note the far greater creep rate of rapidly chilled as compared with slowly cooled polymethyl methacrylate. The greater weakness of chilled glass is in keeping with its lower density. We should be able to characterize the comparative thermal history of given glasses by their density. The small maximum density difference caused by chilling ($<1\%$) and the difficulty of measuring the density of solids very accurately militate against the fulfillment of this need. Measurement of refractive index and its conversion to density through use of molar refractivity may be a solution to this problem.

In all elastic-modulus and relaxation measurements we must therefore specify the rate and amplitude of deformation as well as the thermal history of the sample and the instantaneous temperature T and thus call for definition of T_g as a discontinuity in a multidimensional coordinate system.

Calculation of the temperature at which $\eta = 10^{13}$ P (by extrapolation of the viscosity versus temperature curve) yields the conventional T_g (11) very often, but not always. For such systems T_g in conventional time scale is obviously an isoviscous state.

The extensive work by Litovitz (12) has demonstrated the observability of the glass transition at temperatures far above the "static" values of T_g if the time scale of the deformation is reduced to 10^{-3} sec or less. In the present context, however, we consider only $t \gtrsim 10^2$ sec. An example of the effect of rate of measurement on T_g is given in Table 7.

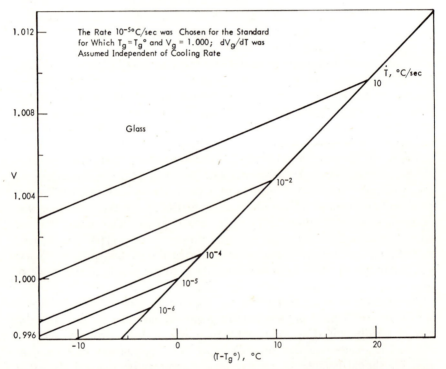

Fig. 1   Effect of cooling rate T on the observed glass-transition temperature and specific volume according to the theory of Saito et al (9). The rate 10^{-5}°C/sec was chosen for the standard, for which $T_g = T_g°$ and $V_g = 1.000$; dV_g/dT was assumed independent of cooling rate.

7. CHANGES IN *PVT* PROPERTIES AT T_g

When one expresses the density of solids and liquids as packing density $\rho^* = \rho V_W/M$—that is, as the ratio of the volume occupied by the molecules (V_W) to the molal volume—one obtains a rather better feel for the meaning of density than from the directly measured numerical value of the density. One finds that the packing density of molecular crystals covers the range expected from crystallography (Table 8) and that the packing density of

Table 7 Time Required for the Density of Polystyrene To Come to within $1/e$ of its Equilibrium Contraction on Quenching to Various Temperatures (4)

$T - T_g$ (°C)	$t(1/e)$ (sec)	$t(1/e)$ (years)
11	0.01	—
6	1	—
2	40	—
1	120	—
0	300	—
−1	1.1×10^3	—
−2.5	3.6×10^3	—
−4	1.8×10^4	—
−7	1.8×10^5	—
−10	5.2×10^6	0.16
−12	3.2×10^7	1
−50	—	10

Table 8 Packing Density of Molecular Crystals at 0°K (4)

Substance	Crystal Type	ρ_0^*	Ref.
F_2	—	0.56	13
Cl_2	Orthorhombic (layer structure)	0.74	14
Br_2	Orthorhombic (layer structure)	0.75	14
I_2	Orthorhombic (layer structure)	0.76	14
CF_4	—	0.69	13
CCl_4	Monoclinic	0.69	15
CBr_4	Monoclinic	0.69	15
CI_4	Monoclinic	0.68	15
O_2	Rhombic, body centered	0.60	13
CO	Cubic (like N_2)	0.60	13
N_2	Face-centered cubic	0.59	16
NO	Cubic	0.73	13
N_2O	Face-centered cubic (like CO_2)	0.70	13
CO_2	Face-centered cubic	0.76	13
CS_2	—	0.69	13
S_8	Rhombic	0.766	13
P_4	Rhombic/monoclinic	0.676	13
CH_4	$N_c > 5$ triclinic	0.67; C_3H_8: 0.695; n-C_5H_{12}: 0.711; n-C_9H_{20}: 0.727	13
C_2H_6	$N_c > 4$ orthorhombic	0.68_4; n-C_4H_{10}: 0.725; n-C_6H_{10}: 0.722; n-C_8H_{10}: 0.735	13
C_2H_4	—	0.670	13
Benzene	Orthorhombic	0.697; φMe: 0.675; 1,3,5-$Me_a\varphi$: 0.736	13
Naphthalene	Monoclinic	0.721	13
Anthracene	Monoclinic	0.733	13
MeOH		0.677; EtOH: 0.698; n-BuOH: 0.720; n-$C_{12}OH$: 0.736	13
Phenol	Monoclinic	0.690; phenethanol: 0.716	17
Cellulose		0.720; $(CH_2)_\infty$: 0.762	13

molecular liquids is essentially a function of a suitably generalized temperature scale, which to a crude approximation amounts to the ratio T/T_b, where T_b is the atmospheric boiling point of the liquid.

It would be very gratifying if the packing density ρ_g^* at T_g had a universal value. Reality does not oblige, however, and Table 9 shows that ρ_g^* covers a range, albeit not a very large one, yet narrower than that of crystalline solids.

Table 9   Comparison of Glass Density (at T_g) with Crystal Density at $0°K$ for Simple Liquids and Polymers (4)

Substance	ρ_g^*	V_g/V_0
Butene-1	0.673	1.07
Isopentane	0.678	1.035
2-Methylpentane	0.688	1.025
3-Methylpentane	0.678	1.04
3-Methylhexane	0.678	1.05
4-Methylnonane	0.670	1.08
Ethyl alcohol	0.648	1.072
1,3,5-Trinaphthylbenzene	0.640	1.14
Polystyrene	0.625	1.142
Polyisobutylene	0.680	1.04[a]

[a] Data from Robbins et al. (18).

Figure 2, a typical graph for packing density versus temperature for a substance in all three condensed states—liquid, glass, and crystal—demonstrates the proximity of the glassy state to the crystalline one. A more accurate estimate of the effect of the time scale (t) of the experiment on the magnitude of T_g can be obtained by using the Williams–Landel–Ferry equation in the form

$$T_g(t) - T_g(t_0) = \frac{2.96}{(1/17.44) + [1/\log(t_0/t)]}$$

where t_0 is the time scale of the "standard" glass-transition experiment, say 10^2 sec.

The thermal expansion passes through a step change at T_g, and, as one might expect from Fig. 2, is of the same magnitude in the glassy state as in the crystalline one. Typical examples are shown in Table 10. Since T_g is very often determined dilatometrically, many thermal expansion data for glassy substances have been published. Should a needed datum not be available,

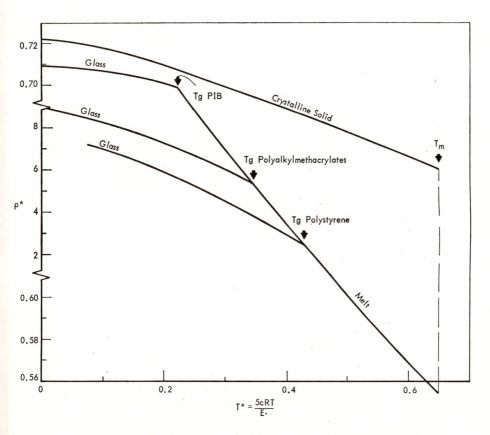

Fig. 2 Schematic representation of diagrams of packing density versus generalized temperature for crystalline and for vitrifying systems.

Table 10 Comparison of Thermal Expansion Coefficients for Crystalline and Glassy Solids (4)

Substance	$10^4 \, \alpha_c$	$10^4 \, \alpha_g$	Ref.
1,3,5-Trinaphthylbenzene	1.25	1.20	19
Glycerol	1.86	2.4	7, 20
Glucose	0.84	0.90	7, 21
Polystyrene	~2.8	2.3	7, 22
Polyethylene terephthalate	1.71	1.87	23
Polyethylene 1,5-naphthalate	1.85	2.14	23

111

however, one could use the estimating methods developed for the corresponding molecular crystal (4).

The compressibility of molecular glasses is of the same order of magnitude as that of molecular crystals of the same or related molecular structure. Although there are no systematic data, it is safe to assume that the compressibility of a glass sample at $T < T_g$ is affected by its thermal history, paralleling its effect on the specific volume. On the other hand, the dependence of compressibility on deformation rate disappears at $(T_g - T) > 20°C$.

The heat capacity C_p (glass) of low-molecular-weight glasses is slightly larger (by about $R/2$) than that of the corresponding crystal. With crystallizable high polymers one finds the glass and crystal heat-capacity data to be in general virtually indistinguishable at $(T_g - T) > 10°C$. Hence the overall change in heat capacity on changing from the liquid to the glassy state can be estimated by the methods developed for calculating the change from the liquid to the crystalline state (4).

Time effects are particularly striking in heat-capacity measurements on glasses. The determining factor is the cooling rate through the glass-transition range and the residence time in the temperature range just below T_g.

8. EFFECTS OF MOLECULAR STRUCTURE ON T_g

There is no lack of "theories" of vitrification. So far none of the proposed conceptual models has accounted successfully for the observed phenomena. A theory or a mathematical model must be consistent with the conclusion of thermodynamic analysis, that the glassy state be characterized by two independent parameters, besides being able to provide quantitative representation of the equilibrium properties and of the time-dependent properties of the system. The most successful qualitative description ascribes the rapid increase of viscosity η as the liquid is cooled toward T_g to the loss of free volume (v_F) such that at some temperature T_0 ($< T_g$), $v_F \to 0$ and $\eta \to \infty$.

The progress of phenomenological treatments has been hampered by the difficulty of finding a conveniently measured, reliable, second, or ordering, parameter that characterizes the "degree of vitrification." Experimental difficulties prevent the adoption of the most logical measure, precision density data. As already mentioned, they could conceivably be replaced by the more easily achieved precision refractive-index measurement.

The common, comparatively crude, formulations of free-volume theory are invalidated by the inconvenient fact that the viscosity of the glass does not approach infinity but rises very smoothly, even linearly, as $\log \eta \sim T^{-1}$ through T_g to temperatures much below T_g. Hence there is no T_0 at which v_F or some important component of v_F disappears.

In the face of this dilemma we have recourse to the familiar powerful tool of the organic chemist—to synthesize a vast number of compounds with carefully planned molecular structures, measure the desired property, and rationalize the empirical dependence of the data on molecular structure in terms of whatever rudimentary theory is available.

Such data have been assembled in Tables 11, 12, and 13. The comparative constancy of the dimensionless ratios T_g/T_b, $RT_b/\Delta H_{vb}$, or $5cRT_g/E^\circ$ (all defined in the table) for liquids composed of rigid molecules shows inter-molecular forces to be the primary factor determining the magnitude of T_g. The magnitude of T_m has then, of course, very little to do with T_g, as is shown by the comparatively wide range of T_g/T_m.

Table 11   Reduced Glass-Transition Temperatures of Monomeric Substances—Aliphatics, Cycloaliphatics, Aromatics, and Esters[a]

Substance	$\dfrac{{}^b 10 RT_g}{\Delta H_{vb}}$	${}^c T_g^*$	θ_L (°K)[c]	$\dfrac{T_g}{T_m}$	$\dfrac{T_g}{T_b}$	T_g (°K)
Aliphatic Substances						
1-Butene	0.229	0.199		0.69	0.23	
trans-2-Hexene	0.244			0.61	0.25	85
1-Heptene	0.244			0.59	0.25	91
				$(0.67)^d$		
cis-2-Heptene	0.250				0.25	
cis-2-Octene	0.248			0.60		101
2-Methylpentane	0.24	0.22		0.67	0.26	
3-Methylpentane	0.21	0.18		—	0.21	
3-Methylhexane	0.233	0.214		—	0.24	
2,3-Dimethylpentane	0.228	0.200		—	0.23	
3-Methylheptane	0.245	0.232	431	0.65	0.25	
4-Methylnonane	0.264	0.254		—	0.28	
$(Et)_2C_2H_4(n\text{-}C_{10}H_{21})_2$		0.299		—	—	
Cycloaliphatic Substances						
Methylcyclohexane	0.228	0.194	437	0.58	0.23	85
Ethylcyclohexane	0.238	0.208	470	0.61	0.24	98
i-Propylcyclohexane	0.25	0.212	508	0.59	0.24	108
n-Butylcyclohexane	0.248	0.236	504	0.60	0.26	119
n-Pentylcyclohexane	0.254	0.238	524	0.58	0.26	125
n-Hexylcyclohexane	0.258	0.249		0.58	0.27	133
9-Cyclohexyleicosane	—	0.292	607	0.67	—	177
⬡—C—C—C(n-C$_8$)$_2$	—	0.272	613	0.70	—	167
(⬡—C—C)$_2$·C·(n-C$_8$)	—	0.282	630	—	—	178
11-α-Decalylheneicosane	—	0.308	606	—	—	187

Table 11—*Continued*

Substance	$\dfrac{^b 10RT_g}{\Delta H_{vb}}$	$^c T_g^*$	$\dfrac{\theta_L}{(°K)^c}$	$\dfrac{T_g}{T_m}$	$\dfrac{T_g}{T_b}$	T_g (°K)
Aromatic Substances						
Methylbenzene	0.282	0.246	460	0.63$_5$	0.29	113
Ethylbenzene	0.258	0.227	488	0.62	0.27	111
n-Propylbenzene	0.280	0.242	504	0.70$_5$	0.28	122
i-Propylbenzene		0.242	508	0.71	0.30	125
n-Butylbenzene		0.242	514	0.68	0.28	125
sec-Butylbenzene		0.246	516	0.67	0.30	127
tert-Butylbenzene		0.272	521	0.66	0.32	142
n-Pentylbenzene		0.243	527	0.66		128
n-Hexylbenzene		0.252	545	0.67		137
o-Terphenyl		0.340	720	0.71		245
1,3,5-Trinaphthylbenzene	—	0.366	950	0.72		348
Distyrene	0.35	0.32				
Esters						
Di-(2-ethylhexyl) sebacate	—	0.270	641			173
Dimethyl phthalate	—	0.279	691	0.69		193
Diethyl phthalate	—	0.266	707			188
Di-*n*-butyl phthalate	—	0.278	635	0.74		176
Di-isobutyl phthalate	—	0.29(?)	622(?)			(188)
Di-(2-ethylhexyl) phthalate	—	0.285	643			184
Tritolyl phosphate	—	0.301	695	0.82		209
Phenyl salicylate (Salol)	—	0.322	673	0.69		216
2-Naphthyl salicylate (Betol)	—	0.326	746	0.66		244
Ethyl malonate				0.68	0.32	

[a] From Chapter 13 of reference 4.

[b] $\Delta H_{vb} = \Delta H_{vap}$ at T_b; hydrocarbon data primarily from *Selected Physical Properties of Principal Hydrocarbons* (24).

[c] $T_g^* = T_g/\theta_L$, $\theta_L (= E°/5cR)$ determined by density method, $E°$ is the energy of vaporization per mole or per mole of repeating units at the temperature at which $M/\rho V_w = 1.70$; c is one-third of the number of external degrees of freedom per molecule or per repeating unit, including those due to internal rotation (see Chapter 14 of reference 4).

[d] The value in parentheses is T_g/T_1, where T_1 is the lowest first-order solid-to-solid transition temperature.

Two significant forms of deviation from the general correlations are noted. The approach of molecular geometry toward spherical symmetry, by introduction of tertiary butyl group(s), raises T_g/T_b, $RT_g/\Delta H_{vb}$, and $5cRT_g/E°$ above the respective "average" level for similarly sized rigid-molecule compounds. The other, surprisingly small, deviation from the general patterns of those generalized vitrification temperatures is associated with hydrogen bonding.

Table 12   Reduced Glass-Transition Temperatures of
Monomeric Substances—Alkyl Halides[a]

Substance	[b]$\dfrac{10RT_g}{\Delta H_{vb}}$	[c]T_g^*	T_g $(\pm 3°\text{K})$	$\dfrac{T_g}{T_m}$	$\dfrac{T_g}{T_b}$
1-Chloropropane		0.248	92	0.64	0.28
1-Chloro-*n*-butane			102	0.58	0.27
1-Chloro-*n*-hexane			110	0.61	0.27
1-Bromopropane			90	0.55	0.27
1-Bromo-*n*-butane			92	0.57	0.26
1-Bromo-*n*-pentane			109	0.59	0.27
1-Bromo-*n*-hexane			115	0.61	0.27
1-Iodopropane		0.25	108	0.60	0.29
1-Iodo-*n*-butane		0.23	105	0.62	0.26
1-Iodo-*n*-pentane			115	0.61	0.26
1-Iodo-*n*-hexane			118	0.60	0.27
1-Chloro-2-methyl-propane	0.24	0.224	88	0.62	0.26
1-Bromo-2-methyl-propane	0.23	0.23	90	0.61	0.25
1-Bromo-3-methyl-*n*-butane	0.26	0.23	101	0.65	0.26
n-Butyl cyanide			110	0.62	0.27

[a] From Chapter 13 of reference 4.
[b] $\Delta H_{vb} = \Delta H_{vap}$ at T_b; hydrocarbon data primarily from *Selected Physical Properties of Principal Hydrocarbons* (24).
[c] $T_g^* = T_g/\theta_L$, θ_L $(= E°/5cR)$ determined by density method, $E°$ is the energy of vaporization per mole or per mole of repeating units at the temperature at which $M/\rho V_w = 1.70$; c is one-third of the number of external degrees of freedom per molecule or per repeating unit, including those due to internal rotation (see Chapter 14 of reference 4).

At a given molecular weight and functional-group content T_g is lower for liquids composed of long flexible molecules than it is for liquids composed of rigid molecules. In the reduced forms of T_g this is not very visible because molecular flexibility also lowers T_b, and it appears explicitly in the form $5cRT_g/E°$, where $3c$ is the number of external degrees of freedom, *including those due to internal rotation*. This factor is particularly important with high polymers, for which there is, of course, no T_b nor ΔH_{vb}. A few typical polymer data have been assembled in Table 14.

The division of the factors determining the magnitude of T_g into a cohesive-energy contribution and an internal rotation (or flexibility) contribution has

Table 13   Reduced Glass-Transition Temperatures of Monomeric Substances—Alcohols[a]

Substance	T_g (1 Hz) (°K)	$^bT_g^*$	$T_g^{(1)}$ (°K)[c]	$\dfrac{T_g}{T_m}$	$\dfrac{T_g}{T_b}$
Methanol	110	0.296	156 (!?) 93	0.63	0.33
Ethanol	100 (95 static)	0.248	131 (!?)	0.64	0.28
n-Propanol	109	0.244	138 (!?) 90	0.74	0.29
i-Propanol	121	0.286	143 (!?)	0.66	0.34
n-Butanol	118 (111 static)	0.245		0.61	0.30
i-Butanol	127	0.270		0.77	0.33
sec-Butanol	127 (125 static)	0.267		0.80	0.34
tert-Butanol[d]	180	0.39–0.406		0.60	0.506
n-Pentanol	124 (128 static)	0.245	93	0.64	0.30
2-Methyl-n-butanol	131		91, 113, (158 !?)		0.33
3-Methyl-n-butanol	125			0.80	0.41
2,2-Dimethyl-n-propanol[d]	166			0.51 0.73	
n-Pentanol	140	0.258		0.36	
1-Ethyl-n-propanol	143		158 (!?)	0.72	
3-Methyl-sec-butanol	149		128		
2-Methyl-sec-butanol[d]	154	0.317		0.59	0.41
Glycerol				0.64	

[a] From Chapter 13 of reference 4.
[b] See footnote c in Table 12.
[c] $T_g^{(1)}$ are other temperatures at which tan δ exhibits discontinuities. That at $T < T_g$ appears to be a secondary relaxation process, whereas that at $T > T_g$ is often, but not always, at a first-order solid-to-solid transition.
[d] By extrapolation.

been undertaken by several authors (4, 31, 32). Whereas the estimation of cohesive energy is comparatively straightforward because of the transferability of heat-of-vaporization data from low-molecular-weight compounds, the choice of a flexibility measure is still a matter of controversy. The constant

**Table 14   Limiting Values for T_g^* at $M \to \infty$, Compared
with the Energy of Rotational Isomerization (ΔE_{iso}) (4)**

Polymer	[a]T_g^*	ΔE_{iso} (kcal/mole)	Ref.
Polyisobutene	0.216	0	26
cis-1,4-Polyisoprene	0.27		
Polyethylene	0.27	0.5–0.6	26
Polydimethylsiloxane	0.30	0.80	27
Polymethyl methacrylate	0.33	0.8	[b]
Poly-n-butyl methacrylate	0.33		
Polypropylene (isotactic)	0.35	0.6–1.5	26
Polymonochlorotrifluoroethylene	0.36		
Polyvinyl acetate	0.39	~1.0	
Polyvinyl chloride	0.42	1.9[c]	30
Polystyrene	0.435	1.0–1.5	26

[a] For T_g data on polymers see *Polymer Handbook* (25).
[b] Quoted by Roetling (28) as unpublished datum of S. Havriliak;
for isotactic PMMA Sakurada et al. (29) recommend $\Delta E_{iso} = 1.2$
kcal/mole.
[c] This value is rather high; 1.2 kcal/mole is a more probable magnitude.

3c mentioned in the preceding paragraph is obtained from bulk-density data,
whereas another author (31) estimates the molecular-flexibility measure from
the Mark–Houwink (33) constant K_θ derived from intrinsic-viscosity (dilute
solution) data determined at the so-called θ-temperature (see Chapter 4).

Qualitatively all enumerated formulations agree with the empirical observation that at constant interaction energy high molecular flexibility decreases
the absolute magnitude of T_g, and conversely that intramolecular stiffness
raises T_g. Polar functional groups in the backbone chain can exhibit both
effects simultaneously, namely, to raise T_g by virtue of raised interaction
energy and to reduce T_g because they are more flexible than the methylene
chain links, which they replace. Attachment of alkyl branch chains to such
functional groups reduces T_g by dilution of the functional-group concentration per unit volume and by raising the concentration of flexible groups per
unit volume.

"Molecular flexibility" is actually an ambiguous term that can refer to a
dynamic phenomenon, the *rate* of internal rotation, governed by the magnitude of a potential energy barrier (V°) hindering internal rotation, or it can
refer to an equilibrium phenomenon, the relative concentration of a molecule

in different geometrical conformations, governed by the sign and magnitude of the energy of rotational isomerization (ΔE_{iso}). Although so far no entirely satisfactory theory of glass transition contains these two molecular properties in the form of mutually independent parameters, as required by the physics of the situation, one can discern their relative importance by similitude analysis. The external degrees of freedom of high polymers c are related primarily to $V°$, so that the dimensionless glass transition $T_g^* = 5cRT_g/E°$ should be a universal number if $V°$ alone mattered. The data in Table 14 show that T_g^* is not a universal number, but seemingly has ΔE_{iso} as an ordering parameter. On the other hand, since ΔE_{iso} of given bonds cannot yet be predicted from data on molecular structure and environment, this correlation is of limited utility for the prediction of the magnitude of T_g in given cases.

A special case of partial immobilization is crosslinking. Small degrees of crosslinking have no significant effect on T_g. However, once the crosslink density reaches a magnitude of >0.001 cm^{-3}, beyond which point the conventional osmotic and elastic measures of crosslink density begin to fail, T_g begins to rise rapidly with crosslink density. This region is of only slight interest in rubber vulcanization, where it represents the range leading toward so-called ebonite, or hard rubber, and where the sulfur content is only a very crude measure of crosslink density because of low crosslink efficiencies. However, the newly developing field of "ionogenic" polymers has revived interest in the quantitative aspects of the effect of "crosslinks" on T_g. In the usual environment of hydrocarbon chain segments one expects that the ionic components, such as metal carboxylate groups and quaternary ammonium salt groups, aggregate into insoluble microdroplets or domains that then act as crosslinks and fillers.

9. MIXTURE RULES FOR T_g

Two types of mixture are of practical significance: mixtures of glass, especially glassy polymer with plasticizer, and the combination of two or more monomers into copolymers. The experimental data of either group show that the T_g of the mixture is rarely, if ever, a linear combination of the T_g of the components, regardless what units of mixture concentration one tries. In the case of copolymers the curve of T_g versus concentration may even pass through a minimum or a maximum (34).

The interactions that lead to the observed behavior are qualitatively understood, but their quantitative formulation suffers from the absence of a truly reliable theory of vitrification. The two mixture-rule models that can deal reasonably effectively with the problem at hand are derived from the free-volume theory of vitrification (4). A direct consequence of both models is that the contribution of the mixture components to the T_g of an "ideal mixture " is dominated by the T_g of the component with more free volume, usually the component with the lower T_g; that is, either the component with appreciably lower molecular weight or with more internal molecular flexibility. With nonideal mixtures one finds that marginal compatibility decreases the mixture $T_g^{(12)}$ below its ideal-mixture value, whereas mutual chemical affinity will raise the $T_g^{(12)}$ above the ideal-mixture value. These minima and maxima are observed more often with copolymer systems than with physical blends. The T_g of incompatible polyblends or composites is essentially that of the pure components, so that if both components differ appreciably in T_g, the final blend has two observable values of T_g. Even massive amounts of finely divided second (solid) phase do not change the T_g of the continuous phase, although the elastic moduli are raised appreciably thereby.

The simpler of the two reasonably intelligible mixture rules has the form

$$T_g^{(12)} = \frac{\phi_1 \, \Delta\alpha_1 T_g^{(1)} + \phi_2 \, \Delta\alpha_2 \, T_g^{(2)} + k\phi_1\phi_2}{\phi_1 \, \Delta\alpha_1 + \phi_2 \, \Delta\alpha_2}$$

where ϕ_1, ϕ_2 are the volume fractions of components 1 and 2, $\Delta\alpha = \alpha_l - \alpha_g$ are the differences in thermal expansion between liquid and glass at T_g, and $-k = v^E/v_{12} \, \phi_1\phi_2$, v^E being the excess volume of mixing, a measure of the interaction between the mixture components near T_g, and v_{12} being the specific volume of the mixture. For many polymer–plasticizer systems $k \approx -0.04$ (4). If one considers the end groups of a polymer molecule as plasticizer and treats them as the dimer of the monomer, one can use the above mixture rule as a means of estimating the effect of molecular weight on T_g.

10. EXPERIMENTAL TECHNIQUES FOR T_g DETERMINATIONS

The change from liquid to glass can be determined by observation of a discontinuity in the change of an equilibrium or transport property with the appropriate state variable, usually the temperature. The time rate of change of the state variable is an important parameter that must be specified, as must the thermal and mechanical history of the sample's conversion from liquid to glass, in order to guarantee reproducibility of the results.

Three types of measurement are commonly employed for the determination of phase transitions, as well as of T_g: dilatometry, calorimetry, and mechanical deformation. They are described in that order.

a. Dilatometry

Dilatometry is only rarely used in first-order phase-transition measurements because the volume changes are so large and abrupt that their precise determination is very difficult. However, dilatometry is the preferred mode of determination for T_g because the discontinuity is in the thermal expansion coefficient rather than in the volume. Recent development of automatic differential dilatometry (35, 36) records the thermal expansivity rather than the volume against temperature and thus provides a very sensitive measure of the onset of vitrification at T_g and other volume-relaxation processes.

b. Calorimetry

Calorimetry in the form of a conventional heat-capacity measurement gives as sensitive an indication of vitrification and of T_g as does thermal expansivity. However, it is a rather more cumbersome measurement, requiring highly skilled operators. Differential calorimetry in the form of differential thermal analysis (DTA), by contrast, is even simpler than dilatometry and is correspondingly popular because it has been automated very successfully. This method, amply described in the literature (37), measures the temperature difference between the sample under consideration and a "nontransiting" sample of similar heat capacity while immersed in a bath whose temperature changes at a constant rate. It is just as well suited for the determination of first-order phase-transition temperatures as for that of T_g.

c. Mechanical Deformation Measurements

Mechanical deformation measurements yield unequivocal data for the temperature of first-order phase transitions because of the very sharp change in elastic, especially shear and tensile, moduli at such transitions. The change from glassy to liquid or rubbery state, however, is accompanied by a rather less sharp change in shear or tensile moduli, so that the location of T_g on the curve for modulus versus temperature requires an arbitrary decision regarding the "reference" modulus level. Deformation rate is also a critical factor that has to be standardized. In shear-stress relaxation moduli Tobolsky and his school specified the reference state as 0.33×10^8 dynes/cm^2 after 10 sec of stress relaxation (38). The resulting temperature is very near the dilatometrically or the calorimetrically determined value of T_g.

A more commonly employed deformation method for detecting phase changes and vitrification is the measurement of mechanical energy absorption in oscillatory deformation at constant frequency versus temperature, especially in shear by means of a torsion pendulum oscillating at 1 Hz. Here a first-order transition is noted as a frequency-independent step change in energy absorption, and vitrification is observed as a comparatively steep maximum in the curve for energy absorption (logarithmic decrement) versus temperature. The temperature corresponding to the maximum at 1 Hz is generally near the T_g measured by dilatometry or calorimetry (39).

d. Spectroscopic Methods

Spectroscopic methods for the determination of first-order solid-to-solid and solid-to-liquid transitions have been used for some time. They usually employ the infrared active splitting of a molecular vibration in the ordered crystal that is absent in the disordered solid and/or in the liquid phase (10). Since state-independent molecular vibrations serve as internal standards, one can often use the intensity of the split band as a measure of the extent of crystallization. Vitrification has recently been observed to change the intensity of some molecular vibrations sufficiently to locate T_g (40).

REFERENCES

1. I. Darmon and C. Brot, *Mol. Crystals* **2**, 301 (1967).

2. E. R. Andrew and P. S. Allen, *J. Chim. Phys.* **85** (1966).

3. K. H. Illers, *Rheol. Acta* **3**, 185, 194, 202 (1964).

4. A. Bondi, *Physical Properties of Molecular Crystals, Liquids, and Glasses*, Wiley, New York, 1968, Chapters 3, 4, 5, 6, and 13.

5. J. G. Aston, in *Physics and Chemistry of the Organic Solid State*, Vol. 1, Wiley, New York, 1963, p. 593.

6. G. Martin, *J. Chim. Phys.* **64**, 347 (1967).

7. A. J. Kovacs, *Fortschr. Hochpolym. Forschg.* **3**, 394 (1963).

8. D. J. Plazek and J. H. Magill, *J. Chem. Phys.* **45**, 3038 (1966).

9. N. Saito et al., *Solid State Physics*, **14**, 344 (1963).

10. J. R. McLoughlin and A. V. Tobolsky, *J. Polym. Sci.* **7**, 658 (1951).

11. M. R. Carpenter, D. B. Davies, and A. J. Matheson, *J. Chem. Phys.* **46**, 2451 (1967).

12. T. A. Litovitz and C. M. Davis, *Physical Acoustics*, W. P. Mason, Ed., *Academic Press*, New York, Vol. 2B, 1965.

13. W. Blitz, *Raumchemie der Festen Stoffe*, Leopold Voss (Leipzig, 1934).

14. R. C. Collin, *Acta Cryst.* **5**, 431 (1952).

15. H. Sackman, *Z. Phys. Chem.* **208**, 235 (1958).

16. L. H. Bolz et al., *Acta Cryst.* **17**, 247 (1959).

17. C. Scheringer, *Z. Krist.* **119**, 273 (1963).

18. R. F. Robbins et al., *Adv. Cryog. Eng.* **8**, 289 (1963).

19. J. H. Magill and A. R. Ubbelohde, *Trans. Faraday Soc.* **54**, 1811 (1958).

20. A. K. Schulz, *J. Chim. Phys.* **51**, 530 (1954).

21. G. S. Parks et al., *J. Phys. Chem.* **32**, 1366 (1928).

22. F. Danusso et al., *Chim. Ind.* **41**, 748 (1959).

23. H. J. Kolb and E. F. Izard, *J. Appl. Phys.* **20**, 564 (1949).

24. *Selected Physical Properties of Principal Hydrocarbons*, American Petroleum Institute.

25. J. Brandrup and E. H. Immergut, Eds., *Polymer Handbook*, Wiley, New York, 1966.

26. T. M. Birshtein and O. B. Ptitsyn, *Conformations of Macromolecules*, Interscience, New York, 1966.

27. P. J. Flory et al., *J. Am. Chem. Soc.* **86**, 146 (1964).

28. J. Roetling, *Polymer* **6**, 311 (1965).

29. I. Sakurada et al., *Kolloid-Z*, **186**, 41 (1962).

30. H. Germar, *Kolloid-Z*, **193**, 25 (1963).

31. D. P. Wyman, *J. Appl. Polymer Sci.* **11**, 1439 (1967).

32. R. A. Hayes, *J. Appl. Polymer Sci.* **5**, 318 (1961).

33. H. Mark and R. Houwink, See References in Chapter 5.

34. G. Kanig, *Kolloid-Z.* **190**, 1 (1963).

35. R. A. Haldon and R. Simha, *J. Appl. Phys.* **39**, 1890 (1968).

36. R. A. Haldon and R. Simha, *Macromolecules* **1**, 340 (1968).

37. "Thermal Analysis," in E. Slade, Jr., and L. T. Jenkins, Eds., *Techniques and Methods of Polymer Evaluation*, Vol. 1, Dekker, New York, 1966.

38. A. V. Tobolsky, *Properties and Structure of Polymers*, Wiley, New York, 1960.

39. K. Schmieder and K. Wolf, *Kolloid-Z.* **134**, 149 (1953).

40. A. Andon, *J. Appl. Polymer Sci.* **12**, 2117 (1968).

The Mesomorphic State: Liquid and Plastic Crystals

7

D. B. DuPré, E. T. Samulski, and A. V. Tobolsky

1. MESOMORPHISM

Liquid crystals are highly associated liquids that occur over various temperature ranges in melts or solutions of certain large organic molecules and some polymers (1–5). Liquid crystallinity is an intermediate state of matter, existing in a peculiar twilight zone between the boundaries of the usual crystalline solids and isotropic liquids. It is one example of mesomorphism, a term denoting intermediate (Greek *mesos*) form (Greek *morphe*).

Conventionally we think of matter existing in one of three distinct states of aggregation: the solid state, where constituent molecules (atoms) execute small vibrations about firmly fixed lattice positions but cannot rotate; the liquid state, characterized by relatively unhindered rotation but no long-range order; and the gas state, where particles are free to roam through the entire volume of their container, under almost no constraint.

The melting of normal solids, occurring abruptly, involves the collapse of the overall positional order of the lattice array and marks the onset of essentially free rotation of the particles. Two intermediate phases of matter have been identified where only one or the other of these two freedoms (freedom of position or freedom of rotation) has been at least partially obtained. These are liquid crystals and plastic crystals. These mesophases do not satisfy all the criteria for either a true solid or a true liquid and, in fact, exhibit many physical properties that are characteristic of both; for example, liquid crystals flow, but their optical properties are those usually expected only of regular solids. Molecules within plastic crystals, on the other hand, have rotational and diffusional mobilities approaching those of the liquid phase although the solid condition is maintained. In liquid crystals molecules are free to move where they please, but their rotational mobility is restricted; in plastic crystals the molecules are free to rotate in place and to some extent change lattice sites, but they still form a regular crystalline superstructure.

The phenomenon of mesomorphism is fundamentally a consequence of molecular shape. A necessary (but not sufficient) condition for the formation of a mesophase is that the molecules comprising the system be either (*a*) elongated and in some cases also flat (possibility of liquid crystallinity) or (*b*) approximately spheroidal (possibility of plastic crystallinity).

A rough determining factor of mesomorphism is therefore the length-to-width ratio of the molecular frame, which we denote by R. If $R \gg 1$, there is a possibility of liquid-crystal formation. If $R \simeq 1$, a plastic-crystal phase might occur. Other qualities enter into or prevent the formation of a mesophase. These are considered in detail later.

a. Liquid Crystals

Compounds capable of forming liquid-crystal mesophases typically have $R \gg 1$. It is clear that all the molecules listed in Table 1 as being liquid crystals are long, relatively rigid, and rod-shaped.† One of the most extensively studied compounds is *p*-azoxyanisole the first compound shown in Table 1, for which $R \simeq 2.6$ (length to width) and $R \simeq 5.2$ (length to thickness) (7). When solid *p*-azoxyanisole is heated to its melting point ($T_m = 116°C$), it becomes a turbid fluid, similar to water in viscosity. At 134°C there is an abrupt change to a clear isotropic phase (clearing point), with all of the properties a normal liquid. The intermediate state is, however, "solidlike" in that, among other things, it is not optically extinct when viewed between crossed Polaroids. The phase is a genuine fluid in that it flows; it exists over a 17°C temperature range.

In a sense the liquid-crystal mesophase represents a stable and protracted condition of incomplete melting. Sufficient thermal energy has not been supplied to allow the individual molecules to rotate (except about their long axes perhaps), and a good deal of the positional order of the parent solid is still present. Significant intermolecular interactions (to be discussed later) exist, preserving extensive solidlike order within the fluid. In the absence

---

† An extensive list of substances forming liquid-crystal mesophases, together with their respective transition temperatures, has been compiled by Kast (6).

of external orienting influences (e.g., electric or magnetic fields) this order does not prevail over the entire bulk of the sample; that is, in moving from region to region within the fluid, one would be just as likely to find a liquid-crystal molecule pointing in one direction as in another. However, within a given region a remarkable degree of parallelism of the major axes of these molecules does occur and gives the phase some of the anisotropic properties of the solid state. At higher temperatures the liquid crystal passes into a normal isotropic fluid as the intermolecular order is finally broken down and the long molecules find room to rotate.

A measure of the degree of internal order (or alignment) within a fluid is given by

$$S = \tfrac{1}{2}\langle 3 \cos^2 \theta - 1 \rangle$$

where θ is the angle between the long axis of a representative molecule and some preferred direction. The angle brackets indicate the average:

$$\langle (\cdots) \rangle = \int_0^\pi (\cdots) f(\theta) \sin \theta \, d\theta \Big/ \int_0^\pi f(\theta) \sin \theta \, d\theta$$

where $f(\theta)$ describes the angular distribution of the molecules. For a perfectly parallel arrangement S would be identically 1.0. For a completely isotropic liquid $f(\theta)$ is a constant and $S = 0$. The order factor for liquid crystals falls somewhere in between, with S as high as 0.9 for the smectic type, where a good deal of the order in two dimensions of the parent crystal still remains. It will be seen later that molecules of the smectic mesophase are arranged by significant lateral associations into loosely connected layers that are free to slide over one another. The values of S are smaller for the inherently less ordered nematic liquid crystals. The molecules of this structure are associated in head-to-tail fashion, forming mobile threads that extend for several thousand molecular lengths. As such the nematic mesophase possesses remnants of crystalline order in only one direction. The highest order values observed here are about 0.8. The values of S near the clearing point of the nematic mesophase is usually between 0.3 and 0.5 (3).

b. Plastic Crystals

In contrast to liquid crystals, which are solidlike liquids, plastic crystals (8–10) are liquidlike solids. Due to their approximately spheroidal shape ($R \simeq 1$), the molecules that make up this intermediate phase (notable examples being camphor and adamantane) are able to attain a large degree of rotational freedom even before passing into a true liquid condition. The phase is not amorphous, however. A definite crystal structure is maintained, although it appears that intersite diffusion occurs rather readily. As a consequence of this high microscopic mobility, these solids are quite easily deformed. They can be

Table 1 Some Liquid-Crystal Compounds: Structures, Melting (T_m) and Clearing (T_c) Points

Compound	Formula	Mesomorphic Structure	T_m (°C)	T_c (°C)
p-Azoxyanisole	$H_3C-O-C_6H_4-N=N^+(O^-)-C_6H_4-O-CH_3$	Nematic	116	133
Anisal-*p*-aminoazobenzene	$H_3C-O-C_6H_4-CH=N-C_6H_4-N=N-C_6H_5$	Nematic	151	185
Ethyl-*p*-azoxybenzoate	$C_2H_5O-CO-C_6H_4-N=N^+(O^-)-C_6H_4-CO-OC_2H_5$	Smectic	114	122
6-Methoxy-2-naphthoic acid	$CH_3O-C_{10}H_6-CO-OH$	Nematic	206	219
trans-*p*-Methoxycinnamic acid	$H_3C-O-C_6H_4-CH=CH-COOH$	Nematic	173.5	190
p-*n*-Propoxybenzoic acid	$C_3H_7O-C_6H_4-CO-OH$	Nematic	145	154

Compound	Structure	Phase		
p-n-Hexylbenzoic acid	C_6H_{13}—C$_6$H$_4$—COOH	Nematic	97.5	114.5
4′-Methoxybiphenyl-4-carboxylic acid	CH_3O—biphenyl—COOH	Nematic	258	300
4-Propyl-4′-ethoxybiphenyl-4-carboxylate	C_2H_5—biphenyl—COOC_3H_7	Smectic	102	103
Ethyl-*p*-azoxycinnamate	C_2H_5O—CO—CH=CH—C$_6$H$_4$—N=N(→O)—C$_6$H$_4$—CH=CH—CO—OC_2H_5	Smectic	141	264
Nona-2,4-dienoic acid	$CH_3(CH_2)_3CH=CHCH=CH$—COOH	Nematic	23	49
4,4′-Di-(benzylideneamino)-biphenyl	C$_6$H$_5$—CH=N—biphenyl—N=CH—C$_6$H$_5$	Nematic	234	260
4-*p*-Methoxybenzylideneamino-biphenyl	CH_3O—C$_6$H$_4$—CH=N—biphenyl	Nematic	161.5	173.5
2-*p*-Methoxybenzylideneamino-phenanthrene	CH_3O—C$_6$H$_4$—CH=N—phenanthrene	Nematic	155	213.5

extruded at relatively low pressures, and some even have the consistency of butter. (Perfluorocyclohexane flows by the weight of its own bulk). For these reasons the mesophase is termed plastic, although the condition has nothing to do with high polymers, for which the label is usually reserved.

Some examples of plastic crystals are listed in Table 2, along with respective temperatures and entropies of transition and fusion (8). The phenomenon is not limited to spherical organic molecules: weakly interacting particles, such as the noble gases, hydrogen, hydrogen sulfide, and metals (including gold, potassium, sodium, and silver), also exhibit a degree of plasticity when cooled below their freezing points.

It is noteworthy that plastic crystals melt to a liquid with only a slight

Table 2   Temperatures and Entropies of Transition and Fusion of Some Plastic Crystals

Substance	T_t (°K)	T_m (°K)	ΔS_t	ΔS_m
Camphor	250	453	7.6	2.8
CH_4	20.5	90.7	0.76	2.48
CCl_4	225.5	250.3	4.86	2.40
CF_4	76.2	89.5	4.64	1.87
CBr_4	320.0	363.3	4.98	2.60
$C(CH_3)_4$	140	256.6	4.4	3.03
$C(CH_2OH)_4$	457	539	22.8	3.16
$C(CH_3)_3Cl$	219	248	8.7	2.0
$C(CH_3)Cl_3$	224	241	7.97	4.5
$C(CH_3)_3 \cdot C_2H_5$	126.8	174.2	10.17	0.80
$C(CH_3)_3SH$	151.6	274.4	6.41	2.16
SiH_4	63.45	88.5	2.61	1.80
GeH_4	73.2	107.3	1.78	1.86
SF_6	94.3	222.5	4.07	5.40
PtF_6	276	334.4	7.7	3.2
C_2Cl_6	344.6	458	5.7	5.5
$Si_2(CH_3)_6$	221.9	287.6	10.5	2.51
$(CH_3)_2CH \cdot CH(CH_3)_2$	136.1	145.2	11.41	1.32
Cyclohexanol	263	297	7.45	1.4
Cyclobutane	145.7	182.4	9.36	1.43
Cyclopentane	122.4	179.7	9.52	0.80
Cyclohexane	186.0	279.8	8.59	2.22
N_2	35.6	63.1	1.54	2.73
Argon	—	83.85	—	3.35
HCl	98.4	158.9	2.89	2.99
H_2S	103.5	187.6	3.55	3.03

change in volume, whereas normal solids typically expand quite considerably at the melting point. This is an indication that the free volume per molecule in liquid and plastic phases is roughly the same. The ΔH_m for these compounds tends to be low, suggesting that there exists very little difference between intermolecular bonding energies in the liquid and plastic–solid phases. Furthermore, the vapor pressure of plastic crystals are abnormally high for solids; for example, plastic neopentane has a vapor pressure of 200 mm Hg, whereas its isomeric counterpart n-pentane, a normal solid, has a vapor pressure of only 1 mm Hg. These facts suggest the substantially liquidlike character of the plastic mesophase.

Direct evidence for rotation within plastic solids has been obtained by study of dielectric and nuclear-magnetic-resonance (NMR) relaxation properties.

For a material to display a dielectric constant much greater than 1.0 its molecules must be both polar and able to reorient in the presence of a static or alternating electric field. The dielectric constants of polar solids (other than ferroelectrics) are usually close to unity, as molecular motion is severely restricted within the dense medium. Only induced electronic distortions act to polarize the sample. The dielectric constants of plastic crystals, on the other hand, are comparable to those of liquids: convincing evidence of the high internal mobility characteristic of this phase of matter. Figure 1 shows the dielectric properties of methyl chloroform through the liquid and three solid phases (11). It is seen that the dielectric constant (static, ε_0, and dynamic, ε') of the plastic mesophase (solid III) is just as high as that of the liquid. The loss factor ε'' is also appreciable in the plastic phase. A sharp decrease in ε_0 and ε' occurs, however, at the plastic-to-solid II transition point as molecular positions and orientations become frozen in.

Dielectric studies demonstrate that many spherical molecules in the camphor series experience about the same low resistence to rotation in the solid phase just below the melting point as they do in the liquid (9). The dielectric constant of the plastic phase of tert-butyl bromide is actually greater than that of the liquid (12).

The degree of molecular mobility is also reflected in NMR line widths. The faster the reorientation of a molecule containing the resonant proton, the more homogeneous the local magnetic environment and the narrower the signal. Nuclear-magnetic-resonance studies of methane (13), methyl chloroform (14), cyclopentane (15), adamantane (16), carbon tetraflouride (17), and neopentane (14) show that these molecules, if not actually in free rotation, are switching rapidly among stable orientations well below the melting temperature. Further measurements demonstrate that considerable diffusional motion is taking place in the solid phase at rates of at least 10^4 displacements per second (18).

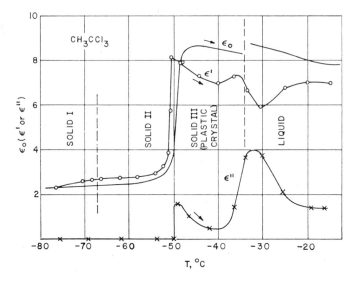

Fig. 1   Dielectric constant (ε_0, ε') and loss (ε'') of methyl chloroform as a function of temperature. The value of ε_0 was measured in a static field; those of ε' and ε'', in a microwave cavity with radiation of 3.22-cm wavelength. Solid state III is a plastic crystal. [Data of Powles et al. (11).]

X-Ray diffraction studies (19) of plastic crystals reveal lattice structures of unexpected simplicity (frequently cubic). Not quite spherical molecules would, at lower temperatures, be obligated to form arrays of less symmetry. Due to high rotational mobility in the plastic phase, geometric anisotropies seem to be averaged out, and the molecules arrange themselves accordingly. X-Ray measurements further show that the molecules of many plastic-crystal compounds do not in fact have sufficient room to rotate freely *in place;* for example, in the case of adamantane the experimental distance between molecular centers is 6.68Å, whereas the minimum estimated diameter of the rotating molecule would be 7.55Å. A fair number of lattice sites must therefore be vacant in these solids to permit the flow behavior mentioned above.

The clear dielectric and NMR evidence for rotational mobility suggests that the molecules are participating in a cooperative effort of rotating together and, frequently, of translating into crystalline defects, or voids. Apparently methane is one of the few plastic crystals in which completely free rotation occurs (8).

We should point out that mesophases are true and distinct states of matter in their own right, with their own peculiar physical properties. It is clear that liquid crystallinity is not due to any sort of colloidal aggregation, nor does it represent a grossly nonequilibrium state of liquid supercooling. Liquid crystals frequently exist over wide temperature ranges and have uniform "solidlike" optical properties. A colloidal suspension of small crystallites, for example, when viewed in the polarization microscope, would appear "speckled" with random bright spots. This is not the case. The optical field is uniform. Plastic crystals, moreover, have a definite crystal lattice. definite melting and transition points, and a certain amount of resiliency, as do true solids. Transitions to the solid and the isotropic liquid in both types of mesophase are perfectly well defined, reproducible, and reversible. The "in-between" phases must therefore be treated on their own terms.

In the following sections we consider in more detail the structure and physical properties of liquid crystals.

2. STRUCTURE OF LIQUID CRYSTALS

Liquid crystals are known to exist in three types of structure—smectic, nematic, and cholesteric—which differ in the nature of local molecular order.

a. Smectic Structure

In the smectic structure long molecules are arranged side by side in layers much like those in soap films (see Fig. 2). The term "smectic" (soaplike) is,

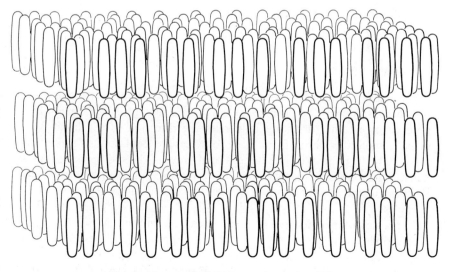

Fig. 2   The smectic liquid-crystal structure (20).

in fact, derived from the Greek word ($\sigma\mu\varepsilon\gamma\mu\alpha$) for grease or slime. The layers are not strictly rigid but form flexible, two-dimensional molecular sheets that can slide past one another. Movement in any direction other than tangent to the layer surfaces is, however, difficult. Molecular motion is rather slow, and smectic mesophases are typically quite viscous.

Smectic liquid crystals arise from organic solids that themselves are stratified. The melting process evidently disrupts end-to-end molecular cohesions, but the temperature at which the mesophase is stable is not sufficient to break apart lateral associations and the layers remain essentially intact. Smectic liquid crystals therefore retain a good deal of two-dimensional solid order; for example, the S-value of smectic 4,4'-bis(n-heptyloxy)azoxy-benzene is 0.9 (21).

The layered nature of this liquid-crystal structure was suggested by early observations that smectogenic solids, when melted on a very clean glass slide, produce a terraced texture. The strata, the thickness of many molecular lengths, are called Grandjean plates (22). Movement of a cover slip demonstrated that the molecular plates easily slide over one another, but displacement in any other direction is resisted. A compelling idea was that the long molecules lie in two-dimensional layers with long axes parallel to one another and perpendicular to the plane of the terraces. All other evidence indicates that this presumption is correct.

Optically smectic layers behave as uniaxial, birefringent crystals. The velocity of light transmitted parallel to the molecular layers is greater than that transmitted perpendicularly. The temperature dependence of this birefringence indicates that very little change of internal order occurs over the smectic interval. The smectic mesophase is the most "solidlike" of all the liquid-crystal modifications.

b. Nematic Structure

Molecules remain parallel to one another in the nematic structure as well, but the positions of their centers of gravity are disorganized (Fig. 3). In the absence of external orienting influences the direction of the long axes of nematic molecules varies from place to place in a continuous fashion. As a result only the parallel or one-dimensional order of the solid persists, and the nematic mesophase is considerably more mobile. Intense movement within the bulk of the nematic liquid is evident by the erratic course of stray dust particles.

Molecules of nematic liquid crystals can be substantially oriented, and their mobility can be reduced by adhesion to supporting surfaces; for example, nematic molecules tend to lie parallel to the rough surface of a glass slide. In thin sections the order pattern established by the surface may extend all the way across the sample from base to cover slip. So persistent is the influence of surface effects that the material can in fact be made optically active by twisting the cover slip relative to the base, generating a helicoidal, or screw, structure within the intervening fluid.

A bright satinlike texture is observed when nematic liquid crystals are viewed between crossed Polaroids. Characteristic dark threads appear at lines of optical discontinuity. These wavering filaments give the mesophase its name, taken from the Greek word *nematos* (νημα), meaning fiber.

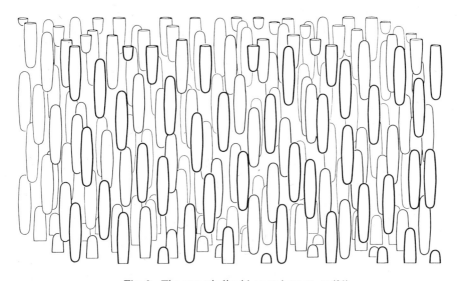

Fig. 3   The nematic liquid-crystal structure (20).

X-Ray studies (23) of the crystal structure of nematogenic solids of *p*-azoxyanisole and *p*-azoxyphenetole show that the molecules within the solid precursor lie parallel to one another but are not organized into layers.

c. Cholesteric Structure

The third type of liquid-crystal modification, the *cholesteric* structure, is so named because many compounds that form this mesophase are derivatives of cholesterol. More generally, however, the classification has to do with the unusual optical properties of the structure. Molecules that bear no relation to cholesterol, such as polybenzylglutamate in concentrated solutions, compounds of other sterol systems, and amyl *p*-(4-cyanobenzylideneamino) cinnamate (1) are "cholesteric." Curiously enough, cholesterol itself is not mesomorphic.

$$NC-\langle \rangle-CH=N-\langle \rangle-CH=CH-CO_2C_5H_{11}$$

1

The distinguishing feature of cholesteric liquid crystals is their iridescent color and extreme optical rotatory power. It was mentioned that the nematic mesophase could be made optically active by mechanically inducing a screw structure with a thin sample. Cholesteric liquid crystals are naturally this way. On inspecting a molecular model of a cholesterol ester (2) one sees that the elongated molecule is essentially flat except for side groups that project out of the plane of the fused rings. It is thought that these molecules arrange themselves parallel, but with the long axes lying in planes. The parallel alignment within each plane layer is like that of nematic liquid crystals, but the fluid is also stratified in the sense of the smectic mesophase. The planes pile on top of one another and are potentially able to glide past one another.

2

Each layer must, however, be slightly twisted with respect to the next to accommodate the functional groups extending out of the molecular plane. The effect (averaging about 15 minutes of arc per layer) is cumulative, and an overall helicoidal architecture results as shown in Fig. 4.

These twisted molecular layers are responsible for the dispersion of reflected white light into brilliant colors and the rotation of the polarization of transmitted light, unique to this mesophase. The regular spacing between layers acts as an internal diffraction grating, breaking up incident white light into its component oscillations. The precise color observed depends on the angle of illumination, the nature of the cholesteric material, and the temperature. The color is further affected by mechanical disturbances (pressure, jarring, or shear) and traces of foreign vapors (20). The helicoidal twist of the structure (of pitch between 0.2 and 20 μ) rotates the electric vector of polarized light as it makes its way through the fluid. So pronounced is the molecular organization that optical rotations on the order of 1000°/mm of material are observed in cholesteric liquid crystals. This should be compared to normal optically active organic liquids, whose optical rotatory powers are seldom much greater than 300°/mm.

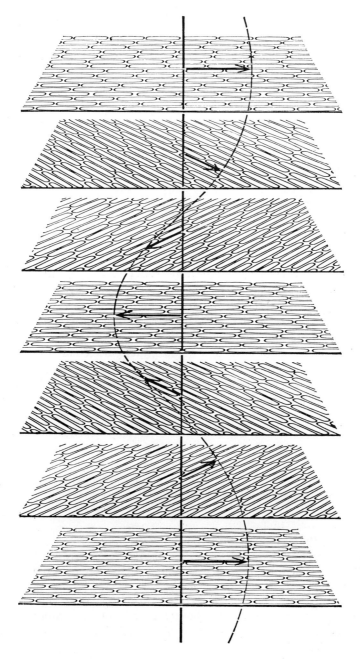

Fig. 4 The cholesteric liquid-crystal structure (20).

The "form" optical rotation of the helicoidal structure originates in the selective reflection of one circularly polarized component of the light. This differs from the more familiar cases of optical rotation encountered in optically active molecules, which originates in the selective absorption of one circularly polarized component of the light. In the first case the wavelength of the reflection band is determined entirely by the pitch of the helicoidal structure, whereas in the second case the wavelength of the absorption band is determined by the chemical constitution of the material.

A few of the more than 1500 known liquid crystals are listed in Table 1 with a notation about their structure: smectic, nematic, or cholesteric.

3. CHEMICAL CONSTITUTION OF LIQUID-CRYSTAL COMPOUNDS

We have indicated that the essential feature necessary for liquid-crystal formation is a large molecular length-to-breadth ratio. It is not true, however, that just because a molecule is elongated that it will engage in a meso-morphic structure; for example, n-paraffins and homologs of acetic acid do not display any properties that are characteristic of liquid crystals, although these molecules are long, narrow, and meet the requirements of geometric anisotropy. The forces of attraction between these molecules are simply not sufficiently strong for an ordered, parallel arrangement to be retained after the melting of the solid. To appreciate what factors support liquid crystallinity one must focus attention on the nature and type of intermolecular inter-actions present in these systems. The particular mesomorphic structure that occurs—smectic, nematic, or cholesteric—not only depends on molecular shape but also is intimately connected with the strength and position of polar groups within the molecule and its overall polarizability.

For the purpose of our discussion molecular interactions that lead to attraction can be separated into four categories:

1. Dipole–dipole interactions.
2. Dipole-induced dipole interactions.
3. Dispersion forces.
4. Hydrogen bonding.

Dispersion forces alone (at least in simple aliphatic compounds) are apparently inadequate to achieve the degree of molecular order necessary for liquid crystallinity, as witness the straight-chain paraffins, which melt to form normal liquids.

In order for type 1 and 2 interactions to be effective the molecule must contain polar groups and/or be highly polarizable. Ease of electronic distortion is favored by the presence of aromatic groups and double or triple bonds. These groups are frequently found in the molecular structure of liquid-crystal compounds. Inspection of Table 1 bears this out. Rodlike *p*-isomers of benzene, 4,4'-disubstituted biphenyls, and 4,4''-disubstituted *p*-terphenyls typically are found to be mesomorphic. The importance of unsaturation is clearly illustrated by the fact that 2,4-nonadienoic acid (see Table 1) forms a mesophase, whereas the *n*-aliphatic carboxylic acids do not. The two double bonds enhance the polarizability of the molecule and bring intermolecular attractions up to a level suitable for mesophase formation.

It should be emphasized that the overall linearity of the molecule must not be sacrificed in potential liquid-crystal candidates. For example, whereas *trans-p-n*-alkoxycinnamic acids (3) are mesomorphic, the *cis* isomers (4) are not, a reflection of the greater breadth of the *cis* molecules.

Freedom of rotation about the double methylene bridge in compound (5) destroys the rod shape of the molecule and prevents liquid-crystal formation. The stilbene derivative (6) however, is essentially linear and mesomorphic.

Bulky, even if highly polarizable, functional groups or atoms placed anywhere but on the end of a rod-shaped molecule usually destroy mesomorphic capabilities. Enhanced intermolecular attractions are more than countered as the molecule deviates from the required linearity. For example, the inclusion of the bromine atom at position 3 of *p-n*-decyloxybenzoic acid (7) prevents mesomorphic behavior.

Hydrogen bonding plays two contrary roles in liquid-crystal formation. In the case of carboxylic acids it may act to induce mesomorphic behavior by lengthening the molecular unit through dimerization, as in compound (8).

$$R-C \begin{matrix} O \cdots H-O \\ \\ O-H \cdots O \end{matrix} C-R$$

8

On the other hand, hydrogen bonding may lead to nonlinear molecular associations that disrupt the parallelism; for example, phenolic compounds are never mesomorphic. Hydrogen-bonding associations may also be so strong that by the time the solid reaches its melting point the thermal energy is too intense to permit substantial order to remain within the fluid. In this case the solid passes directly into the isotropic liquid. Such reasoning could explain the absence of mesomorphism in cholesterol and its presence in the esters of cholesterol. 4-Amino-4″-nitro-p-terphenyl (**9**) melts at 300–301°C and is not mesomorphic. Both amine hydrogens in this compound participate in intermolecular associations. The substituted nitro compound (**10**) however, melts at only 218–219°C and is mesomorphic. One of the hydrogens in this case is tied up in *internal* association with the —NO$_2$ group.

9

10

Although it is difficult to predict exactly which type of mesomorphic structure will be formed by a molecule that meets the general requirements of liquid crystallinity, rough trends can be recognized. The presence of functional groups that lead to strong lateral interactions—such as dipoles operating across the molecule's long axis—favor the layered, smectic structure (see Table 1 for examples). When these structural elements are not present but the molecule is otherwise suitable for mesomorphism (i.e., is long and narrow), the nematic modification is likely. In certain cases interactions between terminal dipolar groups are responsible for the stability of the nematic structure. For example, the methoxy dipoles of neighboring p-azoxyanisole molecules reinforce one another, as shown below, to give the end-to-end threadlike organization of the nematic phase.

A chemical structure with repeating units containing CH_3 groups, ester linkages ($-O-$), aromatic rings, and $-N=N-$ (azo) groups.

As mentioned above, the majority of the compounds that form the cholesteric mesophase are derivations of sterols. At the present time there is no complete understanding of the nature of the forces responsible for the existence of the cholesteric mesophase. Recent studies point out the similarity between the nematic and cholesteric structures, emphasizing the importance of terminal groups on the stability of the mesophases (24). The experimental observation that single isomers of certain optically active compounds exhibit a cholesteric mesophase whereas racemic mixtures of the compounds or mixtures of cholesteric compounds with opposite-form optical rotation are nematic, suggests that the chirality of the dispersion forces in the optically active compounds may be capable of imposing the unidirectional twist found in the cholesteric mesophase (25).

An exhaustive account of the topics of this section may be found in Gray's book (1).

4. POLYMESOMORPHISM AND LYOTROPIC MESOMORPHISM

A liquid-crystal compound can take on more than one type of mesomorphic structure as the temperature is raised. Transitions between these various forms occur at definite temperatures and signal abrupt changes of internal mesophase order. A few examples of polymesomorphic compounds are listed in Table 3 along with their respective transition temperatures. It is apparent that for a compound that displays both a smectic and a nematic mesophase the smectic structure always occurs at the lower temperature. This is of course consistent with the greater degree of order within the smectic arrangement.

Whereas smectic-to-smectic transitions are known to occur, no nematic-to-nematic transformations have been observed. Structural differences between the different smectic modifications in the former case are not understood (see Section 5).

Many open-chain aliphatic esters of cholesterol adopt a smectic structure as well as the cholesteric one. There exists no known case, however, of a compound having both a cholesteric and a nematic mesophase.

Up to this point we have discussed exclusively liquid-crystal mesophases that occur on the melting of solids. The production of the mesomorphic state by heat is referred to as *thermotropic* mesomorphism. Just as the regular lattice of a solid can be disrupted by thermal vibrations, a solvent can also disrupt a crystal lattice. Hence liquid crystals may be produced when certain

Table 3 Examples of Some Polymesomorphic Liquid Crystals

Compound	Formula	Transition Temperature (°C)[a]
4'-n-Pentyloxybiphenyl-4-carboxylic acid	(C$_5$H$_{11}$O-biphenyl-COOH structure)	S $\overset{227.5}{\rightleftharpoons}$ SM $\overset{229.5}{\rightleftharpoons}$ NM $\overset{275}{\rightleftharpoons}$ L
p-Ethoxybenzol-p-azoxybenzoic acid–ethyl ester	(C$_2$H$_5$OCO–N=N$^+$–O$^-$ –OC$_2$H$_5$ structure)	S $\overset{76}{\rightleftharpoons}$ SM $\overset{83}{\rightleftharpoons}$ NM $\overset{112}{\rightleftharpoons}$ L
5-Chloro-6-n-propyloxy-2-naphthoic acid	(Cl, OC$_7$H$_{15}$ naphthalene COOH structure)	S $\overset{165.5}{\rightleftharpoons}$ SM $\overset{176.5}{\rightleftharpoons}$ NM $\overset{201}{\rightleftharpoons}$ L
Ethyl-p-(4-ethoxybenzyl-diamino)-cinnamate	(C$_2$H$_5$O– –CH=N– –CH=CH–CO–OC$_2$H$_5$ structure)	S $\overset{78}{\rightleftharpoons}$ SM I $\overset{110}{\rightleftharpoons}$ SM II $\overset{154}{\rightleftharpoons}$ NM $\overset{158}{\rightleftharpoons}$ L
4-Methylbenzylidene-amino-4'-octoxybiphenyl	(C$_8$H$_{17}$O– –N=CH– –CH$_3$ structure)	S $\overset{155}{\rightleftharpoons}$ SM $\overset{178}{\rightleftharpoons}$ NM $\overset{222}{\rightleftharpoons}$ L
Cholesteryl pelargonate	(CH$_3$(CH$_2$)$_7$CO–O–cholesteryl, C$_8$H$_{17}$ structure)	S $\overset{78}{\rightleftharpoons}$ SM $\overset{79}{\rightleftharpoons}$ CM $\overset{90.5}{\rightleftharpoons}$ L

[a] Abbreviations: S = solid, SM = smectic mesophase, NM = nematic mesophase, CM = cholesteric mesophase, L = isotropic liquid.

compounds are dissolved in appropriate solvents. At low solvent concentrations these substances form an anisotropic phase that is neither a true solid nor a true solution. The phenomenon is termed *lyotropic* mesomorphism. At higher solvent concentrations lyotropic liquid crystals convert into normal isotropic solutions. Water solutions of 9-chloro-phenanthrene-3-sulfonic acid, certain naphthylamine disulfonic acids, and the potassium salts of methyl orange, naphthol yellow, and *p*-aminoazotoluene are anisotropic (26). Many colloidal suspensions of soaps and polypeptides exhibit liquid-crystal properties. Viscous solutions of poly-γ-benzyl-L-glutamate in helicogenic solvents, such as dioxane and CH_2Cl_2, show anisotropic behavior characteristics of the cholesteric mesophase (27) (see Section 7).

5. PHYSICAL PROPERTIES

In this section we discuss some physical aspects of liquid crystals that are peculiarly a result of their anisotropic molecular organization.

a. Viscosity

As already mentioned, the mobility of liquid-crystal molecules in the nematic structure is similar to that in the normal, isotropic liquid phase. Cholesteric and smectic liquid crystals, on the other hand, are typically quite viscous.

In examining a specimen of a smectic liquid crystal prepared in the terraced, Grandjean plate texture, we would find that the material displays a differential viscosity, movement in directions parallel to the smectic layers being very much easier than that in the perpendicular direction, as the strata yield to layer flow. In usual circumstances, however, such a unique arrangement does not obtain, and the mesophase is highly viscous in every direction.

Experiments (28–31) on nematic liquid crystals, such as *p*-azoxyanisole, indicate that the molecules of this structure readily orient themselves with their long axes more or less parallel to the direction of flow. This results in a viscosity lower than that of the isotropic liquid, as the rod-shaped molecules find it easier to stream past one another in the flow-oriented phase.

Flow may be impeded if an electric or magnetic field is applied at right angles to the direction of shear. In this manner Miesowicz (32) was able to raise the viscosity of *p*-azoxyanisole in the nematic state above that of the isotropic liquid. When the mesophase is so aligned, the viscosity of course becomes anisotropic.

A discontinuity in the viscosity of *p*-azoxyanisole and *p*-anisal-*p*-aminoazobenzene occurs at the nematic–isotropic liquid transition point (29). Curious pretransition effects have also been noticed in both viscosity and careful density and volume measurements (29). The phenomenon probably reflects the gradual and cooperative onset of a major reduction in internal molecular order within a few degrees of the transition temperature.

The viscosity of the cholesteric mesophase is much higher than that of the corresponding isotropic liquid and is strongly dependent on shear rate (30). A more complicated flow process for cholesteric liquid crystals is evident.

b. Liquid Crystals in Electric and Magnetic Fields

It is characteristic of liquid crystals that their physical properties are enormously influenced by electric and magnetic fields (as well as surfaces). We have already discussed the effect of external fields on the viscosity of nematic melts.

Liquid crystals are easily and rapidly oriented in the presence of electric and magnetic fields of even small intensity. Early work (33) on the orientational effects of magnetic fields on nematic mesophases demonstrated that the molecules do not react to the disturbance individually but must be associated in "swarms" of approximately 10^5 molecules in size. These swarms are considered to be dynamic molecular bundles that have no permanent integrity but nevertheless respond as a unit on short time scales. Molecules are constantly forming and breaking away from a given swarm cluster. There is a considerable amount of evidence from many physical sources for the existence of these molecular aggregates (34). Dielectric loss measurements (33, 35, 36) indicate that the "fundamental units" in p-azoxyanisole have a relaxation time of 10^{-5} sec as compared with 10^{-11} sec for single molecules. Polarization saturation has been observed in fields of only about 10^4 V/cm in several nematic liquid crystals (37). To obtain this complete alignment in fluids of ordinary polar molecules would require a field of more than 10^7 V/cm.

Recent X-ray studies of Chistyakov and Chaikovskii (38) show that even strong magnetic fields have little effect on the mutual orientation of *individual* molecules. The alignment of small volume elements containing large numbers of molecules, however, occurs readily. Almost complete alignment of these otherwise randomly oriented volume elements was achieved in a magnetic field of only 500 G. Little change in the ordering of neighboring molecules occurred, however, as the field was increased some threefold.

Thin layers of p-azoxyanisole and other nematic liquid crystals have several electrical properties that are characteristic of ferroelectric crystals. Williams (39), for example, has reported the formation of optical domain patterns in the presence of moderate electrical fields. Figure 5 shows quite clearly the difference in the appearance of p-azoxyanisole with and without an electric field.

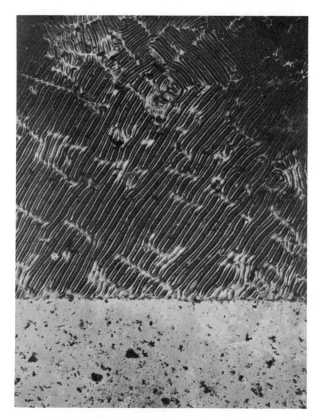

Fig. 5 Optical domain pattern formed in thin section of *p*-azoxyanisole liquid crystal in the presence of an electrical field. No field was applied to the liquid in the relatively clear portion of the photograph. (Reproduced by permission of Dr. Richard Williams, RCA Laboratories, Princeton, N.J.)

Ferroelectric hysteresis loops of polarization versus electric field have also been observed in various nematic liquid crystals (37, 40). The appearance of switching transients and high low-frequency dielectric constants further support the possibility of ferroelectricity in liquid crystals. Williams and Heilmeier (37) notice that all three electrical anomalies persist well into the isotropic phase. Similar hysteresis loops also occur for some liquids that are not mesomorphic.

The ferroelectric nature of liquid crystals is the subject of some debate (41–43). The phenomenon may in fact be explained on the basis of piezoelectric effects induced by strains of elastic curvature within the liquid-crystal structure (43).

The dielectric constant and loss factor of liquid crystals may be made anisotropic by the ordering influence of electric or magnetic fields (44–50) (see Fig. 6). The dielectric anisotropy of nematic 4,4'-di-p-methoxyazoxy-benzene increases rapidly with increasing magnetic-field strength up to approximately 1000 G, where it becomes independent of field strength (44). Differences between the dielectric constant and loss, longitudinal and transverse to applied fields, abruptly disappear at the mesophase–isotropic liquid transition point.

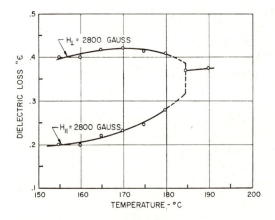

Fig. 6 Temperature dependence of the dielectric loss of anisal-p-aminoazobenzene at a frequency of 24 kMHz (50). An external magnetic field of 2800 G was applied parallel and perpendicular to the microwave electric field. Above the clearing point (185°C) the effect of the magnetic field was negligible. (Reproduced by permission from *Advances in Chemistry Series No. 63*.)

The color and reflectivity of cholesteric liquid crystals are dramatically altered by electric fields (51–53). Electric fields have been shown to produce an increase in the intensity of light reflected from the cholesteric mesophase. The nature of the scattered light depends not only on sample temperature, wavelength of incident light, and observation angle but also on the sample's past thermal and mechanical history. This is consistent with the extreme sensitivity of liquid-crystal order to external stimulus. A hysteresis effect involving the retention of acquired brightness in the absence of the electric field has been observed in cholesterol nonanoate (53).

c. Heat and Volume Changes at Liquid-Crystal Phase Transitions

Transitions from solid to mesophase, from mesophase to isotropic liquid, and between multiple mesophase structures are of the first order (54–56). Pretransition and posttransition effects are observed to occur, however, as the volume begins rapidly changing within a few degrees of the transition temperatures. These anomalies lead to some confusion in the determination of the transition order. Definite volume discontinuities are observed, however, and this is characteristic of first-order phase transitions.

Bauer and Bernamont (57) report a discontinuous volume change amounting to about 0.6% at the nematic-to-isotropic transition of p-azoxyphenetole. An increase in volume some 14 times greater (8.4%) was observed at the solid melting point.

In contradiction, more recent dilatometric work (58) on the structurally similar p-azoxyanisole shows that the density change at the solid-to-nematic transition is no more than 0.1%, whereas the change at the nematic-to-isotropic transition is about 0.36%. Measurements in this study of nematic p-anisal-p-aminoazobenzene and anisaldazine also showed that larger volume changes occur at the transition to the isotropic liquid. The experimenter concluded that the trend is indicative of the substantial residual order within the nematic mesophase.

The thermodynamic behavior of these compounds, however, is just the reverse, as the higher temperature nematic-to-isotropic transitions involve the smaller latent heats (56). p-Azoxyanisole absorbs 28.1 cal/g at the solid-to-nematic transition and only 0.68 cal/g at the clearing point. The corresponding heats of transition (melting, clearing) of anisaldazine and the azobenzene derivative are 26.5 versus 0.59 cal/g and 25.9 versus 0.41 cal/g, respectively.

Smectic ethyl-p-azoxybenzene melts with a heat change of 14.3 cal/g, whereas the smectic-to-isotropic transition occurs with an absorption of only 3.8 cal/g (59).

Pretransition anomalies are also noted in measurements of specific heat, viscosity, dielectric constant, optical transparency and flow and magnetic birefringence (55). Real discontinuities in these physical properties usually can be detected at transition points, which is characteristic of transitions of the first order.

d. X-Ray Diffraction in Liquid Crystals

X-Ray diffraction studies of mesomorphic compounds give information about molecular arrangements in the liquid crystal. The order present in some of the fluid mesomorphic states makes it possible to obtain detailed information about typical spacings between molecules. X-Ray studies of mesomorphic compounds below the temperature of transition from the solid to the liquid crystal (where these substances behave as true crystals) also give important information about the structure of the liquid crystal.

i. Smectic Mesophase

The two-dimensional order characteristic of the smectic mesophase makes this state particularly suitable for X-ray investigations. Between the solid-to-smectic and the smectic-to-isotropic transitions p-azoxybenzoate and ethyl p-azoxycinnamate exhibit diffraction patterns that are indicative of a regular arrangement of the molecules in layers. Above the smectic-to-isotropic transition temperature no diffraction patterns are observed for these compounds. Below the solid-to-smectic transition the diffraction patterns are similar to those of normal three-dimensional crystals.

Substantial differences are apparent, however, when the structural information obtained from the solid state is compared with that found in the smectic liquid-crystal state. The distance between the molecular layers of the smectic state, for example, is roughly equal to the length of the molecule. On the other hand, the distance between the layers in the crystalline solid are some 20 to 30% smaller than the molecular length. These differences can be reconciled by proposing that the long molecular axes are tilted at some angle with respect to the layers in the solid state, whereas they are normal to the layers in the smectic state. Similar observations have been reported for thallium oleates and thallium sterate (60).

These findings have suggested a mechanism to explain how some mesomorphic compounds can exist in more than one smectic phase (smectic polymesomorphism) (see examples in Table 3). Specifically, on passing through successive mesophases, the angle of inclination of the long molecular axes to the plane of stratified layers may increase in stages. Well-defined inclinations correspond to different smectic mesophases.

In general X-ray studies of smectic mesophases confirm the ideas reached on the basis of optical studies regarding the stratified arrangement in this type of mesophase.

ii. Nematic Mesophase

X-Ray diffraction techniques have been less successful in elucidating the details of the molecular arrangement in other mesomorphic states, however. The lower degree of order in the nematic mesophase results in diffraction patterns that are similar to those of isotropic liquids. The amount of structural information can be increased if nematic liquid crystals are examined while oriented by means of an external electric or magnetic field.

Recently Chistyakov and Chaikovskii (7) studied the diffraction pattern of p-azoxyanisole in a magnetic field. The most prominent spacing, 3.5 Å, could be accounted for if the molecules were close-packed, side by side, with the planes of the aromatic rings in neighboring molecules being parallel. Early work by Bernal and Crowfoot (23) on the solid crystal of this compound gives additional information about the molecular arrangement in the nematic state. In the solid state the long axes of the *p*-azoxyanisole molecules are parallel, but the centers of gravity of nearest neighbors are shifted relative to one another by one-half the molecular length; that is, the ether group at the end of a given molecule is positioned very close to the azoxy groups of its nearest neighbors. It is possible that this staggered arrangement persists in the nematic state, although the long molecular axes would be considerably less aligned in the nematic mesophase.

iii. *Cholesteric Mesophase*

The details of the molecular arrangement in mesophases of the cholesteric structure are not known. Very little microscopic information is obtained by applying X-ray studies directly to this mesophase, and little can be inferred from studies of crystalline solids that melt to form the cholesteric mesophase.

iv. *Degree of Intermolecular Ordering*

Small-angle X-ray methods offer a direct means of measuring long-range order in liquids. This method has recently been applied to liquid crystals to measure the degree of intermolecular ordering both in the isotropic and liquid-crystal phases as a function of temperature and in the vicinity of the nematic-to-isotropic transition (61). For *p*-azoxyanisole a correlation length of 300 Å was first observed some 16°C above the nematic-to-isotropic transition temperature. This length increased rapidly to approximately 1850 Å at the transition temperature and remained constant throughout the nematic region. An estimate of the mean number of molecules in the associated regions (swarms) was obtained from the correlation length. The value of 10^6 molecules per swarm is in good agreement with earlier reports (1). As already pointed out, however, the ordered regions must be considered as dynamic entities, with constant changes of molecules between ordered and disordered regions.

6. LIQUID CRYSTALS IN BIOLOGICAL SYSTEMS

It has been recognized that many biological systems exhibit the properties of liquid crystals (62). Considerable concentrations of mesomorphic compounds have been found in many parts of the body, often as sterol derivatives. A liquid-crystal phase has been implicated in at least one major degenerative disease (atherosclerosis) (63). Living tissue, such as muscle, tendon, ovary, adrenal cortex, and nerve, shows the optical-birefringence properties that are characteristic of liquid crystals. The mesomorphic state has been identified in many pathological tissues, particularly in areas of large lipid deposits. Massive deposits of mesomorphic compounds have been found in the kidneys, liver, brain, spleen, marrow, and aorta walls as cholesterol derivatives. Certain living sperms possess a mesomorphic state. Also, solutions of tobacco mosaic virus, collagen, hemoglobin of sickle-cell anemia, native protein, nucleic acid genetic material, and fibrinogen show resemblances to the state. A liquid-crystal condition has also been obtained in concentrated solutions of a synthetic helical polypeptide, polybenzylglutamate (27).

Liquid crystals are not only present in living tissues but also appear to play an important role in its structure and biochemical function. The dynamic plastic quality of the cellular structure (i.e., capacity for change in shape and material interchange) has been attributed to the property of liquid crystal-linity. It may be that catalytic processes at the cellular level find a favorable environment in the mesomorphic structure.

Much more work needs to be done before the true significance of this state in biological systems is understood. However, it should be pointed out that mesomorphic modifications are singularly well suited to provide the delicate balance of organization and lability so characteristic of life processes. It seems probable that liquid crystallinity has important biological conse-quences, both because of its presence in living material and its unusual dependence on slight changes in composition and in the physical and chemical environment for its formation, continuation, or cessation.

The class of mesomorphic compounds that has received the most attention in this context is the phospholipids. This attention is well deserved since phospholipids comprise some 30% of the human brain, are a major component of the myelin sheath (the fatty sheath wrapped around nerve fibers), and are found in blood corpuscles and in the chloroplasts of plants.

Phospholipids are amphiphilic molecules; that is, these large molecules possess two distinct and separate regions: one lipophilic (hydrophobic) in character; the other, hydrophilic. They consist of two hydrocarbon chains, 14 to 18 carbon atoms in length (lipophilic section), attached to a polar group containing phosphorus and nitrogen atoms (hydrophilic section). The structural formula of α-cephalin (phosphatidylethanolamine) is shown below.

The degree of unsaturation and the hydrocarbon chain lengths found in naturally occurring phospholipids are dependent on the type of cell membrane from which the phospholipids have been extracted.

Chapman (64) has investigated the phospholipids by a variety of techniques, including infrared-absorption spectroscopy, X-ray diffraction, optical birefringence, differential thermal analysis (DTA), and nuclear magnetic resonance. One particularly striking observation from the infrared studies is the high degree of mobility in the hydrocarbon section of these molecules. The infrared spectra of the hydrocarbon chains are suggestive of a liquid at room temperature, although the physical appearance of the phospholipids is that of a solid. Since the melting point of the anhydrous phospholipids is above 200°C, these observations suggest that phospholipids exist as liquid crystals at room temperature. It has also been shown (65) that the mesomorphic transition temperature of liquid-crystal phospholipids is dependent on the amount of water in the lipid phase. The degree of unsaturation and its perturbing effect on the conformation of the hydrocarbon chain also influence the mesomorphic transition temperature. As the chains of the phospholipid becomes more unsaturated, the transition temperature is lowered.

Stein (66) has recently reported similar order–disorder transitions among hydrocarbon chains of intact membranes of microorganisms. Furthermore, pure phospholipids extracted from the microorganisms exhibit the same temperature dependence on the degree of unsaturation as that observed in intact membranes. The degree of unsaturation in the latter was varied by controlling the growth medium of the microorganism. These findings are strong evidence for a liquid-crystal organization of the phospholipids in living cells.

The nature of the supramolecular organization found in the liquid-crystal phases of phospholipids has also been investigated. Luzzati and Husson (62) have reported that brain lipids exist in several different types of liquid-crystal structure. Although phospholipids are relatively insoluble in water, the amphiphilic nature of the molecules nevertheless cause them to assume specific arrangements in water. The two main types of structures found by these workers are as follows:

1. A lamellar structure, composed of alternating polar and hydrocarbon layers, similar to the smectic mesophase.

2. A hexagonal structure in which the water molecules lie in the interior of long cylinders constructed in such a manner that the polar groups of the phospholipid molecules are on the inside of the cylinder and the hydrocarbon chains extend outward from the surface.

Cholesterol is not readily dispersed in water or electrolyte solutions under normal conditions. A large amount of cholesterol is found, however, in the brain and is integrated into all cells and most body fluids. It is thought that the liquid-crystal phospholipids may be responsible for the solubilization of cholesterol. With regard to this possibility, NMR measurements indicate a change in the mobility of the phospholipid hydrocarbon chains when cholesterol is added to dispersion of liquid-crystal phospholipids (67). These studies suggest that NMR spectroscopy is a potentially powerful technique for elucidating the molecular details of the associations between phospholipids and other molecules.

Other classes of molecules found in biological systems—for example, helical macromolecules—are known to form lyotropic liquid crystals. These compounds are discussed in the next section.

The unusual combination of lability and lateral cohesion makes the liquid-crystal state biologically useful. Now that structures with the characteristics of liquid crystals have been observed in the extracts of animal tissue, increasing attention has been given to the role of liquid crystals in biological systems (68).

7. POLYMERIC LIQUID CRYSTALS

The term "liquid crystal" has been used by many workers in the field of high polymers to describe supramolecular arrangements in amorphous and semicrystalline polymers. The one- and two-dimensional order associated with the liquid-crystal phases apparently is also characteristic of the type of molecular packing that is found in the amorphous regions of close-packed polymer chains. There are, indeed, well-established incidences of mesomorphism in certain classes of polymers, two examples being amorphous block copolymers and synthetic polypeptides.

a. Block Copolymers

In the preceding section we pointed out that amphiphilic molecules form lyotropic liquid crystals in suitable solvents. This type of mesophase also occurs in aqueous solutions of Arkopals (11) [the condensation product in the reaction of p-nonylphenol with low-molecular-weight polyoxyethyleneglycols (69)].

$$CH_3(CH_2)_8 - \!\!\!\bigotimes\!\!\! - O-(CH_2CH_2-O)_{p-1}-CH_2CH_2OH$$

11

Structurally similar copolymers of polystyrene–polyoxyethylene (**12**) also form liquid-crystal phases in the presence of a good solvent for one of the block units (70).

$$H-(CH_2-CH_2)_{\overline{m}}(CH_2CH_2O)_{p-1}-CH_2CH_2OH$$
$$\underset{C_6H_5}{|}$$

12

The crystallizable block of polyoxyethylene is not essential for the existence of the liquid-crystal phases, however. Several amorphous A–B type block copolymers, including polystyrene–polyisoprene, polystyrene–polybutadiene, and polyisoprene–poly-2-vinylpyridine, are liquid crystals under certain conditions. The liquid-crystal behavior arises from the differences in the solubility of the blocks within the copolymer. When the copolymer is placed in a solvent specific for one block, the insoluble block moieties tend to aggregate in regular arrangements in the gel formed by the soluble portion of the polymer.

The nature of the arrangement depends on the concentration and the type of solvent, the molecular weight of the copolymer, and the precise composition of the blocks. Three types of mesomorphic arrangement have been identified by using X-ray diffraction studies on the gels:

1. *Lamellar structure*: a set of plane, parallel, equidistant sheets; each sheet results from the superposition of two layers, one formed by the insoluble block B, the other by the solution of the soluble block A in the preferential solvent X. Example: A = polystyrene, B = polyisoprene or polybutadiene, X = toluene.

2. *Hexagonal structure*: indefinitely long cylinders arranged in a regular hexagonal two-dimensional array; the long cylinders contain the insoluble block *B* and are separated from one another by the solution of the soluble block A. Example: A = polyisoprene, B = poly-2-vinylpyridine, X = ethyl acetate.

3. *Cubic structure*: spherical particles filled by the insoluble block B packed in a centered cubic lattice and separated from one another by the solution of the soluble block A. Example: A = polyisoprene, B = polystyrene, X = isoprene monomer.

Changing the solvent so that B becomes the soluble block causes an inversion of the liquid-crystal structure; for example, for the hexagonal example above, when X = dioxane, the cylinders are filled with the insoluble polyisoprene block. It is suspected that the nature of the insoluble block is an important factor in determining the type of structure formed by the copolymers. Studies on different copolymers with the same soluble block and A–B–A' block copolymers with two different soluble blocks may clarify this point.

b. Synthetic Polypeptides

Fifteen years ago it was observed that synthetic polypeptides dissolved in certain solvents exist in a rigid rodlike α-helix conformation rather than in the random-coil configuration (71).

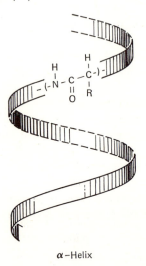

α−Helix

The observation stimulated a large body of investigation of the dilute-solution properties of this class of polymer. In more concentrated solutions (10–50% polymer) the synthetic polypeptide poly-γ-benzyl-L-glutamate (PBLG; $R = CH_2CH_2CO—O—CH_2C_6H_5$) has been observed to form a lyotropic liquid-crystal phase. Robinson and co-workers have extensively characterized the structure of this ordered phase and have found it to be similar in many respects to the cholesteric liquid-crystal structure (27).

The liquid-crystal phase of PBLG can be described in the following manner: In any given volume element or domain the helical axes of the molecules are essentially parallel. In such a domain it is possible to visualize a set of equispaced parallel planes in which the helical axes lie. The optically active molecular conformation causes molecules in a given plane to have a slight twist relative to those in adjacent planes about an axis of torsion perpendicular to the set of planes. The incorporation of successive layers of appropriately oriented molecules results in a helicoidal supramolecular structure. The texture axis (axis of torsion) adopts all orientations throughout the macroscopic fluid, however, and the change in orientation of this axis from one region to another occurs in a continuous fashion. When viewed between crossed Polaroids these birefringent solutions present an image that is very reminiscent of a fingerprint (see Fig. 7a). The spacing between the

(b)

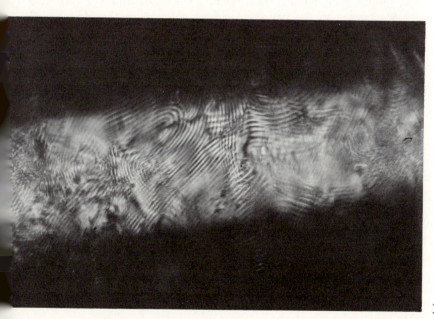

(a)

Fig. 7 Retardation lines characteristic of a helicoidal supramolecular structure are observed in the photomicrographs of both the liquid and solid states of poly-γ-benzyl-L-glutamate (PBLG): (*a*) birefringent fluid liquid-crystal solution of PBLG in dioxane, $S = 50$ microns, 10% PBLG (vol.); (*b*) birefringent solid film of PBLG plasticized by 3,3'-dimethylbiphenyl, $S = 2$ microns, 30% PBLG (vol.).

153

alternating bright and dark lines is equal to one-half the pitch of the heli-coidal structure.

Polypeptide liquid crystals differ from the previously mentioned lyotropic phases in that they depend on no specific interactions between the solvent and the polymer molecule. The mesomorphic behavior appears to be a consequence of the extreme asymmetry of the rodlike macromolecules. The critical volume fraction depends on the axial ratio D/L of the rodlike particle in the following manner (where D is diameter and L is length):

$$V_c = 4(D/L)$$

The relation has been confirmed experimentally by measuring the volume fraction at which the liquid-crystal phase appears in PBLG samples of different molecular weight (72). In fact a statistical mechanical development based on a lattice model for polymer solutions predicts that at a certain volume fraction, V_{rods}, a solution of noninteracting rodlike particles in a solution of inert small solvent molecules should form an ordered aniso-tropic phase.

When solid films of pure PBLG are cast from such solvents as chloroform or methylene dichloride, X-ray evidence (73) and anisotropic swelling characteristics (74) clearly indicate that the rodlike PBLG molecules lie in the plane of the film (parallel to the casting surface). The helical axes are, however, randomly oriented in this plane. These observations suggest that the cast films retained the helicoidal structure found in the fluid liquid-crystal solutions of PBLG.

Carefully prepared plasticized films do in fact exhibit the optical retardation lines characteristic of the helicoidal structure in the liquid-crystal phase of PBLG (see Fig. 7b).

The lateral spacing between PBLG helices in solid films diluted with plasticizer has been examined. X-Ray studies show that this spacing increases continuously with increasing plasticizer concentration until phase separation occurs in the films (50% PBLG) (75). The concentration dependence of this spacing is the same as that observed by Robinson (27) in the fluid liquid-crystal phase of PBLG. The spacing between the optical retardation lines is also dependent on PBLG concentration in the plasticized films. Analogous behavior has been reported for liquid-crystal solutions of PBLG. This is quite convincing evidence that the unusual supramolecular arrangement of the fluid liquid-crystal phase of PBLG does exist in the solid state of mixtures of PBLG plus plasticizer. The PBLG films are solid with regard to mechani-cal properties and possess a thermodynamically stable liquid-crystal molecular arrangement. This property is different from the fluidity previously associated with liquid crystals and probably results from the high molecular weight of the PBLG molecules.

It has been known for some time that magnetic fields on the order of several hundred oersteds cause spontaneous ordering of nematic liquid crystals (33). The local order present in a nematic liquid crystal allows the magnetic field to interact with the diamagnetic anisotropy of large number of molecules (swarms) in a cooperative manner, causing the direction along which the diamagnetism is the smallest to align parallel to the applied magnetic field.

The influence of a magnetic field on the "cholesteric" structure of PBLG solutions has been observed indirectly by using NMR spectroscopy. It was found that the NMR spectrum of methylene dichloride is split into a doublet in the liquid-crystal phase, PBLG + CH_2Cl_2 (74, 76, 77). No spectrum is observed for PBLG in these concentrated solutions because nuclear dipolar interactions broaden the signal of the polymer beyond the limits of detection in high-resolution NMR signals. On the other hand, the methylene dichloride molecules, though partially oriented by interactions with the polymer, retain a high degree of mobility and exhibit a spectrum whose line widths are similar to those observed in ordinary liquids. The partial orientation or anisotropic tumbling of the methylene dichloride molecule in the liquid-crystal matrix does, however, produce a nonzero average of the direct dipole–dipole interactions between the pair of protons on a methylene dichloride molecule, giving a doublet for its spectrum.

The optical retardation lines in "cholesteric" PBLG solutions are no longer visible in solutions subjected to a strong magnetic field. This observation, the X-ray studies of magnetically oriented films of PBLG (described below), and the NMR results indicate that a sufficiently strong magnetic field untwists the "cholesteric" structure and forms an oriented nematic structure with the rodlike PBLG molecule parallel to the applied field. Splitting of the NMR absorption of methylene dichloride due to intramolecular direct dipole–dipole coupling would be expected in this kind of oriented nematic environment.†

When solutions of PBLG in a volatile solvent are evaporated in the presence of a strong magnetic field, highly oriented films are obtained (75). The orientation occurs while the solutions pass through the concentration range in which they are liquid crystalline and becomes permanently locked in when the mixture of solvent plus PBLG becomes solid. The nature of the orientation in the films indicates that the PBLG liquid crystal in a strong magnetic field has an ordered nematic structure. In this structure the PBLG helices are oriented parallel to the direction of the applied magnetic field.

The uniaxial orientation in these films produces X-ray diffraction patterns that are very similar to the fiber patterns obtained from mechanically oriented

† See reference 78 for a review of this phenomenon.

streaks with Bragg reflections occuring on the equator only, implying that the lateral spacing between helices is fairly regular but that there are probably random displacements parallel to the helical axis. The diffraction data obtained from the magnetically oriented films are quite adequate for determining the helical parameters of the polypeptide conformation. It is interesting to point out that magnetically oriented films of PBLG cast from methlene dichloride have the same helical conformation as that proposed in the early 1950s by Pauling et al. (79) for the α-helix (3.6 residues per turn). Films of PBLG cast from chloroform exhibits a distorted helix (3.5 residues per turn) (80). These observations suggest that small differences in polypeptide conformations occurring in different solvents might be detected by this new technique of orienting polypeptides. It is also quite possible that similiar studies could be extended to other rodlike biological macromolecules that have a liquid-crystal phase, particularly nucleic acids.

8. PRACTICAL USES OF LIQUID CRYSTALS

Some of the unusual physical and optical properties of liquid crystals have been put to practical application.

The temperature-dependent variation in the color of cholesteric liquid crystals has led to the use of these substances in the measurement of temperature and temperature gradients. A cholesteric substance or a mixture of cholesteric substances is found to always exhibit the same color at the same temperature. Moreover, the precise color is a very sensitive function of the ambient temperature. Liquid crystals can be color–temperature calibrated and serve as optical registers of absolute temperatures and temperature gradients. Temperature variations of less than 0.1°C can be visually distinguished with liquid-crystal temperature sensors. It is possible to locate "hot spots" and flaws as small as one-thousandth of an inch in microminiature circuits with these substances. In medicine cholesteric liquid-crystal substances applied to the surface of the skin have been used to locate veins, arteries, and tumors, which are warmer than the surrounding tissue.

The pitch of the helical cholesteric structure, and hence its color, is also profoundly affected by the presence of small impurity molecules. The addition of only a few parts per million of impurity will result in an immediate change of color, extending through the medium. This property has been exploited to construct trace-vapor indicators for certain chemicals.

Electrooptical response characteristics of nematic liquid crystals are being examined as the engineering basis of new electrical display devices. When a thin film of a transparent nematic fluid is placed between two conductive glass plates, the substance is observed to turn rapidly opaque at any point where a modest voltage is applied. The opaque areas will in turn reflect incident light and appear bright. If voltages are applied to the liquid-crystal film in some pattern, an image will result. One can envision many technological applications of this effect: electrooptical light shutters, automatic dimming windows, alphanumerical displays, clocks with no moving parts, and thin, wall-mounted television screens. The image devices would not be subject to contrast "washout" under conditions of high light levels. In fact increasing the ambient light level would increase the brightness of the opaque areas, with little effect on the transparent regions. The display of information in direct sunlight becomes feasible. This capability will undoubtedly open up frightening new possibilities to the billboard and television industries.

Nematic liquid crystals have become useful research tools in the field of magnetic resonance. Molecules dissolved in such solvents experience an anisotropic environment and give a very highly resolved NMR spectrum exhibiting intermolecular dipole–dipole fine structure. Analysis of the spectra of molecules in liquid-crystal solvents leads to information on the anisotropy of chemical shifts, direct magnetic dipole–dipole interactions, indirect spin–spin couplings, and bond angles and bond lengths. This new kind of high-resolution NMR spectroscopy on partially oriented molecules has not only contributed to the understanding of NMR spectroscopy but has also provided a new method of determining molecular structure, particularly for small organic molecules (42, 81-83).

Finally some liquid crystals have been used to separate *meta-para* isomers of organic compounds. The chromatographic effect apparently is due to the preferential retention of one geometric isomer over the other in the anisotropic matrix formed by the liquid-crystal molecules (84, 85).

REFERENCES

1. G. W. Gray, *Molecular Structure and the Properties of Liquid Crystals*, Academic Press, New York, 1962.

2. G. H. Brown and W. G. Shaw, *Chem. Rev.* **57**, 1049 (1957).

3. A. Saupe, *Angew. Chem. Intern. Ed.* **7**, 97 (1968).

4. I. G. Chistyakov, *Soviet Physics–Crystallography* **5**, 917 (1961).

5. V. A. Usoltseva and I. G. Chistyakov, *Russian Chem. Rev.* **32**, 495 (1963).

6. W. Kast, in Landolt–Bornstein, 6th ed., Vol. II, Part 2a, Springer, Berlin, 1960, p. 266.

7. I. G. Chistyakov and V. M. Chaikovski, *Soviet Physics–Crystallography* **12**, 770 (1968).

8. L. A. Staveley, *Ann. Rev. Phys. Chem.* **13**, 351 (1962).

9. J. G. Aston, in D. Fox et al., Eds., *Physics and Chemistry of the Organic Solid State*, Vol. I, Interscience, New York, 1963.

10. *The Journal of Physics and Chemistry of Solids* **18**, January 1961 issue.

11. J. G. Powles, D. E. Williams, and C. P. Smyth, *J. Chem. Phys.* **21**, 136 (1953).

12. L. M. Kushner, R. W. Crowe, and C. P. Smyth, *J. Am. Chem. Soc.* **72**, 1091 (1950).

13. J. T. Thomas, N. L. Alpert, and H. C. Torrey, *J. Chem. Phys.* **18**, 1511 (1950).

14. J. G. Powles and H. S. Gutowsky, *J. Chem. Phys.* **21**, 1695 (1953).

15. F. A. Rushworth, *Proc. Roy. Soc.* (*London*) **A222**, 526 (1954).

16. D. W. McCall and D. C. Douglass, *J. Chem. Phys.* **33**, 777 (1960).

17. J. G. Aston, Q. R. Stottlemeyer, and G. R. Murray, *J. Am. Chem. Soc.* **82**, 1281 (1960).

18. F. A. Rushworth, *Phys. Chem. Solids* **18**, 77 (1961).

19. W. J. Dunning, *Phys. Chem. Solids* **18**, 21 (1961).

20. J. L. Fergason, *Scientific American* **211**, 77 (1964).

21. W. Maier and J. Englert, *Z. Elektrochem.* **64**, 689 (1960).

22. F. Grandjean, *Compt. Rend.* **166**, 165 (1917).

23. J. D. Bernal and D. Crowfoot, *Trans. Faraday Soc.* **29**, 1032 (1933).

24. G. W. Gray, *Mol. Crystals Liquid Crystals* **7**, 127 (169).

25. E. Sackmann, S. Meiboom, and L. C. Snyder, *J. Am. Chem. Soc.* **89**, 5981 (1967).

26. G. W. Gray, *Molecular Structure and the Properties of Liquid Crystals*, Academic Press, New York, 1962, p. 13.

27. C. Robinson, *Mol. Crystals* **1**, 467 (1966).

28. R. S. Porter and J. F. Johnson, *J. Phys. Chem.* **66**, 1826 (1962).

29. R. S. Porter and J. F. Johnson, *J. Appl. Phys.* **34**, 51 (1963).

30. R. S. Porter, E. M. Barrall, and J. F. Johnson, *J. Chem. Phys.* **45**, 1452 (1966).

31. R. S. Porter and J. F. Johnson, in F. Eirich, Ed., *Rheology*, Vol. 4, Academic Press, New York.

32. M. Miesowicz, *Nature* **158**, 27 (1946).

33. L. S. Ornstein and W. Kast, *Trans. Faraday Soc.* **29**, 931 (1933).

34. G. W. Gray, *Molecular Structure and the Properties of Liquid Crystals*, Academic Press, New York, 1962, Chapter IV.

35. W. Kast, *Z. Physik* **71**, 39 (1931).

36. W. Kast, *Phys. Z.* **36**, 869 (1935).

37. R. Williams and G. Heilmeier, *J. Chem. Phys.* **44**, 638 (1966).

38. I. G. Chistyakov and V. M. Chiakovskii, *Kristallografiya* **12**, 833 (1967) [*Soviet Physics–Crystallography* **12**, 770 (1968)].

39. R. Williams, *J. Chem. Phys.* **39**, 384 (1963).

40. A. P. Kapustin and L. K. Vistin, *Kristallografiya* **10**, 118 (1965) [*Soviet Physics–Crystallography* **10**, 95 (1965)].

41. R. Williams, *J. Chem. Phys.* **50**, 1324 (1969).

42. A. Saupe, *Angew. Chem. Intern. Ed.* **7**, 97 (1968).

43. R. B. Meyer, *Phys. Rev. Letters* **22**, 918 (1969).

44. M. Jezewski, *Z. Physik* **51**, 159 (1928).

45. A. Maier, G. Barth, and H. E. Wiehl, *Z. Elektrochem.* **58**, 674 (1954).

46. E. F. Carr, *J. Chem. Phys.* **38**, 1536 (1963).

47. E. F. Carr, *J. Chem. Phys.* **39**, 1979 (1963).

48. E. F. Carr, *J. Chem. Phys.* **42**, 738 (1965).

49. E. F. Carr, *J. Chem. Phys.* **43**, 3905 (1965).

50. E. F. Carr, in *Ordered Fluids and Liquid Crystals*, Advances in Chemistry Series, No. 63, American Chemical Society, Washington, D.C., 1967, p. 76.

51. J. H. Muller, *Z. Natürforschung* **20**, 849 (1965).

52. W. J. Harper, *Mol. Crystals* **1**, 325 (1966).

53. J. H. Muller, *Mol. Crystals* **2**, 167 (1966).

54. P. L. Jain, J. C. Lee, and R. D. Spence, *J. Chem. Phys.* **23**, 878 (1955).

55. W. A. Hoyer and A. W. Nolle, *J. Chem. Phys.* **24**, 803 (1956).

56. E. M. Barrall, R. S. Porter, and J. F. Johnson, *J. Phys. Chem.* **68**, 2810 (1964).

57. E. Bauer and J. Bernamont, *J. Phys. Radium* **7**, 19 (1936).

58. R. S. Porter and J. F. Johnson, *J. Appl. Phys.* **34**, 51 (1963).

59. M. E. Spaght, S. B. Thomas, and G. S. Parks, *J. Phys. Chem.* **36**, 882 (1932).

60. K. Herrman, *Trans. Faraday Soc.* **29**, 1032 (1933).

61. C. C. Gravatt and G. W. Brady, *Mol. Crystals Liquid Crystals* **7**, 355 (1969).

62. V. Luzzati and F. Husson, *J. Cell. Biol.* **12**, 207 (1962).

63. G. T. Stewart, in *Ordered Fluids and Liquid Crystals*, Advances in Chemistry Series, No. 63, American Chemical Society, Washington, D.C., 1967.

64. D. Chapman, *The Structure of Lipids*, Wiley, New York, 1965.

65. D. Chapman, *Sci. J.* **1**, 32 (1965).

66. J. M. Stein, in *Molecular Associations in Biological and Related Systems*, Advances in Chemistry Series, No. 84, American Chemical Society, Washington, D.C., 1969.

67. D. Chapman, in *Molecular Associations in Biological and Related Systems*, Advances in Chemistry Series, No. 84, American Chemical Society, Washington, D.C., 1969.

68. G. T. Stewart, *Mol. Crystals Liquid Crystals* **7**, 75 (1969).

69. F. Husson, H. Mustacchi, and V. Luzzati, *Acta Cryst.* **13**, 668 (1960).

70. A. Douy, R. Mayer, J. Rossi, and B. Gallot, *Mol. Crystals Liquid Crystals* **7**, 103 (1969).

71. P. Doty, J. H. Bradbury, and A. M. Holtzer, *J. Am. Chem. Soc.* **78**, 947 (1956).

72. P. J. Flory, *Proc. Roy. Soc.* **A234**, 73 (1956).

73. A. J. McKinnon and A. V. Tobolsky, *J. Phys. Chem.* **70**, 1453 (1966).

74. E. T. Samulski and A. V. Tobolsky, *Macromolecules* **1**, 555 (1968).

75. E. T. Samulski and A. V. Tobolsky, *Mol. Crystals Liquid Crystals* **7**, 433 (1969).

76. S. Sobajima, *J. Phys. Soc. Jap.* **23**, 1070 (1968).

77. M. Panar and W. D. Phillips, *J. Am. Chem. Soc.* **90**, 3880 (1968).

78. S. Meiboom and L. C. Snyder, *Science* **162**, 1337 (1968).

79. L. Pauling, R. B. Corey, and H. R. Branson, *Proc. Natl. Acad. Sci., U.S.* **37**, 205 (1951).

80. E. T. Samulski and A. V. Tobolsky, *Liquid Crystals and Ordered Fluids*, pp. 111–121, Plenum Press, 1970.

81. G. R. Luckhurst, *Quart. Rev.* **22**, 179 (1968).

82. L. C. Snyder and S. Meiboom, *Mol. Crystals Liquid Crystals* **7**, 181 (1969).

83. S. Meiboom and L. C. Snyder, *Acc. Chem. Res.* **4**, 81 (1971).

84. M. J. S. Dewar and J. P. Schroeder, *J. Am. Chem. Soc.* **86**, 5235 (1964).

85. M. J. S. Dewar and J. P. Schroeder, *J. Org. Chem.* **30**, 3485 (1965).

Crystallinity in Polymers

8

R. S. Stein and A. V. Tobolsky

1. X-RAY DIFFRACTION

A characteristic of the crystalline state is the regular ordering of molecules, leading to the diffraction of X-rays at discrete angles, described by the Bragg equation

$$n\lambda = 2d \sin \theta \tag{1}$$

where 2θ is the angle between the incident and the diffracted X-ray beams (Fig. 1), λ is the wavelength of the X-rays, n is the order of diffraction, and d is the interplanar spacing of the crystalline planes involved in the diffraction. There is an infinite set of such planes, and the possible values of d and of diffraction angles θ are determined by the symmetry and dimensions of the unit cell; for example, polyethylene, like the n-paraffins, crystallizes in an orthorhombic unit cell (Fig. 2) with three perpendicular axes, along which the unit cell has the dimensions $a = 7.40$, $b = 4.93$, and $c = 2.54$ Å. The polymer-chain axis corresponds to the c crystal axis. The possible values of d are then given by

$$d = \left(\frac{h^2}{a^2} + \frac{k^2}{b^2} + \frac{l^2}{c^2}\right)^{-1/2} \tag{2}$$

161

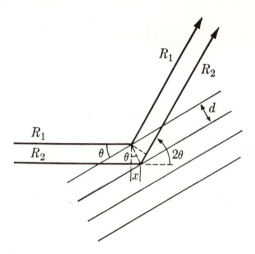

Fig. 1 The geometry of the X-ray diffraction experiment [From Castellan (1)].

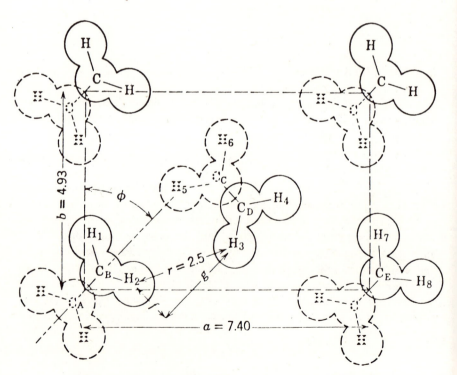

Fig. 2 The unit cell of the polyethylene crystal [From Stein (2)].

where h, k, and l are integers, the Miller indices, that describe the orientation of the crystal plane with respect to the unit-cell axes. The intensity of diffraction is determined by the kind, number, and location of the atoms, within the unit cell, and the fitting of theoretically predicted intensities to measured values is the basis of the art of the determination of crystal structure by X-ray diffraction.

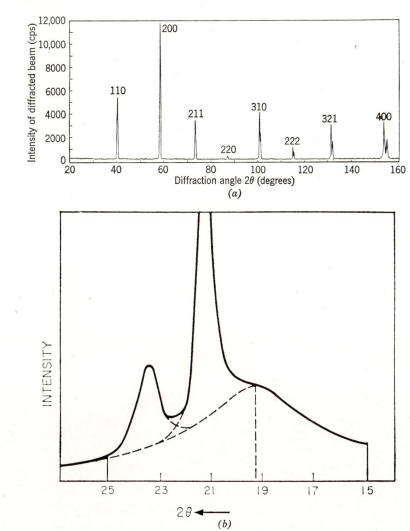

Fig. 3   A comparison of the variation of diffracted intensity with diffraction angle for (a) a large fairly perfect crystal (1) and (b) a semicrystalline polymer (polyethylene) (3).

Large perfect crystals give very narrow diffraction peaks, whereas polymers yield broad diffraction peaks superposed on a diffuse halo (Fig. 3). This broadness is interpreted in terms of the small size and the imperfection of the polymer crystals. If the breadth is interpreted in terms of size, then a characteristic dimension of a polymer crystal, L, perpendicular to a particular diffracting plane, can be obtained from the Scherrer equation (4)

$$L = \frac{\lambda}{B \cos \theta} \tag{3}$$

where B is the half-width of the diffraction peak. In this way one estimates, for example, that the dimension of a typical high-density polyethylene crystal perpendicular to the 110 plane is 500 Å, whereas that of a low-density polyethylene crystal is 300 Å. The crystal's size becomes smaller on stretching and larger with annealing, as might be expected.

Strains and disorder also result in broader diffraction peaks, which can be classified and described according to Hosemann's theories of paracrystallinity (5). Another result of disorder is to increase the rate at which intensity falls off with increasing n, the order of the diffraction. This effect can be used to help separate the contributions to broadening coming from size and from imperfection.

Amorphous polymers, like low-molecular-weight liquids, give broad diffraction peaks associated with the average distances between atoms in the liquid state. The breadth of the diffraction arises because of the statistical distribution of these distances and because of their not being periodic in nature. A classical interpretation of the broad halo part of the crystalline polymer diffraction is in terms of a two-phase model (Fig. 4) in which small

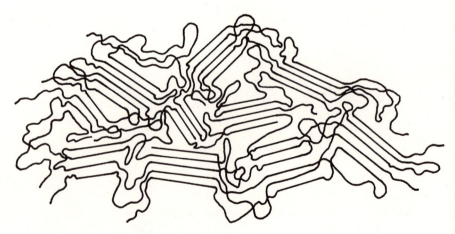

Fig. 4   The two-phase (fringed micelle) model for a semicrystalline polymer (6).

crystals are believed to be interspersed with a liquidlike amorphous phase whose structure is much like that of a completely amorphous polymer (6). The diffraction from the two phases is assumed to be additive, so that the diffracted intensity at a particular angle is assumed to be

$$I = \phi_C I_C + (1 - \phi_C)I_A \tag{4}$$

where ϕ_C is the volume fraction of crystalline material and I_C and I_A are independent of ϕ_C. This is not strictly so, since, for example, we have seen that the width, and hence the intensity, at any point of a diffraction peak is dependent on crystal size and perfection. In consideration of this there is some justification for using the integrated area under a diffraction peak rather than the intensity at a particular angle, since to a first approximation the

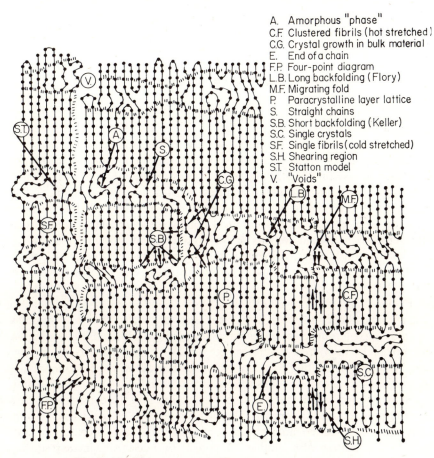

A.   Amorphous "phase"
C.F. Clustered fibrils (hot stretched)
C.G. Crystal growth in bulk material
E.   End of a chain
F.P. Four-point diagram
L.B. Long backfolding (Flory)
M.F. Migrating fold
P.   Paracrystalline layer lattice
S.   Straight chains
S.B. Short backfolding (Keller)
S.C. Single crystals
S.F. Single fibrils (cold stretched)
S.H. Shearing region
S.T. Statton model
V.   "Voids"

Fig. 5   The Hoseman model of a semicrystalline polymer (7).

integrated intensity is believed to depend only on the amount of diffracting material.

Hosemann (7) has shown that a disordered crystalline phase can give rise to diffuse scattering much like the amorphous scattering, so that the interpretation of the diffraction from a semicrystalline polymer in terms of two phases is not unique. An extreme point of view is that of the distributed-defect model in which no amorphous phase at all is postulated but where it is assumed that the diffuse scattering arises entirely from defects within, and on the surface of, the crystals. It is likely that this model is approximated by highly crystalline stereoregular polymers, such as high-pressure crystallized linear polyethylene, and by polyoxymethylene. On the other hand, polymers with relatively low degrees of crystallinity—such as the polyurethanes, polyesters, and polyamides—are better represented by the two-phase model. Most crystalline polymers lie somewhere in between these extremes, with the crystalline regions being of finite size and having disorder, and with the amorphous regions also being of limited size and possessing some order, being restricted by the constraints of the crystalline regions. Hosemann has classified some of the structural features of such a model (Fig. 5).

2. MICROSCOPY

The preparation of single crystals of high polymers from dilute solution has been an exciting new achievement. A typical crystal from a solution of polyethylene is shown in Fig. 6 and has the structure shown in Fig. 7. The morphology is similar to that of an n-paraffin crystal except that its thickness (on the order of 100 Å) is determined not by the molecular weight (except for very low molecular weights or for crystallization at high pressures) but by kinetic considerations. Regular folding with adjacent reentry of chains is shown in Fig. 7, but the validity of this picture has been subject to controversy.

Evidence in favor of regular folding, as shown in Fig. 8a, is (a) the existence of pyramid-shaped crystals with their faces at angles that are crystallographically determined by the regular packing of the folds and (b) the observation of moiré interference patterns between two crystal platelets whose faces are in contact, indicating crystallographic register of their orientations arising from the packing of their fold surfaces. Evidence in favor of irregular folding, as shown in Fig. 8b, includes (a) the observation that the densities of single-crystal preparations are appreciably lower than the unit-cell density, (b) the observation of diffuse amorphous scattering from such preparations, and (c) the observation of a mobile component by nuclear-magnetic-resonance (NMR) spectroscopy. A rational interpretation of these observations is that

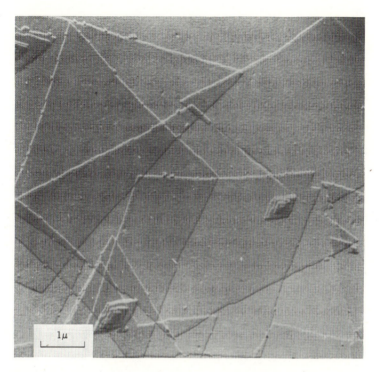

Fig. 6 A single crystal of polyethylene grown from dilute solution (8).

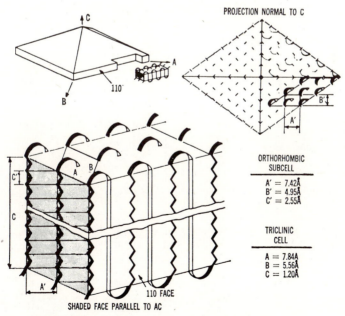

C

PROJECTION NORMAL TO C

A

110°

B

B'

A'

ORTHORHOMBIC
SUBCELL

A' = 7.42Å
B' = 4.95Å
C' = 2.55Å

TRICLINIC
CELL

A = 7.84Å
B = 5.56Å
C = 1.20Å

C'

A B

C

A'

110 FACE

SHADED FACE PARALLEL TO AC

Fig. 7 The structure of single crystals of polyethylene (9).

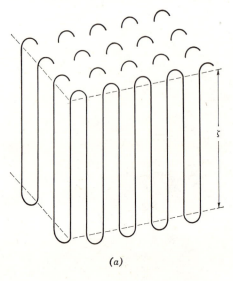

(a)

(b)

Fig. 8 Chain folding with (*a*) adjacent re-entry and (*b*) random reentry (switchboard model) (10).

regular folding can occur, but does not necessarily do so. It is possible that the microscopic evidence arises from selected samples of more perfectly folded crystals.

Recent evidence favoring adjacent rather than random reentry comes from infrared studies of Tasumi and Krimm (11), who have obtained spectra of mixed crystals of ordinary polyethylene and completely deuterated polyethylene. From their analysis it is possible to distinguish the case of random placement of deuterated chains from that shown in Fig. 9, where the chains are arranged in linear arrays. The results prove that ordering prevails and can best be explained on the model of chain folding interconnecting the linear arrays of hydrogenated or deuterated chains. For mixed n-paraffins crystals whose thickness corresponds to the molecular length there is no fold connection among the chains and disordered structures prevail.

The situation for bulk-crystallized polymers is less clear. Electron

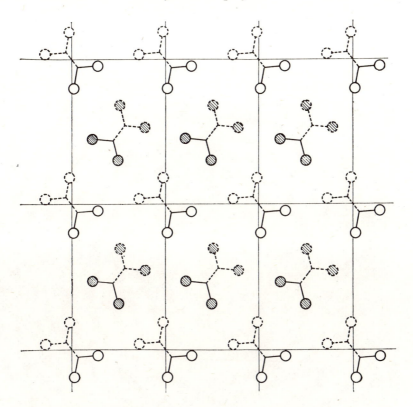

Fig. 9   Ordered structure of a 1:1 mixed crystal of polyethylene (open circles) and deuterated polyethylene (shaded circles). An a—b projection of the unit cell is shown (11).

micrographs reveal free and fracture surface structure interpretable in terms of chain-folded lamellae. The thicknesses of these lamellae are comparable with those of solution-grown single crystals and vary with crystallization temperature in a similar way. The morphology of fragments obtained from nitric acid etching of bulk polymers is similar to that of single crystals. When polyethylene is cocrystallized with *n*-paraffins, after which the *n*-paraffins are extracted out, it yields lamellar structures interconnected with tie chains or extended-chain crystals, whose number increases with increasing concentration of the polymer.

Krimm's infrared analysis favors adjacent reentry of chains in bulk-crystallized polymers. These observations are consistent with the occurrence of folded-chain structures from the melt. It is likely that these structures are interconnected by tie chains that are, in part, responsible for the mechanical integrity of bulk polymer. For polymers with a distribution of crystallizability of molecules, arising from a distribution in molecular weight, branching, or tacticity, segregation will occur in the course of crystallization. The more crystallizable material will form lamellae, and the less crystallizable portions will segregate in interlamellar regions. This material may subsequently crystallize in a secondary crystallization process.

The model for bulk-crystallized polyethylene proposed by Hosemann and

Fig. 10 Photomicrograph of a polyethylene spherulite between crossed Polaroids (8).

shown in Fig. 5 retains aspects of the folded-chain structure of polymer single crystals but contains a variety of defects, such as tie chains, chain ends, and irregular folds.

A characteristic feature of bulk-crystallized polymers is the occurrence of spherulite morphology (Fig. 10). Such spherulites are aggregates of crystals and arise as a consequence of the spherical growth of crystals from hetero-geneous nuclei, as will be discussed later. In cases where the spherulites are volume filling, their diameter is determined by the number of such nuclei per unit volume and may range from a few tenths of a micron up to several centimeters. A characteristic property of such spherulites is the ordered arrangement of their constituent crystals. Usually one crystal axis is oriented along or at some characteristic angle to the radius, giving rise to structures that are anisotropic in their mechanical as well as their optical properties. This can be demonstrated by micro -X-ray diffraction or by selected-area elec-tron diffraction. For example, in polyethylene the b-axis is oriented along the radius and the orientation of the a- and c-axes about the radius varies heli-coidally, as in Fig. 11. This results in the ringed appearance seen when the sample is viewed between crossed Polaroids (Fig. 10). The cause of this twist is not well understood. The period of twist is found to increase with the temperature of crystallization.

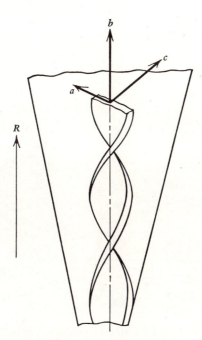

Fig. 11 The orientation of the crystal axes in a polyethylene spherulite.

When closely examined, the spherulites are fibrillar in appearance, and it seems likely that the fibrils, or lamellae, may be identified with chain-folded structures. Electron micrographs (Fig. 12) show that these lamellae are sheet-like and that their twisting is the cause of the helicoidal orientation of the crystal axes. One sees regions in which these lamellae appear to be stacked on their edges, whereas other regions show the lamellae parallel to the plane of the micrograph.

This morphological evidence supports a model in which the radial fibrils of the spherulite are single-crystal-like structures that, in polyethylene, have their *b*-axes parallel to the radius (Fig. 11). These lamellae twist so that the *a*- and *c*-axes rotate about the *b*-axis. In well-developed spherulites the continuity of the extinction bands seen under polarized light (Fig. 10) indicates that such rotation is in phase at a given radial distance from the spherulite center throughout the spherulite. The theory of the polarized-light extinction patterns and of light scattering from such spherulites has been developed and found to agree with experiment.

Spherulites are not completely crystalline. If, as is usually the case, the volume of the polymer is completely filled with spherulites, the degree of crystallinity of the spherulites must be the same as that of the polymer. The amorphous material within the spherulites may be in the form of (a) loose loops of folded-chain crystals, (b) tie chains between crystals, (c) amorphous

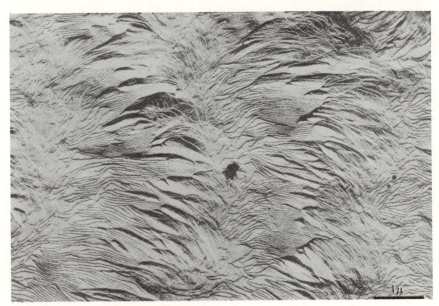

Fig. 12   An electron micrograph of a portion of a polyethylene spherulite showing twisted lamellae [By E. Fischer from Geil [(8)].

material in interlamellar regions, and (d) defects within the crystals. Although the properties of crystalline polymers depend on the kinds of amorphous structures, methods for their determination are not yet available. The measurement of the intensity of fold bands in the infrared spectra may prove a means of determination of the degree of chain folding. The exclusion of poorly crystallizable material into interlamellar regions has been demonstrated by making radioautographs of isotactic polypropylene spherulites cocrystallized with radioactivily tagged atactic polypropylene. The presence of tie chain has been demonstrated by cocrystallizing polyethylene along with an *n*-paraffin, which is later extracted out with a solvent (12). The structures seen (Fig. 13) are not individual tie chains but probably aggregates of tie chains in the form of extended-chain crystals interconnecting the folded-chain lamellae. The number of such tie chains increases with the molecular weight of the polymer and with decreasing temperature of crystallization.

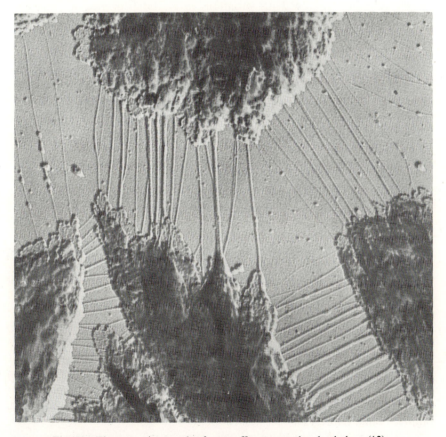

Fig. 13   Electron micrograph of a paraffin extracted polyethylene (12).

The coarseness of the fibrillar structure depends, in part, on this segregation of impurities during crystallization. It is necessary that a crystallizing region have a characteristic thickness δ such that the impurities can diffuse to the surface during crystallization. The order of magnitude of δ is D/G, where D is the diffusion constant of the impurities through the crystallizing matrix and G is the linear growth rate in the direction of the dimension δ. Using reasonable values for D and G, values of δ in agreement with microscopically measured dimensions are obtained. Furthermore, the proper temperature coefficient of δ is obtained from considerations of the temperature coefficients of D and G.

If the degree of crystallinity is to be independent of the distance from the center of the spherulite, it is necessary for the lamella to be branched, so that the concentration of lamellae does not decrease with increasing radius. Such branching will result in a lack of perfect crystallite orientation with respect to the spherulite radius. Thus variations in the birefringence of spherulites with their temperature of growth may be related to variations in the amount of such branching and the angle that the branch makes with the lamella.

Kinetic evidence indicates that not all of the crystallization occurs at the growing spherulite interface (primary crystallization), but some may occur at a later time within the already formed spherulite. Thus "secondary crystallization" involves the formation of crystals between the already formed lamellae, which may come from less crystallizable material rejected during primary crystallization. Such crystals may be differently oriented than the primary crystals.

3. SPECIFIC VOLUME

To a first approximation the specific volumes of the crystalline and amorphous phases, V_C and V_A, are additive, so that

$$V_{sp} = X_C V_C + (1 - X_C)V_A \tag{5}$$

where V_{sp} is the specific volume of the partly crystalline polymer and X_C is the weight fraction of the crystalline phase. Thus it follows that

$$X_C = \frac{V_A - V_{sp}}{V_A - V_C} = \frac{(1/d_A) - (1/d)}{(1/d_A) - (1/d_C)} \tag{6}$$

where d, d_C, and d_A are the densities of the partly crystalline polymer, the crystalline and amorphous phases, respectively. The use of this equation implies that V_A and V_C are constant at a given temperature and independent of X_C. This requires that the structure be best approximated by a two-phase, rather than a paracrystalline, model.

The values of V_C can be calculated from the unit-cell geometry as determined by X-ray diffraction; for example, for an orthorhombic unit cell, such as that of polyethylene, one finds that

$$V_C = \frac{V_{uc}}{W_{uc}} = \frac{abc}{n(M_m/N_A)} \tag{7}$$

where V_{uc} and W_{uc} are the volume and weight of the unit cell, respectively, n is the number of monomer units per unit cell (two for polyethylene), M_m is the molecular weight of the monomer unit, and N_A is Avogadro's number. If the crystal contains defects, an increase in V_C should result and should depend on their number and kind. Neglect of such an increase (as is usual) results in the inclusion of their effect as a decrease in the degree of crystallinity. As V_C will vary slightly with temperature, it is best to either use values of a, b, and c determined at the temperature of measurement of V_{sp} or else to correct V_C for the thermal expansion of the crystal.

It is more difficult to obtain an appropriate value for V_A for use in Eq. 6. One approach is by extrapolation of the value for the molten polymer down to the temperature of measurement of V_{sp}. This assumes that the coefficient of expansion of the amorphous phase is independent of temperature and that its specific volume is not affected by the presence of crystals. The latter requirement is most certainly not true when X_C is high, since one finds from thermal, NMR, and other studies that the amorphous chains are more tightly constrained in a highly crystalline polymer than they are at lower degrees of crystallinity.

For polymers whose glass-transition temperature T_g is higher than the temperature of measurement of X_C it is sometimes possible to quench the polymer from the melt to a temperature below T_g sufficiently rapidly to prevent crystallization. In this way one can obtain an amorphous polymer for which one can measure V_A. It must be assumed that the V_A of this quenched polymer is the same as that of the amorphous regions of the crystalline polymer.

Another way of obtaining an amorphous polymer for measuring V_A is to destroy the crystallites by irradiation. The method has the disadvantage that the density will be affected by the presence of the radiation-induced cross-links.

Densities can be experimentally measured with a pycnometer by measuring the weight loss due to buoyancy or by using a density-gradient column. For routine measurements the latter method is quick and sensitive and requires only very small amounts of sample. For following specific-volume changes during the course of crystallization dilatometers are employed.

One test of the validity of the two-phase model and of Eq. 6 is to compare values of X_C determined by the X-ray and the specific-volume methods. Such a comparison of the variation of X_C during the melting region of polyethylene indicates agreement within experimental error of the methods for this polymer. In other cases, such as polypropylene, a two-phase model is not adequate, and it is necessary to postulate a paracrystalline phase in addition to the crystalline and amorphous phases.

The comparison is often poorest when the degree of crystalline order is low and the crystals are small and imperfect. Under such conditions the small height and the breadth of the crystalline X-ray diffraction make its separation from the amorphous halo and the background difficult. The value of V_c is likely to be higher than that of the ideal crystal. Thus for such polymers as polyvinyl chloride and polyacrylonitrile there is considerable uncertainty as to the value of X_c.

4. INTERACTION WITH ELECTROMAGNETIC RADIATION

It has long been realized that the infrared spectra of crystalline polymers are affected by their crystallinity. Figure 14 compares the infrared spectra in the 14-μ region of polyethylene above and below the melting point (13). The spectra of the crystalline polymer usually exhibit much more fine structure, arising from interactions between vibrations between different molecules within the crystal. For example, the infrared band shown in Fig. 14 at 13.8 μ due to a CH_2 rocking mode of polyethylene in the amorphous state is split into two components at 13.70 and 13.88 μ in the crystalline state. This is due to the in-phase and out-of-phase coupling of vibrations between the two chains

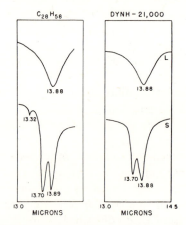

Fig. 14 A comparison of the infrared spectra in the 14-μ region of $C_{28}H_{58}$ and polyethylene (DYNH-21,000) measured above (top curves) and below (bottom curves) their melting points (13).

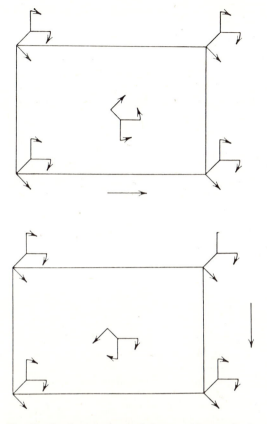

Fig. 15 In-phase and out-of-phase rocking modes in the polyethylene unit cell (14).

in the unit cell (Fig. 15). Such crystalline and amorphous sensitive bands can be used to measure the degree of crystallinity.

As discussed in Chapter 7, protons exhibit characteristic absorption when placed in a magnetic field of strength $\mathbf{H}$ at a frequency v, given by

$$v = C(\mathbf{H} \cdot \boldsymbol{\mu}) \tag{8}$$

where $\boldsymbol{\mu}$ is the magnetic moment of the proton. For a proton in a molecule or in a condensed phase the total magnetic field is

$$\mathbf{H} = \mathbf{H}_0 + \mathbf{H}' \tag{9}$$

where $\mathbf{H}_0$ is the applied magnetic field and $\mathbf{H}'$ is the local field arising from other parts of the molecule or from close-neighbor molecules. In dilute solution, where the molecular conformation and intermolecular distance are continually and rapidly changing, $\mathbf{H}'$ averages to zero, so that the effective field $\mathbf{H}$ is uniform and essentially constant and close to $\mathbf{H}_0$. Thus absorption occurs over a narrow frequency range. In concentrated solutions or amorphous polymers molecular motion is slower, so that in the time for transition between states of orientation of μ in the field $\mathbf{H}'$ will not be uniform and a broadening of the frequency range of absorption will occur. The NMR absorption of a semicrystalline polymer is shown in Fig. 16, where there is a

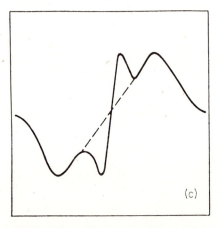

Fig. 16   The NMR absorption of a low-density polyethylene. The derivative of the intensity of absorption with respect to field strength is plotted against the applied magnetic field strength at a constant frequency of radio-frequency field (15).

narrow component arising from the mobile amorphous phase superposed on a broad component from the immobile crystalline phase (15). In this plot use has been made of the conventional procedure, employed for experimental convenience, where the frequency is kept constant and the derivative of absorption with respect to applied field strength is plotted against the strength of the applied field. One may take as a measure of the degree of crystallinity

$$X_C = \frac{A_B}{A_B + A_N} \tag{10}$$

where A_B is the area of the broad component and A_N that of the narrow component.

The NMR results have not been found to correlate well with other methods of measuring X_C. The principal difficulty is that the NMR method is fundamentally a measure of mobility rather than crystallinity. Thus an immobile amorphous phase, such as is found below T_g, also contributes to the broad component. Since the mobility of the amorphous phase and of defects within crystals can be quite dependent on morphology, the technique does not provide a reliable measure of crystallinity. However, when coupled with other methods of measuring crystallinity, it does provide useful information about mobility.

In visible light there are two optical manifestations of crystallinity in polymers: (a) turbidity and (b) unusual light-transmittance patterns in polarized light. Both of these phenomena result from light scattering by crystallites or spherulites whose size is of the order of magnitude of the wavelength of light.

Stein and co-workers (14, 16, 17) have obtained considerable experimental information on transmittance patterns in polarized light and have developed theories of light scattering to explain these unusual patterns. Two types of transmittance pattern have been studied: H_v, where the incident light is vertically polarized and the analyzer is horizontal, and V_v, where the polarizer and analyzer are both vertical. An example of each type of scattering pattern is shown in Fig. 17.

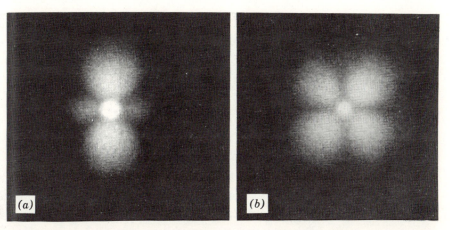

Fig. 17 Measured (a) V_v and (b) H_v light-scattering patterns for a medium-density polyethylene sample (14).

5. THERMODYNAMICS OF CRYSTALLIZATION

The melting of a crystalline polymer can be considered as a first-order phase transition, so that at the melting point T_m° the following relation holds:

$$T_m^\circ = \frac{\Delta H_u}{\Delta S_u} \tag{11}$$

where ΔH_u is the heat of fusion per mole of the repeating structural units of the polymer chain and ΔS_u is the corresponding entropy of fusion.

The heat of fusion of a partly crystalline polymer depends on the amount of crystalline material and can be used to determine the degree of crystallinity. Thus

$$\Delta h = X_C \, \Delta h_0 \tag{12}$$

where Δh is the heat of fusion per gram of polymer and Δh_0 is the heat of fusion per gram of pure crystal. It is assumed that Δh_0 for the crystalline fraction of a polymer is independent of X_C and not appreciably affected by variations in crystal size and morphology. The value of Δh_0 can be obtained from measurements of Δh on a polymer of known X_C by use of the diluent technique to be discussed.

The melting point of a low-molecular-weight solid A is lowered by the introduction of an impurity soluble in the melt according to the formula

$$\frac{1}{T_m} - \frac{1}{T_m^\circ} = -\frac{R}{\Delta H_u} \ln a_A \tag{13}$$

where T_m° is the melting point of the pure solid A, T_m is the melting point in the presence of an impurity, ΔH_u is the heat of fusion, R is the gas constant, and a_A is the activity of A in the melt in the presence of a dissolved impurity. For sufficiently low concentrations of the impurity the activity a_A is equal to the mole fraction of A present in the melt.

Correspondingly the melting point of a crystalline polymer is depressed by "impurities," such as low-molecular-weight diluents, chain ends, and randomly comonomerized units along the chain. The corresponding formulas analagous to Eq. 13 are (18)

$$\frac{1}{T_m} - \frac{1}{T_m^\circ} = \frac{R}{\Delta H_u} \frac{V_u}{V_1} (v_1 - \chi v_1{}^2) \tag{14}$$

$$\frac{1}{T_m} - \frac{1}{T_m^\circ} = \frac{R}{\Delta H_u} \frac{2}{P_n} \tag{15}$$

$$\frac{1}{T_m} - \frac{1}{T_m^\circ} = -\frac{R}{\Delta H_u} \ln X_A \tag{16}$$

Equation 14 describes the effect of a low-molecular-weight liquid (volume fraction v_1, interaction parameter χ) in depressing the melting point of a polymer and is the same as Eq. 42 in Chaper 4. In Eq. 15 P_n is the number-average degree of polymerization of the polymer. In Eq. 16 X_A is the mole fraction of A-units in the copolymer consisting of a few B-units randomly copolymerized with a preponderance of A-units.

Equations 14, 15, and 16 reduce to a common equation at a sufficiently low concentration of impurities:

$$\frac{1}{T_m} - \frac{1}{T_m^\circ} = \frac{R}{\Delta H_u} X_B \tag{17}$$

where X_B is the mole fraction of impurity (i.e., diluent, chain end, or comonomer unit).

The values of ΔH_u for several polymers have been obtained mainly through application of Eq. 14. From Eq. 11 and ΔH_u one can obtain ΔS_u.

Table 1 gives T_m, ΔH_u, and ΔS_u, ΔH_u (per bond), and ΔS_u (per bond) for several polymers. It is particularly interesting to note that ΔS_u per bond is on the order of R cal/°C-mole for many polymers.

Table 1 Thermodynamic Quantities Characterizing the Fusion of Polymers[a]

Polymer	T_m, °C	ΔH_u, cal/mole of Repeating Unit	ΔH_u, per Bond	ΔS_u, per Unit	ΔS_u, per Bond
Polymethylene	137	785	785	1.90	1.90
Polyethylene oxide	66	1980	660	5.85	1.95
Natural rubber	28	1050	350	3.46	1.15
Gutta-percha	74	3040	1013	8.75	2.92
Polychloroprene	80	2000	667	5.7	1.9
Polychlorotrifluoroethylene	210[b]	1200	600	2.50	1.25
Cellulose tributyrate	207	3000	1500	6.2	3.1
Polypropylene	176				
Polytetrafluorethylene	327				
Polyhexamethyleneadipamide	260				
Polyethyleneterephthalate	267				

[a] L. Mandekern, *Chem. Rev.*, **56**, 903 (1956).
[b] Latest value given as 220–225°; J. D. Hoffman and J. J. Weeks, *J. Research Natl. Bureau of Standards*, **60**, 465 (1958).

One might expect that ΔH_u (per bond) would correlate with the strength of molecular cohesion; that is, the solubility parameter of the polymer. Surprisingly this is not necessarily the case; for example, polyesters have larger values of ΔH_u than corresponding polyamides. The higher melting points of polyamides result from low values of ΔS_u. The crystal structures of polyamides seem to allow for more configurational disorder than in corresponding polyesters; this would result in lower values of ΔS_u.

Oriented crystalline polymers will shrink and lose their orientation at the melting point T_m of the crystallites. External tension applied to the oriented polymer will raise the melting temperature (or shrink temperature) quite considerably. They can be treated quite simply by the Maxwell-type equation

$$\left(\frac{\partial X}{\partial T}\right)_p = -\frac{\Delta S}{\Delta L} \tag{18}$$

where X is the tensile force required to maintain equilibrium between the oriented crystalline phase and the shrunk amorphous state of the fiber at temperature T and pressure p; ΔS and ΔL are the latent changes in entropy and length, respectively, associated with the melting transformation of the fiber. Certain rubbers, such as natural rubber, which are amorphous at room temperature when unstretched, develop crystallinity when stretched to high extension ratios. The melting point of the crystallites is obviously raised by increasing the extension ratio. Approximate expressions for this effect have been developed.

First-order crystal-to-crystal phase transitions occur in several polymers, such as polybutene-1, trans-1,4-polybutadiene, and polytetrafluoroethylene. These obey the thermodynamic relations that hold for crystal-to-crystal transitions in low-molecular-weight substances.

6. KINETICS OF CRYSTALLIZATION

The transformation of a supercooled polymer melt into a solid crystalline phase involves two distinct steps: the birth of crystallites, termed primary nucleation; and the propagation of the individual crystallites, called growth. Primary nucleation can occur either heterogeneously or homogeneously. In heterogeneous nucleation adventitious impurities, residual crystalline polymer not completely melted, dispersed small-particle-size solids (such as carbon black, silica, or talc) or the container wall may serve as nucleating agents.

Homogeneous nucleation, sometimes called sporadic nucleation, occurs via thermal fluctuations in the melt, resulting in the continual formation and disappearance of "crystalline clusters" of molecules. Homogeneous nucleation occurs more rapidly at high degrees of supercooling. According to the nucleation theory, clusters equal to or greater than a certain critical size serve as the primary nuclei.

The bulk-crystallization rate of a supercooled melt with spherically symmetrical growth (spherulites) is given by the Avrami equation (19):

$$X_C = X_\infty[1 - \exp(-Kt^n)] \tag{19}$$

where X_C is the weight fraction of crystallinity at time t and X_∞ is the equilibrium crystallinity. The Avrami constant n and the constant K depend on whether nucleation is heterogeneous or homogeneous and on the dimensionality of the growth process. For three-dimensional growth

$$K = \frac{4}{3}\pi \frac{d_c}{d_e}\frac{G^3 N_0}{X_\infty} \qquad n = 3, \text{ heterogeneous} \tag{20}$$

$$K = \frac{\pi}{3}\frac{d_c}{d_e}\frac{G^3 N'(\tau)}{X_\infty} \qquad n = 4, \text{ homogeneous} \tag{21}$$

In Eqs. 20 and 21 d_b and d_e are respectively the crystal- and liquid-phase densities, G is the growth rate, N_0 is the number of heterogeneous nuclei per unit volume, and $N'(\tau)$ is the rate of homogeneous nucleation.

Bulk isothermal crystallization rates are usually obtained dilatometrically. In terms of h_0, $h(t)$, and h_∞, the dilatometer readings at the beginning, intermediate, and end stages of the crystallization process, respectively, Eq. 3 can be cast in the form

$$\ln\left\{-\ln\left[\frac{h(t) - h_\infty}{h_0 - h_\infty}\right]\right\} = \ln K + n \ln t \tag{22}$$

A plot of the left-hand side of Eq. 22 versus $\ln t$ yields, in principle, a straight line whose slope n should be 3 or 4, corresponding to heterogeneous or homogeneous nucleation. The intercept gives the rate constant K, which alternatively can be calculated from the simple relation

$$K = \frac{0.693}{t_{1/2}^n} \tag{23}$$

where $t_{1/2}$ is the time at which X_C/X_∞ equals 0.5.

Equation 22 has been applied to many polymers with varying degrees of success. In many instances the observed n-values are nonintegral. Nonintegral Avrami exponents have no physical basis in the Avrami model as outlined above, and investigators have attributed them to secondary crystallization, time-dependent primary nucleation, and to simultaneously occurring heterogeneous and homogeneous nucleation. These and other complications have been treated in the literature, resulting in modified equations that are far more unwieldy than Eq. 19.

Tables of radial growth rates and Avrami rate constants for many polymers have been tabulated in Brandrup and Immergut's *Polymer Handbook* (20).

It is believed that the nucleation of most polymer crystallization occurs heterogeneously, on container walls, impurities, or structural imperfections. Certain polymers, such as polyethylene, crystallize so rapidly that they are almost impossible to supercool. Other polymers, such as polyethylene terephthalate, crystallize very slowly, so that nucleating agents, such as TiO_2 and talc, are purposefully added. Even so the crystallization is not rapid.

The temperature dependence of the radial growth rate G of a spherulite can be expressed by the relation for both homogeneous and heterogeneous three-dimensional nucleation:

$$G = G_0 \exp\left(-\frac{\Delta F^*}{RT}\right) \exp\left[-\frac{A}{T^2(T_m^\circ - T)}\right] \tag{24}$$

where G_0 and A are constants. The term in brackets is called the transport term, and the second exponential term is called the nucleation term. The quantity ΔF^* is the free energy of activation for transporting a polymer segment to a growing crystal face. The nucleation term relates to the thermodynamic driving force for the nucleation of new layers on the crystal. It is clear that the transport term increases with increasing temperature; the nucleation term, however, increases sharply as the temperature decreases below T_m°, the melting point of the crystal. Because of the contrasting temperature dependence of the transport term and the nucleation term, the maximum rate of crystal growth (for homogeneous nucleation) occurs at about $0.85 T_m^\circ$.

7. ORIENTATION OF CRYSTALLINE POLYMERS

Crystalline polymers often find application for fibers and films, where their orientation is of great importance. By mechanical and thermal treatment it is possible to produce states of orientation in which a given crystal axis is oriented within 95 % of perfect orientation in a given direction. Such orientation produces strength in the orientation directions, which is uniaxial for a fiber but preferably biaxial for a film. It should be noted that amorphous polymers can also be oriented, but much less perfectly.

A very useful method for studying orientation in crystalline polymers is by the X-ray diffraction technique. The photographic method is convenient for qualitative evaluation. Typical X-ray diffraction patterns for unoriented and uniaxially oriented films of polyethylene are shown in Fig. 18 (21). It is seen that the diffraction from the unoriented film is cylindrically symmetrical about the incident beam, whereas that from the drawn film is highly dependent on the azimuthal angle of diffraction. This can be understood in that diffraction can be regarded as a kind of reflection of crystal planes. Cylindrically symmetrical diffraction corresponds to random orientation of such planes, whereas diffraction at specific orientation angles corresponds to the orientation of crystals at particular angles. For a single crystal oriented at specific angles a given crystal plane would diffract as a single spot on the photograph.

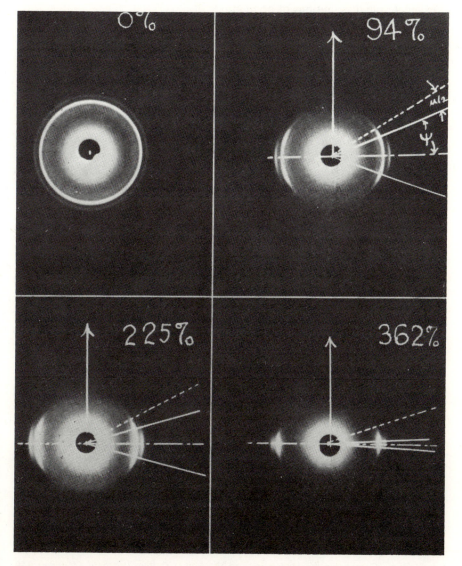

Fig. 18 X-ray diffraction patterns for unoriented and cold-drawn low-density polyethylene samples drawn to different elongations (21).

From the density of diffraction spots as a function of azimuthal angle, obtained from such X-ray diagrams as Fig. 18, it is, in principle, possible to calculate the distributions of orientations of the crystal planes.

In many more cases one must be satisfied with a single numerical value of the orientation rather than a three-dimensional contour map of the distribution function for orientation. Measurements of the birefringence and infrared dichroism give the second-moment orientation function f_2, defined as

$$f_2 = \frac{3\langle \cos^2 \theta \rangle - 1}{2} \tag{25}$$

where θ is the angle that a given crystallite makes with the direction of orientation. The second-moment orientation function f_2 obviously varies between $-1/2$ and unity. Birefringence measures f_2 because it depends on refractive-index anisotropy in oriented samples; infrared dichroism measures f_2 because it depends on infrared-absorption anisotropy in oriented samples.

It is noted that birefringence is related to the total orientation of both crystalline and amorphous regions, whereas the orientation function calculated from X-ray diffraction is that of the crystals. Hence a combination of the two techniques permits the separate determination of the orientation of crystalline and amorphous regions. The orientation function determined from infrared dichroism is that of the transition moment of the particular mode of vibration that is active at the frequency of measurement, which may be characteristic of the orientation of either the crystalline or the amorphous phase and may lie at some particular angle to the crystalline or molecular axis. Other techniques, such as fluorescence polarization, permit the determination of fourth-moment orientation functions.

REFERENCES

1. G. W. Castellan, *Physical Chemistry*, Addison-Wesley, Reading, Mass., 1964.
2. R. S. Stein, *Polymer Eng. Sci.* **9**, 320 (1969).
3. P. H. Hermans and A. Weidinger, *J. Polmyer Sci.* **4**, 709 (1949).
4. P. Scherrer, Göttinger *Nachrichten*, **2**, 98 (1918).
5. R. Hosemann and S. N. Bagchi, *Direct Analysis of Diffraction by Matter*, N. Holland, Amsterdam, 1962.
6. O. K. Gorngross, K. Horrman, and W. Agitz, *Z. Physik. Chem.* (*Leipzig*) **B10**, 371 (1930).
7. R. Hosemann, *Polymer* **3**, 349 (1962).
8. P. Geil, *Polymer Single Crystals*, Interscience, New York, 1963.
9. W. D. Niegisch and P. R. Swan, *J. Appl. Phys.* **31**, 1906 (1960).

10. P. J. Flory, *J. Am. Chem. Soc.* **84**, 2857 (1962).
11. M. Tasumi and S. Krimm, *J. Polmer Sci.* **A2, 6**, 995 (1968).
12. H. D. Keith, F. J. Padden, Jr., and R. G. Vadimsky, *Science* (1955).
13. F. M. Rugg, J. J. Smith, and L. H. Wartman, *J. Polymer Sci.* **11**, 1 (1953).
14. R. S. Stein, in B. Ke, *Newer Methods of Polymer Characterization*, Interscience, New York, 1964.
15. W. P. Slichter and D. W. McCall, *J. Polymer Sci.* **25**, 230 (1957).
16. R. S. Stein, *Proceedings of the R.A. Welch Foundation*, Vol. 10, 1966, p. 207.
17. R. S. Stein, *J. Polymer Sci.* **C15**, 185 (1966).
18. P. J. Flory, *Principles of Polymer Chemistry*, Cornell University Press, Ithaca, 1953, Chapter 13.
19. M. Avrami, *J. Chem. Phys.* **7**, 1103 (1939); **8**, 212 (1940); **9**, 177 (1941).
20. J. Brandrup and E. H. Immergut, Eds., *Polymer Handbook*, Interscience, New York, 1966, Chapter III.
21. R. S. Stein and F. H. Norris, *J. Polymer Sci.* **21**, 381 (1956).

BIBLIOGRAPHY

P. H. Geil, *Polymer Single Crystals*, Vol. 5 of Polymer Reviews, Interscience, New York, 1963.

F. Gornick and J. D. Hoffman, *Ind. Eng. Chem.* **58**, 41 (1966).

J. D. Hoffman and J. I. Lauritzen, *J. Res. Natl. Bur. Stds.* **65A**, 297 (1961).

L. Mandelkern, *Crystallization of Polymers*, McGraw-Hill, New York, 1964.

P. Meares, *Polymers: Structure and Bulk Properties*, Van Nostrand, Princeton, N.J., 1965, Chapters 4 and 5.

M. L. Miller, *The Structure of Polymers*, Rheinhold, New York, 1966, Chapter 10.

F. P. Price, Ed., *The Meaning of Crystallinity in Polymers*, *J. Polymer Sci.*, Part C, Polymer Symposia No. 18, Interscience, New York, 1967.

A. Sharples, *Introduction to Polymer Crystallization*, Arnold, 1966.

B. K. Vainshtein, *Diffraction of X-Rays by Chain Molecules*, Elsevier, New York, 1966.

"Crystallization," in *Encyclopedia of Polymer Science and Technology*, Vol. 4.

Rubber Elasticity

9

A. V. Tobolsky

The physical properties of high polymers are by far their most important attribute in regard to their practical utilization. Natural and synthetic polymers are used as lubricating oils, adhesives, waxes, surface coatings, films, rubbers, plastics, papers, and textiles, and in all these cases some physical characteristics of the substance, such as viscosity or elasticity, are of paramount importance. An understanding of the relationship between the physical properties and molecular structure of polymers is essential to the polymer scientist who desires to synthesize new materials or to modify and process existing polymers.

Perhaps the most striking physical characteristic of polymers compared with the solid state of low-molecular-weight compounds is the property of high elasticity. Whereas metals and other structural materials can be elastically deformed for only a fraction of a percent, rubberlike polymers are capable of reversible elastic extensions of several hundred percent. Other important attributes of certain polymers are their plastic flow and delayed elasticity. The phenomena all have a direct interpretation in terms of the structural characteristics of polymers.

1. CONCEPTS OF STRESS AND STRAIN

When matter is subjected to certain geometrical constraints—for example, when it is confined to a fixed volume—it responds by exerting forces on the bodies imposing the restraints. Stated alternatively, when a fixed portion of matter is subjected to external forces under such conditions that the inertial motion of the body as a whole is prevented, it responds by a change of shape or dimension. The change in geometrical state of the body is described as the *strain*; the balanced system of forces in the interior of the body is known as the *stress*.

There are three particularly elementary types of strain for isotropic materials for which the stress is simply related to the external forces: simple tension, simple shear, and uniform (hydrostatic) compression. These are illustrated in Fig. 1.

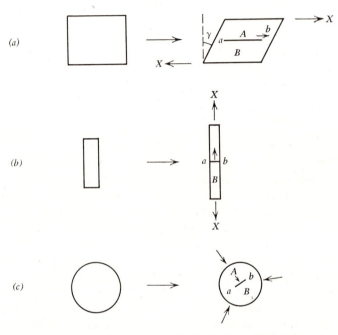

Fig. 1 Elementary types of strain: (*a*) simple shear, (*b*) simple tension, and (*c*) uniform hydrostatic compression.

The shear strain is defined as the tangent of the angle of shear (for sufficiently low shear $\tan \gamma = \gamma$); the tensile strain is defined as $\Delta l/l$, where Δl is the increase in length. For uniform (hydrostatic) compression the strain is defined as $\Delta V/V$, where ΔV is the decrease in volume. The strains discussed in this section are restricted to very small values.

The stresses in a strained substance are defined by calculating the force per unit area exerted by the molecules on one side of a small planar slit on the molecules on the other side in the limit as the area of the slit approaches zero.

In Fig. 1 we consider the forces exerted by the molecules on side A on the molecules on side B across the slit ab. At each point of the body such a slit ab may be oriented to an infinite number of directions. Each direction is defined by the normal to the slit, the direction of the normal being taken from side B to side A in Fig. 1. Even at a given point in the body the stress varies with the direction of the normal. The relation between the stress vectors and the corresponding normal vectors is best described by tensor calculus.

For simple shear the stress associated with the slit ab (Fig. 1) is perpendicular to the normal to ab. For simple tension the stress is in the direction of the normal to ab. For hydrostatic compression the stress is always in the direction opposite to the normal to the slit ab, independently of the direction of ab. In these three cases the stresses defined as above are equal to the force per unit area on the surfaces of the body on which they are applied.

The classical theory of elasticity treats ideally elastic solids, for which, according to the famous dictum of Hooke, the stress is proportional to the strain. In the most general case of a nonisotropic body there are six components of stress and six components of strain. Hooke's law can then be expressed in terms of six linear equations relating each of the components of stress with each of the six components of strain. For isotropic bodies, on the other hand, these relations are very considerably simplified. The elastic behavior in these cases can be described in terms of a shear modulus G, a Young's modulus E, a compression, or bulk, modulus B, and Poisson's ratio σ, only two of which are independent. Poisson's ratio is defined as the ratio between the relative lateral contraction and the relative longitudinal extension in a stretched elastic body. For substances that do not change in volume on stretching Poisson's ratio is 0.5.

The Hooke's law relations for the isotropic elastic solid are as follows:

$$f_\gamma = G\gamma = Gs_\gamma \tag{1}$$

$$f_l = \frac{E\,\Delta l}{l} = Es_l \tag{2}$$

$$f_p = \frac{B\,\Delta V}{V} = Bs_p \tag{3}$$

where f_γ, f_l, and f_p refer to the shear stress, tensile stress, and pressure, respectively; s_γ, s_l, and s_p refer to the corresponding strains, and G, E, and B are shear modulus, tensile (Young's) modulus, and bulk modulus, respectively. In the subsequent discussion we shall often find it convenient to omit the subscripts in writing the strains and stresses, since in most cases it will be obvious which components are under consideration.

The bulk modulus, Young's modulus, and shear modulus are related through Poisson's ratio by the following equations:

$$E = 2(1 + \sigma)G \tag{4}$$

$$B = \frac{E}{3(1 - 2\sigma)} \tag{5}$$

In contrast to elastic solids, ideal fluids, such as liquids or gases, will not support fixed shearing strains or tensile strains. In other words, these substances will show no resistance to infinitely slow changes in shape. They will, however, respond in elastic fashion to changes in volume. Also, if changes in shape are imposed with a finite velocity, shearing stresses will develop and will be proportional to the rate of shear in accordance with Newton's viscosity law:

$$f = \eta \frac{ds}{dt} \tag{6}$$

where η is the coefficient of viscosity. The behavior of most polymeric solids is intermediate between that of an elastic solid and that of an ideal fluid.

2. EQUATION OF STATE FOR IDEAL GASES AND CRYSTALS

The equation of state for an ideal gas is given by the familiar equation

$$p = \frac{NkT}{V} = \frac{nRT}{V} \tag{7}$$

where p is the pressure, N is the number of gas molecules confined in a volume V, T is the absolute temperature, k is Boltzmann's constant (equal to 1.38 $\times$ 10^{-16} erg/°C), and R is the gas constant (equal to 8.3142 $\times$ 10^7 ergs/°C-mole). The quantity n is the number of moles of gas, equal to N/N^*, where N^* is Avogadro's number, equal to 6.023 $\times$ 10^{23}.

The pressure can be interpreted in terms of the internal energy E and entropy S changes with volume, by the following fundamental thermodynamic equation:

$$dE = T\,dS - p\,dV \tag{8}$$

Therefore

$$p = T\left(\frac{\partial S}{\partial V}\right)_T - \left(\frac{\partial E}{\partial V}\right)_T \tag{9}$$

The first term on the right-hand side in Eq. 9 is called the kinetic pressure, the second term is the internal pressure. For an ideal gas the ratio $(\partial E/\partial V)_T$ is zero and the pressure is strictly a kinetic pressure, arising from a change in entropy with volume.

In fact the ideal-gas law can be derived very simply from the ratio $(\partial S/\partial V)_T$. The entropy S and the thermodynamic probability Ω are related by the Boltzmann equation

$$S = k \ln \Omega \tag{10}$$

For a perfect gas consisting of N particles in a volume V the probability of each particle is proportional to the volume. For N independent particles in volume V

$$\Omega = (cV)^N \tag{11}$$

where c is a function of temperature alone.

For an ideal gas therefore

$$p = T\left(\frac{\partial S}{\partial V}\right)_T = kT\left(\frac{\partial \ln \Omega}{\partial V}\right)_T = \frac{NkT}{V} \tag{12}$$

The bulk modulus of the ideal gas is given by

$$B = -V\left(\frac{\partial p}{\partial V}\right)_T = -TV\left(\frac{\partial^2 S}{\partial V^2}\right)_T = p \tag{13}$$

For an ideal gas at 1 atm pressure $B = 1$ atm $= 1.013 \times 10^6$ dynes/cm^2.

If one considers a crystal lattice, especially at low temperatures, the preponderant term in Eq. 9 is the internal pressure, $-(\partial E/\partial V)_T$. The major component of the internal energy is the lattice energy E_L. The bulk modulus is therefore given by

$$B = -V\left(\frac{\partial p}{\partial V}\right)_T = V\left(\frac{\partial^2 E_L}{\partial V^2}\right)_T \tag{14}$$

If the atoms of the crystal lattice are regarded as being connected by springs of force constant k', the bulk modulus is given by the equation of Einstein and Debye

$$B \approx \frac{k'}{a} \tag{15}$$

where a is the lattice constant. The force constant k' is related to the characteristic frequency v of the solid as follows:

$$2\pi v = \left(\frac{k'}{m}\right)^{1/2} \tag{16}$$

The bulk modulus of solids varies between 10^{10} and 10^{12} dynes/cm^2. The contrast between the elastic properties of the solid and the gas is shown by considering that for a gas at 1 atm pressure the bulk modulus is at least 10^4 times smaller than the bulk modulus for solids. This arises from the fundamental difference in molecular mechanism, sharply revealed in Eqs. 13 and 14.

We present the following highly idealized model of a fibrous solid: First consider a one-dimensional array of $N + 1$ atoms vibrating around positions of equilibrium with a lattice distance equal to a_0. For small displacements of an atom from its equilibrium position we assume that the restoring force is proportional to its displacement from equilibrium:

52868—9—1

The equilibrium length L_0 of the lattice is clearly

$$L_0 = Na_0 \tag{17}$$

If the system is subjected to a homogeneous extension, so that the lattice distance is now a, the new length is

$$L = Na \tag{18}$$

If the force constant of the "springs" connecting the atoms is k', the potential energy of each spring, stretched by an amount $a - a_0$, is $\frac{1}{2}k'(a - a_0)^2$. The potential energy E_L of the strained lattice is therefore

$$E_L = \frac{1}{2}Nk'(a - a_0)^2 \tag{19}$$

The tensile force on the strained lattice is given by

$$Z = \frac{dE}{dL} = \frac{1}{N}\frac{dE}{da} = k' \, da = ka_0 \frac{dL}{L_0} \tag{20}$$

For an idealized fiber consisting of parallel chains, each one behaving as described above, Young's modulus E would be

$$E = k' \frac{a_0}{A} = 4\pi^2 v^2 m \frac{a_0}{A} \tag{21}$$

where A is the effective cross-sectional area of each chain.

3. RUBBERLIKE BEHAVIOR

Crystalline solids, such as metals, can be extended reversibly to something under 1 % strain and have a tensile modulus of approximately 10^{12} dynes/cm^2. By contrast many rubbers have a tensile modulus of about 10^7 dynes/cm^2 and can be stretched 500 % or more with nearly complete reversibility. This astonishing phenomenon stems from the molecular architecture of rubber-like substances.

Rubbers are generally made from linear polymers to which are admixed small amounts of potential crosslinking agents, such as sulfur or peroxides. The mixture is forced into a suitable mold, and the crosslinking reaction is

made to occur by using high temperatures. This produces a network structure, which is the basis of rubbery behavior. High-molecular-weight *linear*
polymers may sometimes behave in rubbery fashion, but this is because the
molecules are entangled, giving rise to quasi-crosslinks and a temporary
network structure.

It will be shown that the equation of state for a rectangular strip of rubber
in tension is

$$Z = \frac{N_0 kT}{L_0} \, \Phi \left[\left(\frac{L}{L_0} \right) - \left(\frac{L_0}{L} \right)^2 \left(\frac{V}{V_0} \right) \right] \tag{22}$$

where Z is the force required to stretch the rubber from an initial length L_0
to the final length L, N_0 is the total number of network chains in the sample,
k is Boltzmann's constant, T is the absolute temperature, and Φ is a correction
factor close to unity; V is the volume in the final (stretched) state and V_0
is the volume in the initial (unstretched) state. The ratio V/V_0 is very close to
unity if the rubber has not been swollen by solvent before stretching.

Equation 22, without the correction factor Φ, was derived in the late 1930s
and early 1940s mainly through the work of Kuhn, Guth, Mark, James,
Wall, and Treloar, as described in Treloar's *The Physics of Rubber Elasticity*
(1). The correction factor Φ (equal to r_i^2/r_f^2, as discussed later) was introduced by Tobolsky (2–4) and indirectly by Guth and James (5).

Equation 22 is also valid for a rubber that is swollen before stretching,
provided that L_0 and V_0 are taken to be the dry, unstretched length and
volume, and L and V are taken to be the swollen, stretched length and volume.

Very often Eq. 22 is written in the following form:

$$F = NkT \, \Phi \left[\left(\frac{L}{L_0} \right) - \left(\frac{L_0}{L} \right)^2 \left(\frac{V}{V_0} \right) \right] \tag{23}$$

where F is the force per unit area based on the original (unstretched) cross-
sectional area and N is the number of network chains per cubic centimeter.

For rubber in shear the equation of state is

$$f = NkT\Phi\gamma \tag{24}$$

where f is the shear stress, γ is the shear strain, and N is the number of network chains per unit volume.

The phenomenon of rubberlike elasticity arises because the polymer
chains between network junctures are in a state of segmental Brownian
motion, much more rapid than the normal rate of application of external
force. These motions are partly restricted by the crosslinks. When a strip of
rubber is stretched from rest length L_0 to extended length L, the number of
conformations that the network chains have available at L is smaller than

at L_0. In this sense the force of stretching is largely an entropic force, like the force required for isothermal compression of gases.

The number N of network chains per cubic centimeter is equal to $N*d/M_c$, where d is the density of the rubber and M_c is the number-average molecular weight of the network chain between crosslinks. Equations 23 and 24 can be rewritten as follows:

$$F = \frac{dRT}{M_c} \Phi \left[\left(\frac{L}{L_0} \right) - \left(\frac{L_0}{L} \right)^2 \left(\frac{V}{V_0} \right) \right] \tag{25}$$

$$f = \frac{dRT}{M_c} \Phi \gamma \tag{26}$$

For most rubbers the density is near unity and M_c ranges between 2000 and 10,000.

The chemical structures of a few typical rubbers are shown in Table 1.

Table 1   Some Typical Rubbers

Rubber	Formula
1,4-*cis*-Polybutadiene	$\begin{array}{cc} \text{H} & \text{H} \qquad\qquad \text{H} & \text{H} \\ \vert & \vert \qquad\qquad \vert & \vert \end{array}$ $-CH_2C=CCH_2CH_2C=CCH_2-$
1,4-*cis*-Polyisoprene (natural rubber)	$\begin{array}{cc} CH_3 & \text{H} \qquad\quad CH_3 & \text{H} \\ \vert & \vert \qquad\quad \vert & \vert \end{array}$ $-CH_2C{=}{=}CCH_2CH_2C{=}{=}CCH_2-$
Butadiene–styrene (75 : 25) (SBR rubber)	$-CH_2CH=CHCH_2CH_2CH-$ $\qquad\qquad\qquad\qquad\quad \vert$ $\qquad\qquad\qquad\qquad\ C_6H_5$
Polyisobutylene (base of butyl rubber)	$-CH_2C(CH_3)_2CH_2C(CH_3)_2CH_2C(CH_3)_2-$
Ethylene–propylene rubber	$-CH_2CH_2CH_2CH(CH_3)(CH_2)_5CH(CH_3)-$
Silicone rubber	$\begin{array}{cccc} CH_3 & CH_3 & CH_3 & CH_3 \\ \vert & \vert & \vert & \vert \end{array}$ $-SiO-SiO-SiO-SiO-$ $\begin{array}{cccc} \vert & \vert & \vert & \vert \\ CH_3 & CH_3 & CH_3 & CH_3 \end{array}$

Those without unsaturation along the main chain usually are copolymerized with a few mole-percent of another monomer which leaves residual unsaturation along the chain or pendant to the chain. A small amount of unsaturation is very helpful for crosslinking by sulfur or peroxides.

4. EQUATION OF STATE FOR AN ISOLATED CHAIN

Consider a freely rotating chain of n segments each of length l_0, in the presence of a force field Z acting in the z-direction. The average projected length L of the random chain in the z-direction is

$$L = n \frac{\int_0^\pi l_0 \cos\theta \exp(l_0 Z \cos\theta/kT) 2\pi \sin\theta \, d\theta}{\int_0^\pi \exp(l_0 Z \cos\theta/kT) 2\pi \sin\theta \, d\theta}$$

$$= nl_0 \left(\coth \frac{l_0 Z}{kT} - \frac{kT}{l_0 Z}\right)$$

$$= nl_0 \, \mathcal{L} \frac{l_0 Z}{kT}$$

(27)

The problem solved above is identical to the problem of orientation of gas dipoles in an electric field, and the function $\mathcal{L}$ in Eq. 27 is the famous Langevin function. For small values of y, $\mathcal{L}(y)$ is equal to $y/3$; for large values of y, $\mathcal{L}(y)$ is equal to unity. Therefore for values of $l_0 Z/kT$ reasonably smaller than unity we obtain

$$L = \frac{nl_0^2 Z}{3kT}$$

(28)

$$Z = \frac{3kT}{r_f^2} L$$

(29)

where r_f^2, the mean-square end-to-end length, has been substituted for nl_0^2. It can be shown that Eq. 29 is also valid for a real chain containing energy differences between *trans* and *gauche* configurations; one merely has to use the correct value of r_f^2 rather than nl_0^2. The spring constant in Eq. 29 is $3kT/r_f^2$.

Consider a right-angled chain with $3n$ linkages, each of length $l_0/\sqrt{3}$, successively pointed in the directions $\pm x$, $\pm y$, $\pm z$. The equation of state for this chain acted on by a force Z in the z-direction is

$$L = \frac{\exp(l_0 Z/3^{1/2}kT) - \exp(-l_0 Z/3^{1/2}kT)}{\exp(l_0 Z/3^{1/2}kT) + \exp(-l_0 Z/3^{1/2}kT)} \frac{nl_0}{3^{1/2}}$$

$$= \frac{nl_0}{3^{1/2}} \tanh \frac{l_0 Z}{3^{1/2}kT}$$

(30)

This model corresponds more nearly to a chain with fixed bond angles. The equation of state for this chain for values of $l_0 Z/3^{1/2}kT$ smaller than unity is also given by Eq. 29.

The configurational Helmholtz free energy A_c of a chain obeying Eq. 29 is

$$A_c = \int_0^\infty Z \, dL = \frac{3kT}{2r_f^2} L^2 \tag{31}$$

From Eq. 31 it will be shown in the next section that the equation of state for a network is

$$\frac{F}{NkT\Phi} = \alpha - \frac{V}{V_0} \alpha^{-2} \tag{31'}$$

where α is L/L_0, Φ is the correction factor, and V/V_0 is the ratio of volumes in the final and initial states.

From Eq. 27 one can derive [see texts by Bueche (6) and Treloar (1)]

$$\frac{F}{NkT\Phi} = \tfrac{1}{3}n^{1/2}\left(\mathcal{L}^{-1} \frac{\alpha}{n^{1/2}}\right) - \frac{V}{V_0} \alpha^{-2} \tag{27'}$$

where $\mathcal{L}^{-1}$ is the inverse Langevin function and n is the number of freely rotating segments per network chain. Equation 27' reduces to Eq. 31' when $\alpha/n^{1/2}$ is much less than unity (i.e., at moderately low strains).

From Eq. 30 one can derive the following equation (7):

$$\frac{F}{NkT\Phi} = n^{1/2} \tanh^{-1} \frac{\alpha}{n^{1/2}} - \frac{V}{V_0} \alpha^{-2} \tag{30'}$$

where $\tanh^{-1}$ is the inverse hyperbolic tangent. Equation 30' has an advantage over Eq. 25, in that the work of stretching per cubic centimeter, namely $\int_0^{n^{1/2}} F \, d\alpha$, is finite (equal to $nNkT\Phi \ln 2$) in the latter equation but infinite in the former. For a hydrocarbon chain with three (equal-energy) rotational conformers per bond the work of stretching per cubic centimeter is $nNkT\Phi \ln 3$.

5. EQUATION OF STATE FOR NETWORKS

It is postulated that the Helmholtz free energy for a network is

$$A(T, L_0, V_0) = A_0(T, V_0) + \sum_{i=1}^{N_0} \frac{3kT}{2r_{f,i}^2} (\mathbf{r}_i)^2 \tag{32}$$

where $A_0(T, V_0)$ is the free energy of the network exclusive of the conformational free energy, which is represented by the second term on the right-hand side. In the summation $\mathbf{r}_i$ represents the network-chain vector of the ith chain, replacing L in Eq. 31. The term N_0 represents the total number of network chains. The second term on the right-hand side arises from Eq. 31.

The task of obtaining actual values of $\mathbf{r}_i$ in the unstrained state is an impossible one, since it would involve a detailed knowledge of $A_0(T, V_0)$. It is clear that $\langle r_i{}^2 \rangle$ in the unstrained state is not necessarily the same as $\langle r_{f,i}^2 \rangle$ (2–4). (Angle brackets represent average values.)

We now consider the equation of state for tension, treating both dry and postswollen rubber networks. In both cases the rubber strip goes from an initial state $L_0 V_0 T$ to a final state L, V, T. Let us first apply a hydrostatic pressure (positive or negative) to the network, so that it attains the final volume V. This state is isotropic and has a length L'. The equation for the free energy is

$$A'(T, V) = A_0(T, V) + \sum_{i=1}^{N_0} \frac{3\,kT}{2r_{f,\,i}^2} (\mathbf{r}_i')^2 \tag{33}$$

We now consider a process of stretching whereby the hydrostatic pressure (positive or negative) is continually varied so that the volume remains constant. In the final state the network vectors are $\mathbf{r}^*$ and the length is L. The free energy is

$$A^*(T, L, V) = A_0(T, V) + \sum_{i=1}^{N_0} \frac{3\,kT}{2r_{f,\,i}^2} (\mathbf{r}_i^*)^2 \tag{34}$$

Now the vectors $\mathbf{r}_i'$ have undergone an affine transformation to $\mathbf{r}_i^*$, such that

$$X_i^* = \frac{L}{L'} X_i'; \qquad Y_i^* = \left(\frac{L'}{L}\right)^{1/2} Y_i; \qquad Z_i^* = \left(\frac{L'}{L}\right)^{1/2} Z_i \tag{35}$$

An original spherical distribution of vectors $\mathbf{r}'$ is transformed to an ellipsoidal distribution $\mathbf{r}^*$ such that

$$(\mathbf{r}^*)^2 = (X^*)^2 + (Y^*)^2 + (Z^*)^2$$

$$= \left(X' \frac{L}{L'}\right)^2 + \left[Y'\left(\frac{L'}{L}\right)^{1/2}\right]^2 + \left[Z'\left(\frac{L'}{L}\right)^{1/2}\right]^2$$

$$= \tfrac{1}{3}(\mathbf{r}')^2 \left[\left(\frac{L}{L'}\right)^2 + 2\left(\frac{L'}{L}\right)\right] \tag{36}$$

The free energy in the strained state is therefore

$$A^*(T, L, V) = A_0(T, V) + \frac{N_0 kT}{2} \left\langle \frac{r_i^2}{r_{f,\,i}^2} \right\rangle_{V,\,T} \left[\left(\frac{L}{L'}\right)^2 + 2\left(\frac{L'}{L}\right)\right] \tag{37}$$

where the average value is in the unstrained state. We approximate the average of a quotient as the quotient of the averages, giving

$$A^*(T, L, V) = A_0(T, V) + \frac{N_0 kT}{2} \frac{\langle r_i^2 \rangle_{V,\,T}}{\langle r_{fi}^2 \rangle_{V,\,T}} \left[\left(\frac{L}{L'}\right)^2 + 2\left(\frac{L'}{L}\right)\right] \tag{38}$$

The tensile force is obtained by differentiating A^* with respect to L:

$$Z = \left(\frac{\partial A^*}{\partial L}\right)_{T,V} = \frac{N_0 kT}{L'} \frac{\langle \mathbf{r}_i^2 \rangle_{V,T}}{\langle r_{f,i}^2 \rangle_{V,T}} \left[\left(\frac{L}{L'}\right) - \left(\frac{L'}{L}\right)^2\right] \qquad (39)$$

In order to transform to the quantities L_0 and V_0, the volume and length in the initial state (also unswollen state if postswelling is considered), note that

$$\frac{L'}{L_0} = \frac{V^{1/3}}{V_0^{1/3}} \qquad (40)$$

Inserting Eq. 40 into Eq. 39, we obtain

$$Z = \frac{V_0^{2/3}}{V^{2/3}} \frac{N_0 kT}{L_0} \frac{\langle r_i^2 \rangle_{V,T}}{\langle r_{f,i}^2 \rangle_{V,T}} \left[\left(\frac{L}{L_0}\right) - \left(\frac{L_0}{L}\right)^2 \left(\frac{V}{V_0}\right)\right] \qquad (41)$$

Consider that

$$\frac{\langle r_i^2 \rangle_{V,T}}{\langle r_i^2 \rangle_{V_0,T}} = \left(\frac{V}{V_0}\right)^{2/3}$$

The final equation of state is

$$Z = \frac{N_0 kT}{L_0} \left\{\frac{\langle r_i^2 \rangle_0}{\langle r_f^2 \rangle}\right\} \left[\left(\frac{L}{L_0}\right) - \left(\frac{L_0}{L}\right)^2 \left(\frac{V}{V_0}\right)\right] \qquad (42)$$

The term in the braces is the correction factor Φ of Eq. 22. Experimental results indicate that the order of magnitude of this quantity is unity, though it is not exactly unity, as will be discussed in Section 6. The temperature and pressure (volume) dependence of Φ reveals very interesting information concerning the quantity $\langle r_{f,i}^2 \rangle$, which we have abbreviated to $\langle r_f^2 \rangle$. This quantity is the mean-square end-to-end distance of the network chains in free space, which we take to be the mean-square end-to-end distance the network chains would assume if the network junctures were cut. The subscript on $\langle r_i^2 \rangle_0$ denotes the reference, unstretched, condition of V_0, T. The quantity $\langle r_f^2 \rangle$ is evaluated in the *stretched* state at T, V and is therefore a function of temperature and volume.

6. NETWORK TOPOLOGY

One way of producing rubber networks is by random crosslinking of linear chains of number-average molecular weight M_n. Consider a chemical crosslinking process producing c moles of tetrafunctional crosslinks per cubic centimeter. If the value of M_n approaches infinity, the moles of network chains per cubic centimeter is $2c$ and the average molecular weight M_c of the network chains is $d/2c$. We retain this definition even if M_n is finite. In this case a certain fraction of the original chains (sol fraction) may not be tied to the infinite network, and some portion of the network consists of "terminal chains" tied to the network only at one end (8). These terminal chains and the sol fraction are ineffective in supporting stress. This produces a correction to Eq. 22 as follows:

$$Z = \frac{N_0 kT}{L_0} \left(1 - \frac{xM_c}{M_n}\right) \Phi \left[\left(\frac{L}{L_0}\right) - \left(\frac{L_0}{L}\right)^2 \left(\frac{V}{V_0}\right)\right] \qquad (43)$$

The quantity x varies between 1.0 and 2.0, being equal to 1.0 when $M_c/M_n \ll 1$ and close to 2.0 when $M_c/M_n \sim 0.5$ (9).

The quantity Φ can be evaluated from Eq. 22 if Z, L/L_0, and the other experimental quantities, such as T, V_0, and V, are measured. The quantity N_0 must also be determined, but this is equal to $2V_0 c$, where c is the moles of crosslinks per cubic centimeter. In certain cases c can be determined by special methods, so that an independent evaluation of N_0 can be made. Sometimes c can be determined by delicate chemical analysis. In cases where the network is polymerized directly from monomers (e.g., ethyl acrylate and ethylene glycol dimethacrylate) the stoichiometry of the polymerizing mixture defines c (10). Finally, if the sample is crosslinked by high-energy radiation, the crosslink density can sometimes be obtained by knowing the total energy incident on the sample. If N_0 is accurately known, the numerical value of Φ can be determined from experiment.

By copolymerization of known amounts of ethylene glycol dimethacrylate with various acrylates and methacrylates, Φ values were obtained for these networks (10). It was found that Φ was close to unity for ethyl acrylate but decreased systematically to about 0.3 as one proceeded from ethyl to butyl, to hexyl, to octyl acrylate. Similar results were found with the methacrylates. A similar decrease in Φ was found when these networks were polymerized in the presence of an inert diluent. It was inferred that many of the supposed crosslinks formed intramolecular loops instead.

The values of Φ must be intimately related to network topology. There are at least three different ways of producing networks.

1. Random crosslinking of linear chains.
2. Copolymerizations, such as ethyl acrylate with ethylene glycol dimethacrylate.
3. Tying of functional end groups, such as triimine cure of carboxyl-terminated polybutadiene or triol cure of NCO-terminated polyesters or polyethers.

Under case 3 the molecular-weight distribution of the initial linear molecules can be heterodisperse or monodisperse. In the formation of the network in case 1 we must consider the following problems:

1. The spatial distribution of the crosslinking molecules (e.g., sulfur or peroxide).

2. Interpenetration of polymer chains and network chains as affected by the thickness of the individual polymer chains and by the steric factors produced by previous crosslinks.

3. The distribution in sizes of the network chains as affected by the two preceding conditions.

It will clearly be very difficult to produce a satisfactory mathematical treatment of the topological situation, and hence many more experiments on the numerical value of Φ for different systems are very desirable.

Another aspect of network topology is that certain curing agents, such as peroxide or radiation, form carbon-to-carbon crosslinks which are irreversible at the curing temperature. On the other hand, curing with sulfur and certain accelerators produces polysulfide linkages which break or remake at the curing temperature. The latter gives rise to internally relaxed networks of higher tensile strength (19).

7. THERMOELASTICITY

The basic thermodynamic equation for the stretching of rubber networks is

$$dE = T\,dS - p\,dV + Z\,dL \tag{44}$$

The $p\,dV$ term cannot be entirely neglected because experiments are carried out at atmospheric pressure and the volume does change slightly with temperature and with length.

Much of the force required to stretch a rubber network is entropic in origin. However, some portion of the force, Z_e, is energetic in origin:

$$Z_e = \left(\frac{\partial E}{\partial L}\right)_{T,V} = Z - T\left(\frac{\partial Z}{\partial T}\right)_{V,L} \tag{45}$$

$$\frac{Z_e}{Z} = -T\left[\frac{\partial \ln(Z/T)}{\partial T}\right]_{V,L} \tag{46}$$

By combining Eq. 46 and the equation of state (Eq. 42), it becomes clear that

$$\frac{Z_e}{Z} = T\left(\frac{\partial \ln\langle r_f^2\rangle}{\partial T}\right)_V \tag{47}$$

The magnitude of Z_e/Z is related to the change of $\langle r_f^2\rangle_V$ with temperature. This relates in turn to such important matters as the energy difference between *trans* and *gauche* states of the polymer chain.

Experimentally Z_e/Z can be measured by stretching the network to a fixed length L and then measuring the change of force with temperature at 1-atm pressure. The requisite thermodynamic equation (based on the restrictive assumption that $\langle r_f^2 \rangle$ is independent of volume) is (11, 12)

$$\frac{Z_e}{Z} = -T\left[\frac{\partial \ln(Z/T)}{\partial T}\right]_{p,L} - \frac{\beta T}{(L/L_0)^3 - 1} \tag{48}$$

where β is the coefficient of thermal expansion.

For large values of L/L_0 the second term on the right-hand side becomes negligible. If under these conditions the force at constant length and pressure is proportional to T, then Z_e/Z becomes zero. This means, of course, that in these cases the force is entirely entropic in origin, and, referring to Eq. 47, $\langle r_f^2 \rangle$ is independent of temperature. This condition is analogous to Charles' law for ideal gases; when Charles' law is valid, the pressure on the gas is entirely entropic in origin. This analogy was made in the early days of polymer science and underlies our present sophisticated theory of rubber elasticity, still widely known as the kinetic theory of rubber elasticity based on the early analogy drawn with gas theory.

From Eq. 47

$$\frac{\partial \ln\langle r_f^2 \rangle}{dT} = -\left[\frac{\partial \ln(Z/T)}{\partial T}\right]_{p,L} - \frac{\beta}{(L/L_0)^3 - 1} \tag{49}$$

Consider the following definition of ε', with dimensions of energy per mole:

$$\varepsilon' = -RT^2 \frac{d \ln\langle r_f^2 \rangle}{dT} = -R\left[\frac{\partial \ln(Z/T)}{\partial(1/T)}\right] + \frac{\beta RT^2}{(L/L_0)^3 - 1} \tag{50}$$

The second term on the right-hand side can often be neglected if L/L_0 is larger than 1.5.

A simplified Ising–model treatment of chain dimensions gives the following result (13):

$$\langle r_f^2 \rangle = nl_0^2 e^{\varepsilon'/RT} \tag{51}$$

where ε' is the *gauche–trans* energy difference (per mole), with the *trans* state as the state of zero energy.

To the extent that Eq. 51 represents a fair approximation, ε' from Eq. 50 should be constant with temperature.

Values of $d \ln\langle r_f^2 \rangle/dT$ and ε' from independent authors have been tabulated by Ciferri (14) for various polymers.

A critical discussion of these quantities has been given by Dusek and Prins (15), who claim that they depend on crosslink density and perhaps on extension ratio; other authors find independence of extension ratio (18).

On the assumption that $\langle r_f^2 \rangle$ is independent of volume, one can show from the equation of state that

$$\left(\frac{\partial Z}{\partial p}\right)_{L, T} = \left(\frac{\partial V}{\partial L}\right)_{p, T} = \kappa Z \frac{1}{(L/L_0)^3 - 1} \tag{52}$$

where κ is the compressibility.

Recently Christensen and Hoeve (16) showed that Eq. 52 is not obeyed experimentally for natural rubber.

The restrictive assumption that $\langle r_f^2 \rangle$ is independent of volume is easily removed. Tobolsky, Shen, and Sperling (17, 18) showed that Eqs. 48 and 52 must be modified as follows:

$$\frac{Z_e}{Z} = -T\left[\frac{\partial \ln(Z/T)}{\partial T}\right] - \frac{\beta T}{(L/L_0)^3 - 1} - \beta T\left(\frac{\partial \ln\langle r_f^2 \rangle}{\partial \ln V}\right)_T \tag{48'}$$

$$\left(\frac{\partial V}{\partial L}\right)_{p, T} = \left(\frac{\partial Z}{\partial p}\right)_{L, T} = \kappa Z \left[\frac{1}{(L/L_0)^3 - 1} + \left(\frac{\partial \ln\langle r_f^2 \rangle}{\partial \ln V}\right)_T\right] \tag{52'}$$

Similar modifications hold for Eqs. 49 and 50.

The deviation from Eq. 52 found by Christensen and Hoeve (16) can be fully accounted for by use of Eq. 52' with a value of $\partial \ln\langle r_f^2 \rangle/\partial \ln V$ equal to 0.27 (20).

It is reasonable to assume that $\langle r_f^2 \rangle$ evaluated in the stretched state is a function of the volume of the stretched state. This means that the spring constant of the network chains, $3kT/r_f^2$, changes with extension because the volume is changing with extension.

The basic equation of state, Eq. 42, is applicable to postswollen stretched networks as well as to nonswollen (dry) networks. For swollen networks $\langle r_f^2 \rangle$ is evaluated in the stretched and swollen condition. By measurements of Z, L, V, L_0, and V_0 in a dry silicone network and in the same network swollen by a variety of solvents it was possible to obtain $\langle r_f^2 \rangle_s/\langle r_f^2 \rangle_d$ in a number of solvents (21). The subscript s refers to the swollen network and the subscript d to the dry network. In general it was found that the value of $\langle r_f^2 \rangle_s/\langle r_f^2 \rangle_d$ is similar to that obtained from intrinsic-viscosity measurements on linear polymers in dilute solution.

REFERENCES

1. L. R. G. Treloar, *The Physics of Rubber Elasticity*, 2nd ed., Clarendon Press, Oxford, England, 1958.
2. A. V. Tobolsky, Ph.D. Thesis, Princeton University, 1944.
3. M. S. Green and A. V. Tobolsky, *J. Chem. Phys.* **14**, 80 (1946).

4. A. V. Tobolsky, D. W. Carlson, and N. Indictor, *J. Polymer Sci.* **45**, 175 (1961).

5. H. M. James and E. Guth, *J. Polymer Sci.* **4**, 153 (1949); *J. Chem. Phys.* **15**, 669 (1947); *J. Appl. Phys.* **15**, 294 (1944).

6. F. Bueche, *Physical Properties of Polymers*, Interscience, New York, 1962, Chapter I.

7. Z. Soos and A. V. Tobolsky, *J. Phys. Chem.* **73**, 2864 (1969).

8. P. J. Flory, *Chem. Rev.* **35**, 51 (1944).

9. J. Scanlan, *J. Polymer Sci.* **43**, 501 (1960).

10. D. Katz, M. Shen, and A. V. Tobolsky, *J. Polymer Sci.* **2**, 1513, 1595, and 2749 (1964).

11. T. N. Khasanovich, *J. Appl. Phys.* **30**, 948 (1959).

12. P. J. Flory, A. Ciferri, and C. A. J. Hoeve, *J. Polymer Sci.* **45**, 235 (1960).

13. K. W. Scott and A. V. Tobolsky, *J. Colloid Sci.* **8**, 465 (1953).

14. A. Ciferri, *J. Polymer Sci.* **A2**, 3089 (1964).

15. K. Dusek and W. J. Prins, *Adv. Polymer Sci.* **6**, 1 (1969).

16. R. G. Christensen and C. A. J. Hoeve, *J. Polymer Sci.* **A1, 8**, 1503 (1970).

17. A. V. Tobolsky and M. Shen, *J. Appl. Phys.* **37**, 1952 (1966).

18. A. V. Tobolsky and L. H. Sperling, *J. Phys. Chem.* **72**, 345 (1968).

19. P. F. Lyons and A. V. Tobolsky, *J. Polymer Sci.* **A26**, 1561 (1968).

20. J. C. Goebel and A. V. Tobolsky, *Macromolecules* **4**, 208 (1971).

21. A. V. Tobolsky and J. C. Goebel, *Macromolecules* **3**, 556 (1970).

Viscoelastic Properties of Polymers

10

A. V. Tobolsky

1. MODULUS-VERSUS-TEMPERATURE CURVES FOR LINEAR AMORPHOUS POLYMERS

Polymers can be divided into two categories: the semicrystalline and the wholly amorphous. Semicrystalline polymers are amorphous at temperatures above their melting point and sometimes remain amorphous if quickly quenched from their molten condition. In this section we discuss linear polymers that are wholly amorphous under *all* conditions. Since these polymers are generally stereochemically or structurally irregular, they cannot be made to crystallize. A plot of specific volume versus temperature for an amorphous polymer shows a characteristic change of slope at a particular temperature, denoted as the glass-transition temperature T_g. This is one of the most important of the parameters that define the properties and behavior of the polymer. Below T_g the polymer is hard and glassy. Above T_g a polymer of sufficiently high molecular weight successively behaves as a leather, a rubber, and finally, as the temperature increases, as a liquid.

Pictorially speaking, the structure of an amorphous polymer resembles the contents of a bowl of cooked spaghetti. Above the T_g of the polymer the spaghetti must be thought of as being in constant wriggling motion, more like an entangled mass of live earthworms. A quasi-network is formed by the occasional entanglements between the molecules (or the spaghetti, or the earthworms!).

The simplest way of characterizing the elastic properties of a polymer is to measure its elastic modulus as a function of temperature. Since polymers are viscoelastic, the modulus will depend on the time and the method of measurement. For the moment we discuss the tensile-relaxation modulus $E_r(t)$, which is obtained by measuring stress as a function of time in a sample maintained at constant extension and constant temperature.

For an introductory appraisal of viscoelastic behavior as a function of temperature the time of measurement is fixed and standardized at 10 sec. The quantity being examined as a function of temperature is therefore $E_r(10)$.

Other methods of measuring the modulus will give qualitatively similar results if the time of measurement is the same; for example, measurements of torsional shear creep modulus made at 10 sec can be utilized just as well. The shear modulus is approximately one-third of the tensile modulus, this relation being nearly exact for values of tensile modulus below 10^9 dynes/cm^2. The moduli measured in creep and in relaxation are nearly the same in most cases if measured at the same time. Mathematical transformations can be used to convert relaxation to creep, and vice versa.

The results obtained from $E_r(10)$ with a typical linear amorphous polymer, atactic polystyrene, are shown in Fig. 1. One of the samples shown in this figure is denoted as polystyrene sample C and has a weight-average molecular weight of 325,000 and a number-average molecular weight of 217,000. The other sample shown in this figure is polystyrene sample A, which has a weight-average molecular weight of 210,000 and a number-average molecular weight of 140,000.

Let us consider first polystyrene sample C. The modulus-versus-temperature curve shows five regions of viscoelastic behavior (1). The first region, below 97°C, is the glassy region, in which $E_r(10)$ is between $10^{10.5}$ and $10^{10.0}$ dynes/cm^2. In this region the polymer is glassy, hard, and brittle when examined by hand. In the second region, the transition region, $E_r(10)$ varies between $10^{10.0}$ and $10^{6.7}$ dynes/cm^2 in the temperature interval between 97 and 120°C. In this region of rapidly changing modulus the physical properties are best described as leathery. The onset of this region starts very close to T_g, which is 100°C for atactic polystyrene.

In the third region, known as the rubbery plateau region, $E_r(10)$ remains fairly constant with temperature, at a value between $10^{6.7}$ and $10^{6.4}$ dynes/cm^2. The width of the temperature interval for this region depends on the molecular weight of the polymer. For polymer C this interval is 120 to 150°C. In this interval the polymer, when examined by hand, is quite rubbery, with

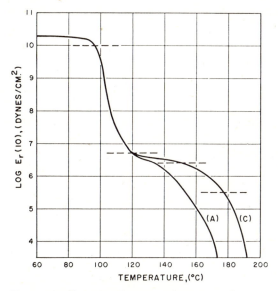

Fig. 1 Tensile relaxation modulus as a function of temperature for amorphous polystyrene samples A (weight-average and number-average molecular weights- of 210,000 and 140,000 respectively) and C (weight-average and number-average molecular weights of 325,000 and 217,000, respectively).

long-range reversible elastic behavior if it is not deformed for prolonged periods of time.

In the fourth region, known as the rubbery flow region, $E_r(10)$ varies from $10^{6.4}$ to $10^{5.5}$ dynes/cm^2. Here the polymer is elastic and rubbery, but it also has a marked component of flow when deformed for periods on the order of 10 sec. For sample C this region extends from 150 to 177°C.

Finally, in the fifth region, with $E_r(10)$ values below $10^{5.5}$ dynes/cm^2, the polymer exhibits very little elastic recovery when deformed for 10 sec and manifests an apparent state of liquid flow. Above 177°C the maximum relaxation time of polystyrene sample C is less than the time of measurement, namely, 10 sec.

The temperature intervals for these regions of viscoelastic behavior depend on the arbitrarily selected reference time of 10 sec. This is why in Fig. 1 the various regions of viscoelastic behavior are indicated in terms of modulus values. A discussion of time effects is presented in Section 2.

Figure 1 shows the modulus-versus-temperature curves of two atactic polystyrene samples of similar molecular-weight distributions but of different average molecular weight. It is clear that the glassy region and the transition region are unaffected by the molecular weight of the sample (for sufficiently large molecular weights). The value of the modulus in the region of the rubbery plateau is likewise independent of molecular weight. The regions of rubbery flow and liquid flow, however, are markedly dependent on molecular weight of chain length. Actually chain length is a more significant quantity than molecular weight because it is more meaningfully transferable from one polymer to another.

In the glassy region the segments of the polymer chain are frozen in fixed positions on the sites of a completely disordered quasi-lattice. They vibrate about these fixed positions just as do the molecules of a true crystalline molecular lattice. However, they undergo little, if any, diffusional motion from one lattice position to another in intervals of less than 10 sec. This type of diffusional motion, characteristic of liquids, sets in only above T_g.

In the transition region the segments of the polymer chain are undergoing short-range diffusional motion. The time for diffusion from one "lattice site" to another is on the order of 10 sec (the arbitrarily chosen reference time). This type of diffusion of small polymer segments is independent of molecular weight. In the transition region the modulus changes very rapidly with time as well as with temperature.

In the quasi-static rubbery region the short-range diffusional motions of the polymer segments are very rapid. However, the motions of the molecules as a whole, involving the cooperative movement of many chain segments, are retarded, particularly by entanglements between the chains, which act as temporary crosslinks. One can in fact compute the molecular weight of the network chain between entanglements from the value of the rubbery plateau modulus.

In the region of rubbery flow the motion of molecules as a whole becomes important. Major long-range configurational changes of the entire molecule, including the slippage of long-range entanglements, take place in times on the order of 10 sec.

Finally in the region of liquid flow the long-range configurational changes of the molecule occur in less than 10 sec. Elastic recovery is nearly completely negligible here for stresses or strains that are maintained for longer than 10 sec.

The regions of rubbery flow and liquid flow can be completely suppressed if chemical crosslinks are introduced to serve as permanent network junctures.

For other amorphous polymers the main features of the above discussion based on polystyrene are still applicable. The major changes from one polymer to the other are the changes in T_g and chain length.

In Fig. 2 we show modulus-versus-temperature curves for three other linear amorphous polymers, with emphasis on the glassy, transition, and

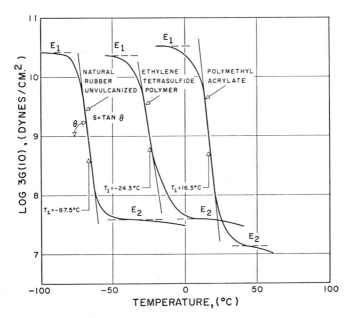

Fig. 2 Modulus-versus-temperature curves for some linear amorphous polymers.

rubbery plateau regions. We also indicate on the graph how we obtain four characteristic parameters—T_i, E_1, E_2, and s—for the polymer from the modulus-versus-temperature curve.

The most important parameter is T_i, the temperature at which $E_r(10)$ attains the value of 10^9 dynes/cm². Although arbitrarily chosen, this modulus value lies more or less in the center of the transition region.

The value of T_i obtained in this way is generally very close to, perhaps 2 to 5°C higher than, the value of T_g obtained from the familiar plot of specific volume versus temperature.

The values of these characteristic parameters for some amorphous polymers are shown in Table 1.† It is interesting to note that the variation in E_1, E_2, and s for amorphous polymers of very different structures are not very great.

---

† The measurements reported here were made in torsion rather than tension. This type of measurement yields a shear modulus G. The shear data are reported as $3G$ in order to be of the same order of magnitude as the tensile modulus E. For values of E lower than 10^9 dynes/cm², E and $3G$ are essentially equal. For higher values of E, approaching the glassy region, the relation between E and G depends on Poisson's ratio, but E must lie between $2G$ and $3G$.

Table 1   Characteristic Parameters for Modulus-versus-Temperature Curves[a]

Polymer	$3G_1$ [(dynes/cm²) × 10⁻¹⁰]	$3G_2$ [(dynes/cm²) × 10⁻⁷]	T_i (°C)	s
Linear Polymers				
Polyisobutylene	3.39	0.892	−62	0.13
Natural rubber (unvulcanized)	2.51	3.98	−67	0.21
Polystyrene	1.82	0.47	101	0.21
Polymethyl acrylate	3.16	1.35	16	0.23
Polymethyl methacrylate	1.35	1.86	107	0.13
Polybutyl acrylate	1.51	0.562	−51	0.21
Polybutyl methacrylate	0.89	1.12	31	0.14
cis-Polybutadiene	2.00	1.07	−106	0.19
Ethylene tetrasulfide polymer	2.19	3.81	−24	0.15
Atactic polypropylene	1.70	2.51	−16	0.16
Ethylene–propylene copolymer (2:1 mole ratio)	1.23	7.08	−58.8	0.16
Slightly Crosslinked Polymers				
Natural rubber (vulcanized)	3.39	4.27	−56	0.20
Tetrahydrofuran polymer	3.17	10.0	−73	0.14
Butadiene–styrene (75:25)	2.09	4.57	−48	0.16

[a] A. V. Tobolsky and M. Takahashi, J. Appl. Polymer Sci., **7**, 1341 (1963).

2. TRANSITION REGION FOR AMORPHOUS POLYMERS

Modulus-versus-temperature curves for amorphous polymers in the transition region can be approximated by the following equation:

$$E_r(10) = \frac{10^9}{10^{s(T-T_i)} + 10^9/E_1} + E_2 \tag{1}$$

where E_1, E_2, T_i, and s are the characteristic parameters discussed in the preceding section and tabulated in Table 1. Inasmuch as E_1, E_2, and s do not show great variations from one amorphous polymer to another, the major parameter is T_i, which is usually given by $T_i = T_g + 4.4°C$.

Figure 3 shows the modulus-versus-temperature curves of a series of butadiene–styrene copolymers, with weight ratios of 100 : 0, 75 : 25, 30 : 70, 10 : 90, and 0 : 100. The T_i values are respectively −90, −60, 15, 65, and 100°C, and these polymers are used respectively for arctic rubbers, general-purpose rubbers, latex paints, plastic sheet, and all-purpose thermoplastics. It is important to note that the T_i values of other general-purpose rubbers

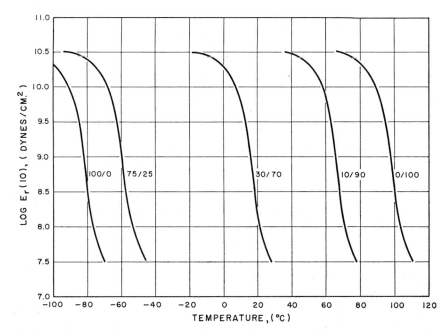

Fig. 3 Curves for E_r (10) versus temperature for butadiene–styrene random copolymers, (weight ratios shown next to each curve).

(natural rubber, *cis*-polybutadiene, butyl, neoprene, nitrile rubbers) range from -90 to $-60°C$. The T_i values of other widely used latex paints (polyvinyl acetate copolymers, acrylic copolymers) are close to $15°C$. Polymethyl methacrylate, with a T_i of $110°C$, is closely matched in physical properties with polystyrene. To a first approximation the physical properties of amorphous polymers of the same T_i and the same chain length are closely matched. For such polymers, whose chemical composition can be very different, the solubility properties and aging properties can vary greatly despite the close match in physical properties.

If the standard time of measurement is changed from 10 sec to a new reference time t_r, the modulus-versus-temperature curves are displaced roughly parallel to each other along the temperature axis. The amount of displacement ΔT is given by the equation

$$\Delta T = -\frac{56(\log t_r - 1)}{\log t_r + 15.1} \tag{2}$$

The derivation (2) of Eq. 2 starts from a modification of the Williams–Landel–Ferry equation (see Section 2 in Chapter 5).

For linear amorphous polymers we can define the molecular weight M_e between entanglements by means of the equation

$$E_2 = \frac{3dRT}{M_e} \tag{3}$$

where E_2 is the tensile modulus of the rubbery plateau and d is the density.

3. CRYSTALLINITY AND CROSSLINKING

Changes in the structural features of polymers show up very clearly in modulus-versus-temperature curves. Figure 4 shows the modulus-versus-temperature curves for four polystyrene samples: two linear amorphous samples (A and C) of differing molecular weights (see Fig. 1), a very slightly crosslinked sample, and a semicrystalline sample based on isotactic polystyrene (the other three samples were based on atactic polystyrene).

It is clear that the major effect of a slight degree of crosslinking is to suppress the region of rubbery flow and liquid flow.

The effect of crystallinity is to enhance the modulus between T_g and the melting temperature T_m. (The value of T_m for isotactic polystyrene is about 235°C.) The exact form of modulus enhancement depends on the degree of

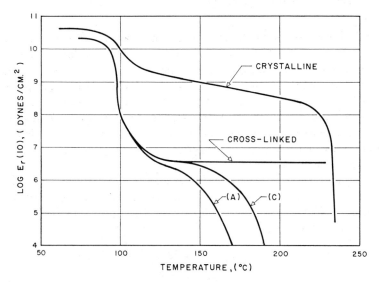

Fig. 4 Curves for E_r (10) versus temperature for crystalline isotactic polystyrene, slightly crosslinked atactic polystyrene, and polystyrene samples A and C, see Fig. 1).

crystallinity and the size of the crystallites or spherulites. For low degrees of crystallinity the crystals act as meltable filler particles and crosslinks. The modulus enhancement resulting from hard-sphere filler particles embedded in a rubbery matrix is given by the equation

$$E = E_0(1 + 2.5\phi + 14.1\phi^2 + \cdots) \qquad (4)$$

where E is the modulus of the filled polymer, E_0 is the modulus of the unfilled polymer, and ϕ is the volume fraction of the filler.† A modulus enhancement also arises from the "crosslinking" characteristic of the crystallites.

At high degrees of crystallinity the above approach is inadequate: it is perhaps better to think of the crystalline phase as continuous with amorphous defects on boundaries of crystallites.

The variety of products that can be obtained by controlling crystallinity is best illustrated by Fig. 5, which shows modulus-versus-temperature curves for polymers of ethylene and propylene—namely, isotactic polypropylene, high-density polyethylene, low-density polyethylene, atactic polypropylene, and amorphous ethylene–propylene copolymer rubber. For polymethylene

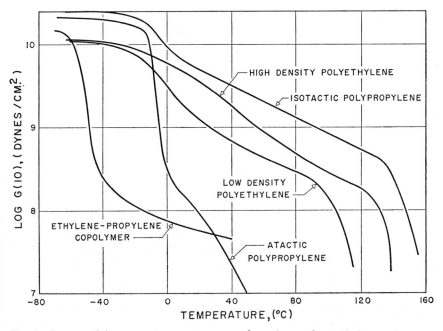

Fig. 5 Some modulus-versus-temperature curves for polymers from ethylene and propylene.

† Note the analogy with the Einstein equation (Eq. 12 in Chapter 5) for hydrodynamics of polymer solutions.

T_m and T_g are 137 and $-80°C$, respectively; for polypropylene T_m and T_g are 175 and $-15°C$, respectively. The five substances shown are rigid plastics, semirigid plastics, and rubbers, depending on their crystallinity.

Figure 6 shows the effect of crosslinking over a wide range of crosslink density. The polymers in question were made by copolymerizing ethyl

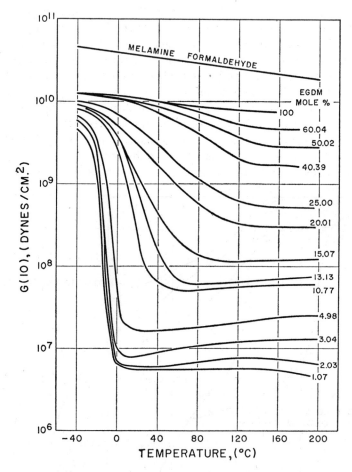

Fig. 6 Modulus-versus-temperature curves for copolymers of ethyl acrylate–ethylene glycol dimethacrylate (EGDM) and for melamine–formaldehyde polymer.

acrylate with ethylene glycol dimethacrylate. From the chemical composition and from the volume shrinkage during polymerization it is possible to determine in an exact calculation the moles of crosslinks per cubic centimeter.

For the polymers shown in Fig. 6 the equation from rubber-elasticity theory was applied to the plateau modulus:

$$G = 2\Phi cRT \tag{5}$$

where G is the shear modulus, Φ is the correction factor, c is the concentration of crosslinks in moles per cubic centimeter, R is the gas constant, and T is the absolute temperature. The correction factor Φ turned out to be unity for mole-percent values of ethylene glycol dimethacrylate below 15% (i.e., for 10 or more C—C main-chain linkages between crosslinks). For higher degrees of crosslinkage Eq. 5 is theoretically and experimentally invalid.

4. BLENDS, GRAFTS, AND BLOCKS

a. Blends

Useful composite structures can often be obtained by blending two polymers with different T_g (or T_i) values. For the blending to be satisfactory one desires (a) that the polymers be sufficiently incompatible to form two distinct phases and (b) that there be satisfactory wetting or adhesion between the two phases.

Impact-resistant polystyrene plastics can be made by blending polystyrene resin with 5 to 10% of a rubbery butadiene–styrene copolymer. The polymers can be mixed in emulsion form, or the styrene can be grafted onto the copolymer emulsion by ordinary vinyl-polymerization techniques. Finally the two polymers can be blended together by mechanical mixing; no matter how the polymers are initially admixed, a severe but controlled mechanical mixing is desirable.

The morphological structure of the impact-resistant polystyrene blends is a continuous matrix of polystyrene in which small rubbery spheres of copolymer are finely and thoroughly dispersed. The impact-resistant quality of these dispersions appears to be related to the initiation and termination of cracks and possibly void formation.

The modulus-versus-temperature curves for polyblends frequently show two distinct transitions, one associated with each phase, as shown in Fig. 7. Sometimes, however, especially if there is compositional heterogeneity (and this can even happen in certain copolymerizations), the modulus-versus-temperature curve will show not two transitions but rather a very broad transition region.

It is of course possible to reverse the phases by changing the weight ratios of the rubbery and hard phases. In an 80 : 20 rubbery–hard polyblend the rubbery phase will be continuous and the hard phase will act as a filler.

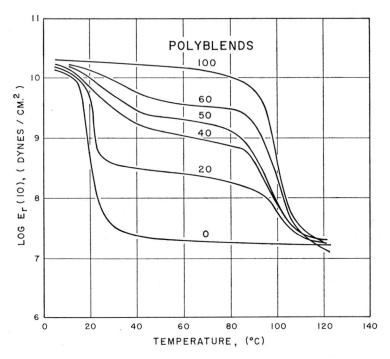

Fig. 7 Curves for E_r (10) versus temperature for polyblends of polystyrene and a 30:70 butadiene–styrene copolymer. Numbers on the curves refer to the weight percent of polystyrene in the blend.

b. Grafts

Grafting ensures good adhesion between phases. In acrylonitrile–butadiene-styrene (ABS) resins a mixture of styrene and acrylonitrile is polymerized around a preformed polybutadiene emulsion. The styrene–acrylonitrile copolymer molecules make occasional chemical attachments to the surface of the polybutadiene emulsion. In the final product the size and shape of the original polybutadiene emulsion particles are fairly well preserved.

c. Blocks

An extremely interesting new development relates to block polymers of the type ABA or ABABABA $\cdots$, etc. Of the first type, the styrene–butadiene-styrene block polymers are the best known. In a typical polymer of this type the molecular weight of each styrene block will be 30,000 and the molecular weight of each butadiene block will be 80,000. The styrene blocks from several molecules aggregate into spherical clusters within a rubbery matrix of poly-butadiene. The polystyrene clusters act as quasi-filler and quasi-crosslink up to 100°C, the softening point of polystyrene. This type of rubber can be injection-molded, but below 100°C it acts like a reinforced, vulcanized elastomer.

The ABABAB ··· type of block polymer is best exemplified by the polyure-thane elastomers. The soft segments are polyester or polyether blocks whose molecular weight is 2000 to 5000. The hard blocks are aromatic urethane or aromatic urea segments whose molecular weight is about 500. The hard-segment clusters may be crystalline in certain cases.

Modulus-versus-temperature curves for some block polymers are shown in Fig. 8.

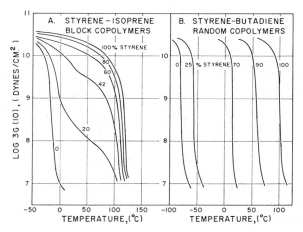

Fig. 8 Modulus-versus-temperature curves for styrene–isoprene block copolymers and for styrene–butadiene random copolymers.

5. TAN δ MEASUREMENTS

Consider a sample of polymer subject to a forced harmonic oscillation of the type

$$s(t) = s_0 \cos \omega t \tag{6}$$

where s is the strain, s_0 is the strain amplitude, and ω is the angular frequency of vibration. If the sample is viscoelastic, the stress will be out of phase with the strain and will be of the form

$$f(t) = f_0 \cos(\omega t + \delta) \tag{7}$$

where δ is the phase difference. In actual measurements both the stress and the strain can be recorded on an oscilloscope, where the phase difference δ is made visual.

It is convenient for dynamic (oscillatory) measurements to describe the dynamic modulus as a complex number G^*(3)

$$G^* = G' + iG'' \tag{8}$$

where G' is the real part of the dynamic modulus and G'' is the imaginary component of the dynamic modulus, both being functions of the angular frequency of measurement. In terms of $s(t)$ and $f(t)$ described above G' and G'' relate to the observables, f_0, s_0, and δ as follows:

$$\frac{G''}{G'} = \tan \delta \tag{9}$$

$$[(G')^2 + (G'')^2]^{1/2} = \frac{f_0}{s_0} \tag{10}$$

The quantities $\tan \delta$, G', and G'' can also be obtained from free-vibration experiments. A rectangular strip of polymer is used as the twisting component of a torsional pendulum whose torsion wheel has a moment of inertia I. After an initial angular twist of θ_0 degrees, the pendulum oscillations are of the form (4, 5)

$$\theta(t) = \theta_0 e^{-\alpha t} \cos \omega t \tag{11}$$

where ω is the observable natural angular frequency of the system and α is an observed damping constant. The quantities G' and G'' are given by the following equations (for $\alpha < \omega$):

$$G' = \frac{I\omega^2}{K} \tag{12}$$

$$G'' = \frac{2\alpha I\omega}{K} \tag{13}$$

where K is a shape factor that depends on the length, width, and thickness of the polymer sample. It follows that

$$\tan \delta = \frac{G''}{G'} = \frac{2\alpha}{\omega} \tag{14}$$

Both G' and G'' are functions of the frequency ω.

Let Δ equal the logarithm of the ratio of successive amplitudes of vibration. From Eq. 11

$$\Delta = \ln \frac{\theta_i}{\theta_{i+1}} = \frac{2\pi\alpha}{\omega} = \pi \tan \delta \tag{15}$$

Equation 15 provides a very direct way of obtaining tan δ from experiment.

Free-vibration experiments are carried out as a function of temperature, modifying the moment of inertia of the torsional wheel so that the natural frequency at all temperatures is about 1 Hz. Forced-vibration experiments are also carried out at fixed frequencies ranging between 1 and 10^4 Hz. Here one can make measurements of G', G'', and tan δ as a function of frequency at constant temperature or as a function of temperature at fixed frequency.

Measurements of tan δ have been carried out on both crystalline and amorphous polymers (5, 6). In amorphous polymers tan δ measurements are especially significant in the glassy state. Sometimes tan δ peaks are observed below 80°K; these are generally associated with side-chain motions.

A very significant tan δ peak occurs in the polycarbonate polymer derived from bisphenol and phosgene (Lexan). Though its T_g is 150°C, this polymer has very high impact strength in comparison with other glassy amorphous polymers, such as polystyrene and polymethyl methacrylate. This high impact strength has been associated with a large tan δ peak at −90°C. The motion responsible for this tan δ maximum has been associated with very-short-range intrachain crankshaft movement.

Many tan δ peaks have been observed in crystalline polymers. For example, in polyethylene there are three transitions α, β, and γ at 90, −30, and −120°C. The exact nature of the molecular motions for these transitions has not yet been established.

The curve of tan δ versus temperature for amorphous polystyrene (at 1 Hz) is shown in Fig. 9. The α-peak at 390°K is associated with T_g and long-range cooperative chain motions. The β-peak at 325°K is associated with the torsional vibration of phenyl groups. The γ-peak at 130°K is associated with the motions of CH_2 moieties. The δ-peak at 38°K is associated with the oscillation, or wagging, of phenyl groups.

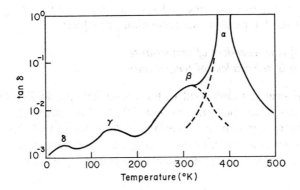

Fig. 9 Tan δ versus temperature for polystyrene. Measurements were made at 1 Hz (After M. Shen).

6. STRESS RELAXATION

Stress-relaxation experiments are carried out by deforming a sample to a fixed strain s_0 at constant temperature and measuring the stress required to maintain the fixed strain as a function of time. For amorphous polymers the effects of time and strain are separable:

$$\frac{f(t)}{s_0} = \Phi(s_0)E_r(t) \qquad \text{tension}$$

$$\frac{f(t)}{s_0} = \Psi(s_0)G_r(t) \qquad \text{shear} \tag{16}$$

where $\Phi(s_0)$ and $\Psi(s_0)$ approach unity for small values of s_0. The terms $E_r(t)$ and $G_r(t)$ are the tensile relaxation and shear relaxation moduli respectively.

For higher values of strain $\Phi(s_0)$ and $\Psi(s_0)$ are approximated as follows (from rubber-elasticity theory):

$$\Phi(s_0) = \frac{(1 + s_0)^2}{3s_0} - \frac{1}{3s_0(1 + s_0)} \tag{17}$$

$$\Psi(s_0) = 1$$

The effect of temperature on $E_r(t)$ and $G_r(t)$ is simply to change the time scale. Plots of $\log E_r(t)$ versus $\log t$ or $\log G_r(t)$ versus $\log(t)$ obtained at different temperatures can be superposed by horizontal translation along the $\log t$ axis. This permits construction of "master curves" at any one temperature with enormously expanded time scales, as shown in Figs. 10 and 11. Equivalently let us define $K(T)$ as the time at temperature T required for $3G_r(t)$ or $E_r(t)$ to attain a value of 10^9 dynes/cm^2. The following law can be stated (the time–temperature superposition principle):

$E_r[t/K(T)]$ *is independent of temperature.*
$G_r[t/K(T)]$ *is independent of temperature.*

The simplifications embodied in the above law and in Eq. 16 were first shown for *amorphous* polymers by Tobolsky and Andrews (6).

The following *approximate* equation has been found to hold for linear or *very slightly* crosslinked amorphous polymers, independently of molecular weight or crosslink density [for theory see reference (9)]:

$$E_r(t) = \frac{E_1}{\{1 + (t/\tau_1)\}^n} + E_2 \tag{18}$$

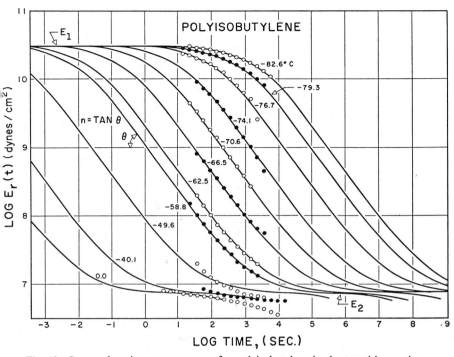

Fig. 10 Stress-relaxation master curves for polyisobutylene in the transition region.

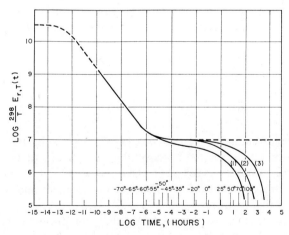

Fig. 11 Idealized master relaxation curve log $E_r(t)$ versus log t for butyl rubber and polyisobutylene of three different average molecular weights:

(1) $M_v = 1.36 \times 10^6$; (2) $M_v = 2.80 \times 10^6$;

(3) $M_v = 6.60 \times 10^6$.

223

where E_1 is the glassy modulus, E_2 the rubbery plateau modulus, and n varies between 0.5 and 1.5. The quantity τ_1 is a characteristic relaxation time of the transition region. The temperature dependence of τ_1 obeys the WLF equation

$$\log \frac{\tau_1(T)}{\tau_1(T_g)} = \frac{-17.44\,(T - T_g)}{51.6 + T - T_g} \tag{19}$$

The value of $\tau_1(T_g)$ is of the order of magnitude of one second.

For the rubbery flow region of linear *monodisperse* polystyrene (115°C) (7)

$$E_r(t) = E_2 \exp[-(t/\tau)^{0.74}]$$
$$\tau = 2.5 \times 10^{-14} M^{3.4} \tag{20}$$
$$E_2 = 4.4 \times 10^6$$

Measurements of $E_r(t)$ for polydisperse polymers provide a method of obtaining the molecular-weight distribution (7). As the distribution broadens, the power of (t/τ) in the exponential decreases.

7. CREEP

Creep measurements are carried out by subjecting a sample to a fixed stress f_0 (in tension or shear) and measuring the strain as a function of time. Tensile-creep tests are generally carried out at fixed load rather than at fixed stress, but the two measurements are equivalent for very small strains. For small shear strains we can define a time-dependent compliance $J(t)$ as follows:

$$\frac{s(t)}{f_0} = J(t) \tag{21}$$

$$J(t) = J_0 + (J_e - J_0)g(t) + \frac{t}{\eta} \tag{22}$$

where J_0 is the instantaneous compliance, $(J_e - J_0)g(t)$ is the delayed recoverable compliance, and t/η is the irrecoverable compliance that could be obtained in a creep-recovery experiment. Such an experiment is carried out by removing the stress f_0 after a long-time creep experiment and following the strain $s(t)$ from this point on.

Inasmuch as amorphous polymers obey laws of linear viscoelasticity at low strains, the quantities $J(t)$ and $G_r(t)$ are related; for example, η and J_e are calculable from $G_r(t)$ as follows:

$$\eta = \int_0^\infty G_r(t)\,dt \tag{23}$$

$$\eta^2 J_e = \int_0^\infty t \, G_r(t) \, dt \tag{24}$$

Furthermore,

$$t = \int_0^t G_r(\tau) J(t - \tau) \, d\tau \tag{25}$$

If either $G_r(t)$ or $J(t)$ is known, the other can be calculated from Eq. 25.

If a crosslinked amorphous polymer is considered, the third term on the right-hand side of Eq. 22 vanishes because η is infinite.

Equation 25 has a simple approximation. If the slope of $\log G_r(t)$ versus $\log t$ is $(-m)$, then

$$G_r(t) J(t) = \frac{\sin(m\pi)}{m\pi} \tag{26}$$

This approximation is valid if $m < 0.8$. If m is zero, $G_r(t)$ and $J(t)$ are reciprocal, as expected. Simple relations between $G'(1/\omega)$ and $G_r(t)$ can also be derived (1).

8. MOLECULAR MODELS

Suppose that we idealize a polymer molecule as being a linear lattice of n balls of mass m connected by springs of force constant c with an equilibrium distance a between balls. The equation of motion for the jth ball is

$$m\ddot{x}_j + f\dot{x}_j + c(2x_j - x_{j-1} - x_{j+1}) = F \qquad j = 1, 2, 3, \ldots, n \tag{27}$$

The term $f\dot{x}_j$ represents a frictional term, to be discussed later. The term on the right-hand side represents the external forces.

A normal coordinate transformation can be used to convert Eq. 27 as follows:

$$m\ddot{q}_p + f\dot{q}_p + K_p q_p = F \qquad p = 1, 2, 3, \ldots, n \tag{28}$$

The value of K_p is

$$K_p = 4c \sin^2\left(\frac{\pi p}{2n}\right) \approx \frac{\pi^2 c p^2}{n^2} \tag{29}$$

Associated with Eq. 28 are n normal mode frequencies and n relaxation times, as follows:

$$v_p = \frac{1}{2\pi} \left(\frac{K_p}{m}\right)^{1/2} \approx \frac{p}{2n} \left(\frac{c}{m}\right)^{1/2} \qquad p = 1, 2, 3, \ldots, n \tag{30}$$

$$\tau_p = \frac{f}{K_p} \approx \frac{fn^2}{\pi^2 cp^2} = \frac{f}{4\pi^2 v_p^2 m} \tag{31}$$

The classical velocity of sound, v, of the linear lattice is $a(c/m)^{1/2}$. The standing waves are

$$\lambda_p = \frac{2na}{p} \qquad p = 1, 2, 3, \ldots, n \tag{32}$$

$$v = a\left(\frac{c}{m}\right)^{1/2} \tag{33}$$

Equation 30 for v_p can in this way be derived from $\lambda_p v_p = v$ without a normal mode treatment. The relation between τ_p and v_p in Eq. (31) is a general one (9).

If we now consider N polymer molecules per cubic centimeter, each idealized as the linear lattice considered above, it can be shown that

$$E_r(t) = Nca^2 \sum_{p=1}^{n} e^{-t/\tau_p} \tag{34}$$

where τ_p is given by Eq. 31.

There are two ways of interpreting the meaning of the constants c and f in Eqs. 27 through 34. The first molecular model, due to Rouse and Bueche (8), considers the polymer molecule as being subdivided into z statistical segments of the Kuhn type. The tensile-force constant c of the Rouse–Bueche model is equal to $3kT/\sigma^2$, where σ^2 is the mean-square end-to-end distance of the segment. This value is an entropic force constant from the rubber-elasticity theory. The friction factor f according to this model is $6\pi\eta\sigma$, where η is the viscosity of the solvent in which the macromolecules are dissolved or the viscosity of the polymer itself if we are considering undiluted polymer.

The second molecular model is the damped torsional oscillator (DTO) model proposed by Tobolsky and coworkers (9). In this interpretation each unit of the linear lattice corresponds to a triad of chain atoms along the polymer chain which is twisting and undergoing occasional hindered rotational transitions. In the DTO model the constant c represents a force constant for torsional oscillation that is 50 times larger than kT/σ^2.

For undiluted amorphous polymers it has been proposed that the DTO model is applicable to the transition region of viscoelastic behavior, whereas the Rouse–Bueche model (suitably modified to account for molecular entanglement) is applicable to the rubbery flow region (9).

For dilute polymer solutions the Rouse–Bueche theory seems to be applicable to the long-time end of the viscoelastic spectrum. The extra viscosity imparted by the polymer molecules to the solvent can be shown to be

$$\eta - \eta_s = \frac{fN\sigma^2 z^2}{36} \tag{35}$$

where z is the number of Kuhn segments and σ^2 is the mean-square length of the segments. This gives rise to exactly the same expression for intrinsic viscosity as does the free-draining model of Huggins and Debye (see Chapter 5).

A modification of the DTO model has been proposed by Zimm and Clark (10). Suppose that in the equation of motion the damping term is considered proportional to the *relative* velocity of the jth segment with respect to the $(j-1)$th and $(j+1)$th segments, rather than proportional to the absolute velocity of the jth segment. The equations of motion become

$$m\ddot{x}_j + f'(2\dot{x}_j - \dot{x}_{j-1} - \dot{x}_{j+1}) + c(2x_j - x_{j-1} - x_{j+1}) = F$$
$$j = 1, 2, \ldots, n \tag{36}$$

By normal coordinate treatment this set of equations becomes

$$m\ddot{q}_p + 4f' \sin^2\left(\frac{\pi p}{2n}\right) + 4c \sin^2\left(\frac{\pi p}{2n}\right) = F \qquad p = 1, 2, \ldots, n \tag{37}$$

The relaxation times associated with Eq. 37 are all the same, all equal to f'/c and quite obviously independent of chain length. The quantity f' in Eq. 36 can be regarded as a coupled frictional factor, where f in Eq. 27 can be regarded as an uncoupled frictional factor (also true for the original DTO model).

9. CHEMICAL STRESS RELAXATION

At high temperatures stress relaxation at constant extension in rubber networks can occur because of the cleavage of chemical bonds either along the main chain or at crosslinks (1).

Figure 12 shows stress decay at constant extension plotted as $\log[f(t)/f(0)]$ versus time for (a) peroxide-cured EPT rubber in vacuum at 250°C, (b) polysulfide rubbers in air or vacuum at 110°C, (c) sulfur-cured EPT rubber in air or vacuum at 130°C, and (d) peroxide-cured EPT rubber in air at 130°C.

Stress decay in these several cases arises from the following chemical mechanisms:

1. Carbon–carbon bond cleavage along the main chain or at crosslinks.
2. Polysulfide bond cleavage along the main chain or at crosslinks.

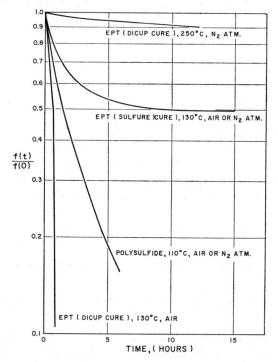

Fig. 12 Chemical stress relaxation (dicup: dicumyl peroxide).

3. Polysulfide link cleavage (trisulfide or tetrasulfide) at the crosslink (disulfide and monosulfide crosslinks are more stable than polysulfide links).

4. Oxidative cleavage of hydrocarbon chains.

The stress at constant extension is proportional to the number of chains per cubic centimeter still uncut. If the cleavage is at the crosslink, we can write the following equation:

$$-\frac{dN(t)}{dt} = k'N(t) \tag{38}$$

$$\frac{N(t)}{N(0)} = \frac{f(t)}{f(0)} = e^{-k't}$$

where $N(t)$ is the number of network chains per cubic centimeter still uncut, $N(0)$ is the initial concentration of network chains, and k' is the specific rate constant for cleavage.

If the cleavage occurs along the main chain, we can write

$$-\frac{dN(t)}{dt} = k'm\,N(t) \tag{39}$$

where m is the number of sites per chain available for cleavage. If m is the same for all network chains, which is true if the network chains are of uniform length, then

$$\frac{N(t)}{N(0)} = \frac{f(t)}{f(0)} = e^{-k'mt} \tag{40}$$

The assumption that gives rise to Eq. 40 also yields the following relation:

$$q(t) = -N(0)\ln\frac{f(t)}{f(0)} \tag{41}$$

where $q(t)$ is the number of cleavages per cubic centimeter that have occurred up to time t. This relation enables one to calculate bond cleavage from stress-relaxation studies. Chemical-stress-decay laws have also been derived for other cases, including networks of nonuniform length.

REFERENCES

1. A. V. Tobolsky, *Properties and Structure of Polymers*, Interscience, New York, 1960.
2. J. B. Yannas and A. V. Tobolsky, *J. Macromol. Chem.* 1, 399 (1966).
3. J. D. Ferry, *Viscoelastic Properties of Polymers*, John Wiley, New York, 1961.
4. L. E. Nielsen, *Mechanical Properties of Polymers*, Reinhold, New York, 1962.
5. P. Meares, *Polymers: Structure and Bulk Properties*, Van Nostrand Ltd., London, 1965.
6. A. V. Tobolsky and R. D. Andrews, *J. Chem. Phys.* 13, 3 (1945).
7. W. Knoff, I. L. Hopkins, and A. V. Tobolsky, *Macromolecules* (in press).
8. F. Bueche, *Physical Properties of Polymers*, Interscience, New York, 1962.
9. A. V. Tobolsky and D. B. DuPré, *Adv. Polymer Sci.* 6, 103 (1969).
10. B. H. Zimm and M. B. Clark, *Polymer Preprints A.C.S.* 12, 116 (1971).

Strength of Polymers

11

H. Mark

1. MECHANICAL PROPERTIES OF POLYMERS

Natural polymers are used and synthetic polymers are made mainly because of their outstanding mechanical properties. All their industrial importance as fibers, plastics, rubbers, adhesives, and coatings is ultimately based on strength, high elongation, high modulus, toughness, abrasion resistance, and thermal and chemical durability.

One of the most important and intriguing questions is therefore the following: Why do polymeric materials of all kinds have such useful mechanical qualities and how can their outstanding mechanical, thermal, and chemical performance be correlated with their molecular structure? In this chapter we describe briefly the mechanical characteristics of the most important polymers and discuss some of their present theoretical aspects. This will not only explain what we possess today but will also suggest the way to further improvements.

Let us begin with Table 1, which places organic polymers as structural materials in the well-known frame of metallic and ceramic systems. If we have at our disposal 10 kg (22 lb) of a number of materials and are to pull with them a certain load from street level to the top of a skyscraper 100 meters (325 ft) high, which material permits us to lift the heaviest load? Evidently, we must first make a 100-meter rod or rope of each material and then compute its tensile strength; lighter substances will have a larger cross section and therefore a higher effective strength. This is already a point in favor of organic polymers, whose specific gravities are between 0.9 and 1.5. Table 1

231

**Table 1   Maximal Tensile Load That Can Be Supported by a
Fixed-Weight Cross Section of Various Materials**

Material	Density (g/cm³)	Cross Section (cm²)	Tensile Strength (kg/cm²)	Maximal Load (kg)
Polypropylene, isotactic	0.90	1.10	20,000	22,000
Polyethylene, linear	0.95	1.00	20,000	20,000
Nylon 66	1.12	0.85	18,000	15,500
Polyester	1.3	0.75	16,000	12,000
Graphite yarn	2.2	0.44	33,000	14,500
Glass	2.4	0.41	25,000	10,000
Aluminum	2.7	0.38	15,000	5,700
Titanium	4.5	0.22	35,000	7,700
Steel	7.5	0.13	50,000	6,500

contains the best presently achieved tensile-strength values and shows that
the heaviest load would be lifted by a rope made from highly oriented filaments of isotactic polypropylene; a steel wire would lift only about one-third
of this load.

The strength figures used in Table 1 are the highest values ever observed
for the individual materials, and it is therefore appropriate to add Table 2,

**Table 2   Tensile Strength and Young's Modulus of
Commercial Structural Materials**

Material	Young's Modulus (kg/cm²)	Tensile Strength (kg/cm²)
Cast iron	1,100,000	4,000
Steel:		
0.60% carbon	1,600,000	8,000
0.8% carbon	1,800,000	12,000
1.0% carbon	2,100,000	14,500
Nickel (hard sheet)	2,000,000	5,500
Aluminum		
(hard sheet)	600,000	1,800
Glass	300,000	5,600
Polyimide (drawn)	70,000	11,500
Polyamide (drawn)	20,000	10,500
Phenolic resin	20,000	1,100
Polystyrene	20,000	800

which shows the characteristic mechanical properties of commercially available structural materials of different origin.

2. COHESIVE FORCES IN POLYMERS

Considering the importance of strength and the large number of organic polymers that are used as fibers, rubbers, and structural plastics, it is tempting to explore what one would expect for the maximal *theoretical* values of the cohesive forces in solid polymeric materials and compare them with the corresponding values for metals and ceramics. It has been long known that the actual strength of metals and ceramics is much lower than what one should expect from the average interatomic bonding forces, which are known from other physical properties. This fact has led to the concept that the failure of materials in mechanical deformation is brought about by small and large flaws. In order to arrive at a valid comparison with organic polymers it is convenient to start with the following considerations.

Suppose we could place a single molecule of linear polyethylene between the clamps of a tensile tester, what force would be required to break one of the C—C bonds of the chain? If instead of polyethylene we were to deal with polyesters, polyamides, or cellulosics, we would have to ask the same question for the C—O and C—N bonds.

In general we do not know this force from any direct measurement, but we have reliable information concerning the energy necessary to dissociate covalent bonds of this type; it is approximately 80 to 90 kcal/mole, or 5 to 6 $\times 10^{-12}$ erg/bond. To a rough approximation this energy E can be expressed as the product of the force f that is applied and the distance d over which f has to act in order to separate the two atoms completely:

$$E = fd \tag{1}$$

We know that the effective distance of covalent bonding forces is very short, so that d is of the order of 1.5 Å, or 1.5×10^{-8} cm. Hence the force that suffices to rupture a C—C, C—O, or C—N bond is

$$f = \frac{E}{d} = 3 \text{ to } 4 \times 10^{-4} \text{ dyne/bond} \tag{2}$$

A better approximation is reached if one considers in more detail the actual shape of the potential energy curve $V(r)$ that characterizes a covalent bond. It was first proposed by Morse that this function has the general form

$$V(r) = D\{(\exp[-2\alpha(r - r_0)] - 2\exp[-\alpha(r - r_0)]\} \tag{3}$$

where D is the dissociation energy, r_0 is the distance of the two atoms in the equilibrium position, and the force constant α is given by

$$\alpha = 2\pi v \left(\frac{\mu}{2D}\right)^{1/2} \tag{4}$$

where v is the frequency with which the two atoms vibrate and the reduced mass μ is

$$\frac{1}{\mu} = \frac{1}{m_1} + \frac{1}{m_2} \tag{5}$$

the terms m_1 and m_2 being the masses of the two atoms.

The Morse curve for a covalent bond with a dissociation energy of 50 kcal/mole (Fig. 1) is an example and indicates the general shape of the energy valley in which the two bonded atoms vibrate against each other; at $r = 7.0$ Å the force between them becomes negligible.

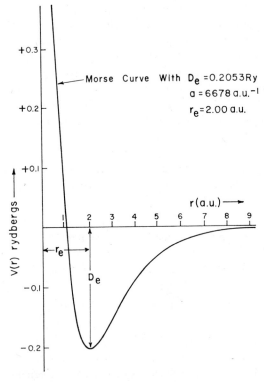

Fig. 1 Morse curve for a covalent bond with a dissociation energy of 50 kcal/mole.

In order to arrive at the force that is necessary to lift the atoms from the depth of this energy valley one must consider the *steepest slope* at the right-hand side of it. The best available knowledge of D, r_0, and α leads to values between 2×10^{-4} and 5×10^{-4} dyne/bond, which justifies the former crude estimate as a reasonable approximation.

A similar consideration can also be applied to *weaker* interactions, such as hydrogen bonds and van der Waals forces, that exist between the individual macromolecules and are responsible for important properties, such as melting and solubility. The energies that are necessary to open these bonds are between 1 and 5 kcal/mole, which corresponds to 0.6 to 3.0×10^{-13} erg/bond.

Hydrogen bonds have a dissociation energy of 5 kcal/mole and a range of action of about 2.8 Å; weaker van der Waals bonds have a dissociation energy of 2 kcal/mole, with a range of action of about 4.0 Å; the forces that are necessary to break these bonds are approximately 10^{-5} and 2×10^{-6} dyne, respectively.

If we compare the two orders of magnitude (about 5×10^{-12} erg for inter-atomic bonds and 10^{-13} erg for intermolecular bonds) with the average kinetic energy of one degree of freedom at room temperature—2×10^{-14} erg—it becomes clear that the strong chemical bonds will not be weakened or broken by thermal motions in the range of ordinary temperatures, whereas the forces responsible for the heats of melting or vaporization of organic compounds can be overcome by the thermally induced molecular vibrations and rotations.

If materials of this type are exposed to mechanical action—such as cutting, tearing, or grinding—all cohesive forces will be taxed, and it is obvious that the weaker van der Waals (secondary) bonds will be severed easier than the chemical (primary) bonds. This leads to the question: Can chemical bonds in polymeric materials at all be ruptured by mechanical action?

For a certain period of time it was not possible to give a clear-cut answer to this question, but recently experimental evidence has been accumulating that mechanical forces can break chemical bonds if they are concentrated in very small volume elements to provide force concentrations greater than 5×10^{-4} dyne/bond. The first indication of this fact was obtained through reducing the average molecular weight of polystyrene melts by forcing them through narrow capillaries in the absence of oxygen at 150°C. Later the breaking of chemical bonds by large shearing forces in calendering and ex-truding was established not only by the reduction of the molecular weight but also by the production of free radicals at the freshly produced chain ends. The most quantitative information on the mechanical severing of chemical bonds is available today as the result of systematic studies of the progressive degradation of macromolecules in ultrasonic fields; it confirms that the strength of the single bonds between carbon, oxygen, and nitrogen atoms are about 5×10^{-4} dyne/bond, and this permits one to describe the breaking process with the methods of molecular statistics.

Let us now assume that we take a bundle of very long polyethylene chains that has a 1-cm^2 cross section, put it in a tensile tester, and stretch it; what force will we need to start stretching and eventually breaking this bundle?

In order to find this out we have to know how many chains of linear polyethylene can be packed parallel to each other in a 1-cm^2 area. The crystal lattice of polyethylene and of many other n-paraffins is known, and it turns out that an area of about 20 Å^2, or 20×10^{-16} cm^2, is needed to accommodate one chain. The number of chains is therefore 5×10^{14} per square centimeter, and, if we take the force to break one chain as 5×10^{-4} dyne, we arrive at a theoretical tensile strength of 25×10^{10} dynes/cm^2, which corresponds to 25×10^4 kg/cm^2, or 3.7×10^6 psi. This is, of course, a maximum value, which would be realized only if the packing of the molecules were completely uniform and regularly arranged without any flaws and discontinuities. In reality the highest values obtained for the tensile strength of highly oriented and crystalline polyethylene filaments are between 200,000 and 240,000 psi, which shows that the actual strength is smaller by a factor of 15 to 20. To the best of our present knowledge this discrepancy is caused by the fact that no specimen is ever so structured that all macromolecular chains are breaking simultaneously over the same cross section. Because of the finite length of the chains and because of both large and small imperfections in the structure, there will always be an initial breaking of many *intermolecular* bonds until finally the stress is concentrated on a statistically favored few chemical bonds that are then ruptured. The crack thereby produced is propagated through the sample by the catastrophic severing of all chemical and intermolecular bonds in the crack path. This leads to the concept of *weak spots* as the precursors of the actual fracture of all real materials. This was recognized many years ago and was put on a quantitative theoretical basis by a series of fundamental and systematic studies. These results made it clear that not only *cohesive* but also *adhesive* bonds are essentially determined by weak spots and weak layers.

Considering the rupture of hydrogen bonds, we can again arrive at a rough estimate if we can assume an average dissociation energy of 5 kcal/mole and a breaking distance of 2.5 Å. Using the same approximation as before, we find that the force necessary to break one hydrogen bond is about 10^{-5} dyne, and the corresponding force to break an average van der Waals bond is about 2×10^{-6} dyne.

If a specimen is essentially held together by hydrogen bonds (such as, for example, cellulose, nylons, or proteins) and if its lateral packing is completely regular so that optimal orientation of the molecules is obtained, its breaking strength should be around 5000 kg/cm^2, or 60,000 psi. This is of the same order of magnitude as the actually measured values of highly oriented fibers of these materials and shows that during the normal breaking of fibers, films, or other test pieces the essential initial contribution to the rupture of the sample occurs by the severing of intermolecular bonds of the hydrogen bridge or van der Waals type. As a consequence there results a force accumulation on the covalent bonds of some chains, which will also break despite the fact that they are 10 to 20 times stronger than the van der Waals cohesion between the chains.

3. GENERAL BEHAVIOR OF NONCRYSTALLINE THERMOPLASTICS

Figures 2 and 3 and Table 3 are presented to identify the amorphous, uncrosslinked polymers as a family of materials that includes noncrystalline resins, raw rubbers, and some semisolids and glasses.

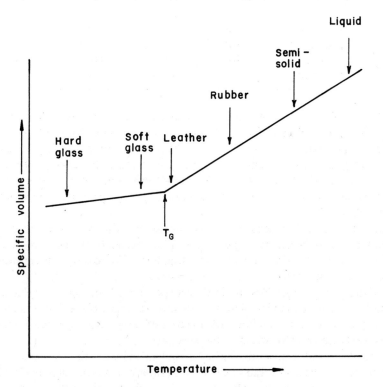

Fig. 2 The specific volume of glasses, and supercooled liquids as a function of temperature. (T_G = glass points).

Figure 2 provides a schematic plot of the temperature dependence of the specific volume or reciprocal density of glasses and supercooled liquids. Without modifying the molecular structure of the polymer, one type of a noncrystalline thermoplastic material can be converted reversibly into another type by dilating or contracting it on heating or cooling and on adding or extracting liquid components (i.e., plasticizers).

Table 3 contains approximations of the Young's moduli, Poisson's ratios, and maximum elongations of these same materials.

Table 3   Approximate Young's Modulus E, Poisson's
Ratio μ, and Maximum Elongation of Amorphous
Polymers

Material	E (dynes/cm^2)	μ	Maximum Elongation (%)
Hard glass	$>10^{11.0}$	<0.30	0.2
Soft glass	$\sim 10^{10.5}$	~ 0.33	2.0
Leather	$\sim 10^{9.0}$	~ 0.38	100
Rubber	$\sim 10^{6.5}$	~ 0.49	1000
Semisolid	$<10^{6.0}$	>0.49	500

The trend for increasing plasticizer concentrations also is indicated in Fig. 3, from which it can be inferred that Young's moduli for the "glassy" and "rubbery" states of these materials, although greatly different in magnitude, are sensibly independent of time and temperature; hence both of these materials are *elastic*. However, the "leathery" and "semisolid" states of the same polymers are strongly dependent on time and temperature; hence both are *viscoelastic*.

Figure 4 illustrates schematically the stress-versus-strain curves of these materials and also the general appearances of the specimens at, and immediately after, rupture. It also indicates the trends for increasing temperature, plasticizer concentration, and rate of extension.

The appearance of the glassy and rubbery specimens after rupture is similar; both fail in an *elastic* manner; that is, both have a fracture plane that is normal to the principal tension, and both, for the case of sufficiently rapid tensioning, have sensibly no residual strain in the specimen.

The leathers and semisolids, however, fail in manners that indicate that pronounced *flow* has occurred in directions that are somewhat inclined to the axis of the principal tension. This flow is most favored along the 45° planes since these planes include the maximum shearing-force components of the

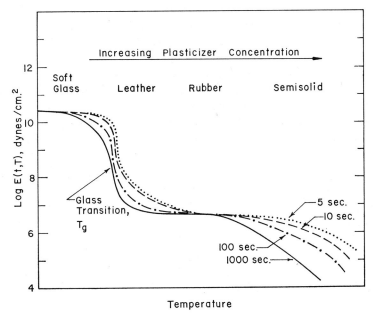

Fig. 3

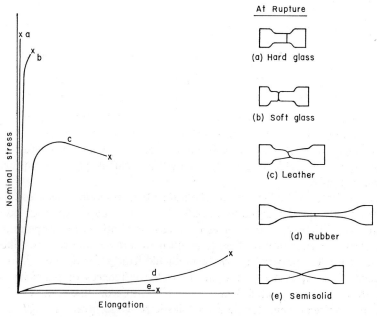

Fig. 4   Stress–strain curves for noncrystalline thermoplastics.

system of forces that are applied longitudinally to the specimen, as shown in Fig. 5. In effect the *viscoelastic* leathers and semisolids exhibit more *ductile* responses at high strains, which result from flow, creating much residual strain in the specimen.

The "soft glasses" show a limited tendency for specimen necking and residual strain, and generally provide a transition from elastic to ductile ruptures. The glass-transition temperature T_g can be crudely regarded as the temperature of transition between the quasi-elastic and quasi-ductile modes of rupture.

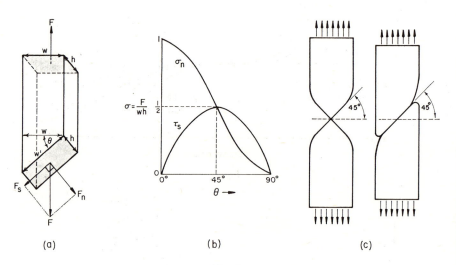

(a)   (b)   (c)

Fig. 5

Although samples of "hard glass" and "rubber" may appear similar after breaking, their rupture processes differ considerably not only in regard to the magnitudes of the strain required but also in regard to the overall shapes of the specimens and their stress-versus-strain curves prior to rupture, as shown in Fig. 4.

It is possible to predict (by interpolation) the probable influences of time or strain rate on the ultimate stresses and strains by comparing the contributions of time with the corresponding contributions of temperature or plasticizer concentration to these same ultimate stresses and strains. Likewise the influences of temperature can be predicted from prior time or plasticizer-concentration experiments, and the probable influences of plasticizer concentration can be predicted from prior temperature or time studies.

Such empirical time–temperature–plasticizer-concentration *shifting procedures* for estimating rupture stresses and strains (and even yield stresses and strains) are very practical and often can be used to forecast the behavior of some materials without direct experimentation.

It has already been mentioned that the discrepancy between the theoretical and actual tensile strengths of polymers indicates the existence of force accumulations in certain restricted areas. This fact is strongly supported by the microscopic and submicroscopic appearance of the specimens before and after rupture. In most cases it is observed that the specimen behaves differently (i.e., it is much softer or more ductile) at a localized region of damage than anticipated from the mechanical behavior of the whole specimen. A glassy specimen may appear to have deformed as if it were a rubber *in the immediate region of damage.* Hence a region of damage is, in a sense, a unique site in which the levels of stress and strain may be up to 10^2 or 10^3 times higher than those in the rest of the sample.

Frequently transparent, noncrystalline thermoplastics lose most of their optical clarity in regions of damage, such as in necking zones or at micro-fractured surfaces. Some of these areas become birefringent and reflect or scatter incident light, depending on the specific circumstances. When the incident light is predominantly reflected, the phenomenon is called " crazing," when it is principally scattered, it is called " blushing," which in noncrystalline specimens is associated either with pronounced birefringence or with gross submicroscopic damage. In the case of submicroscopic damage there could be as many as a trillion cavities per cubic centimeter, with an average diameter in the range of 600 to 1100 Å (see Fig. 6).

Figure 7 is a photomicrograph of a microfractured, or " crazed," surface of polymethyl methacrylate, a soft glass.

4. ASPECTS OF MOLECULAR AND SUPERMOLECULAR STRUCTURE

Having briefly discussed the macroscopic and microscopic phenomena of rupture, we now consider the influences of the macromolecules themselves.

a. Molecular Weight

There exists a characteristic trend of Young's modulus with increasing molecular weight and with increasing temperature of the specimen. It has been found that the semisolid state of the *low polymer* is achieved at a lower temperature than that of the *high polymer*; hence flow by interchain slip increases with decreasing molecular weight. Moreover, the *rubbery state* does not exist in a noncrystalline thermoplastic of low molecular weight; hence an almost direct transition from the leathery to the semisolid state is experienced

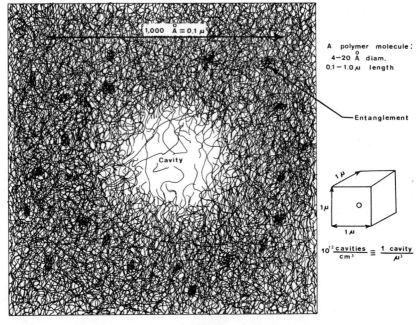

Fig. 6

in short-chain polymers, thereby giving to T_g a more reliable role as a brittle–ductile transition temperature.

Below T_g the tensile strength is affected by the molecular weight in two ways:

1. The higher probability of a low polymer experiencing interchain slip.
2. The higher probability of finding a chain end on the fracture surface of the low polymer.

The first influence encourages a more efficient stress dissipation in the region of macroscopic discontinuities and a more efficient stress-redistribution; the second influence causes a lower fracture-surface energy and with it a weakening in tensile loading.

b. Molecular Orientation

In amorphous polymers strain-induced molecular orientation affects the tensile strength, the mechanism of rupture, and the initiation and population density of microfractures, such as crazes.

For most specimens one can crudely anticipate with proper orientation an enhancement in tensile strength by a factor of 3, since the chains would all be aligned parallel to one direction rather than parallel to *all three*. In practice

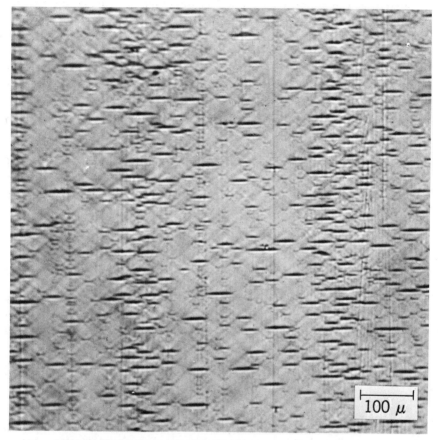

Fig. 7 Photomicrograph of a microfractured surface of Plexiglas.

factors slightly greater than 2 are commonly experienced in amorphous polymers. It can also be anticipated that the tensile strength normal to the axis of molecular orientation will be very much lower than that of an isotropic specimen, since secondary, rather than primary, bonds would predominate along the fracture plane. In fact tensile-strength-reduction factors of about 5 are observed in noncrystalline thermoplastics.

Oriented-chain molecules are rather effective strong points in time-dependent ruptures during which localized forces are continually being redistributed. It is apparently more difficult to realign an oriented polymer molecule in a submicroscopic region of flow, and this may be the reason why uniaxially, and particularly biaxially, oriented specimens do not readily experience as dense a microfracture population as isotropic specimens of the same molecular size and constitution.

Interesting also are the changes in the rupture mechanism and the appearance of the fracture surfaces in oriented samples. It is common for amorphous polymers to show the phenomenon of "cold-drawing," which, however, is not nearly as pronounced as it is in crystalline polymers. This mild degree of "strain-hardening," accompanied by a "uniform drawing" of the material at temperatures approaching T_g, is apparently caused by the uniaxial molecular orientation taking place in the necking zone. The ductile failure otherwise anticipated in this zone is displaced by the orientation process, which curtails large-scale flow along the planes of maximum shearing stress.

The quasi-brittle fracture surfaces of a *biaxially* oriented material are markedly affected by the planar orientation, frequently showing a series of parallel ridges projecting normal to the fracture surface when the direction of applied uniaxial tension is in the plane of the biaxial orientation of the specimen.

c. Crosslinking

As we depart from the *thermoplastics* and move toward the *thermosets*, we discover a completely new scheme of things pertinent to this chapter.

First, a ductile mode of specimen rupture cannot take place in sufficiently well-crosslinked specimens because polymers can experience only a large-scale flow when chain scissions and interchain slips are freely occurring, and this is impossible if the specimen is still well crosslinked at the time of rupture. This means that the *semisolid state cannot exist* for well-crosslinked polymers.

Second, with the elimination of the semisolid state, the highest thermal state possible for crosslinked polymers is the *rubbery state*.

Third, from crosslinking of the polymer the specimen derives considerable *toughness* at temperatures above T_g, and it displays higher tensile strengths than those of glassy, amorphous thermoplastics of similar molecular structure at comparable temperature levels.

The value of strength for crosslinked polymers of organic constitution approximates 15,000 psi; the rupture modes vary from brittle to rubberlike-elastic as the temperature is increased. There is essentially no tendency for cold-drawing since all large-scale flow processes are restricted by the cross-links.

As the degree of crosslinking is increased, the material *becomes harder*. In fact crosslinking can be extended sufficiently to alter the state of the material all the way from a rubber to a hard glass, thus shifting it through the entire spectrum of rupture mechanisms and tensile-strength values.

d. Crystallinity

In this section we deal with polymeric materials that differ from noncrystalline varieties in regard to the following:

Table 4 Mechanical Properties of Several Commercial Fibers

Property	Fiber[a]								
	Nylon 66	Nylon 6	PACM-9	PACM-12	Nylon 4	Nylon 7	Nylon 12	2GT	Nylon 11
Melting point (°C)	260	220	240	205	262	235	175	265	192
Glass-transition temperature (°C)	42	37	140	120	72	60	40	80	47
Tenacity (g/denier)	3.5–5.5	3.5–5.5	2.0–4.0	2.0–3.0	2.5–4.5	3.0–4.5	2.5–4.5	3–5	3–5
Elongation (%)	35–50	35–50	18–40	18–45	30–50	30–50	30–50	25–45	30–50
Tensile modulus (g/denier):									
Dry at 65% relative humidity and 22°C	35	33	45	40	33	30	33	40	30
Wet at 23°C	12	10	35	35	8	9	11	25	12
Wet at 95°C	5	5	15	15	2	5	6	12	10
Moisture regain (%)	4.0	4.5	3.2	3	7.3	2.8	2.0	0.6	1.3

[a] The fiber-forming compounds are as follows:

PACM-9: poly-bis-p-aminocyclohexyl methane azelamide.
PACM-12: poly-bis-p-aminocyclohexyl methane dodecane diamide.
Nylon 4: poly-4-aminobutyramide.
Nylon 7: poly-7-aminoenanthamide.
Nylon 12: poly-12-aminolauramide.
2GT: polyethylene terephthalate.
Nylon 11: poly-11-aminoundecanamide.

1. Physical state
2. Fracture mechanisms
3. Microfracture and submicrofracture habits
4. Tensile-stress-at-rupture values.

Semicrystalline thermoplastics can be divided into hard and soft glass, and hard and soft leather. The soft-glassy and hard-leathery semicrystalline polymers show distinct "cold-drawing," a yielding in which the necking zone experiences a cold-flow-induced, uniaxial recrystallization, such as that employed for preparing high-tensile-strength synthetic fibers.

Rubbers, whether or not crosslinked, do not exhibit irreversible cold-drawing at high extension but may experience a reversible strain-hardening that is produced by an enhanced, uniaxial crystallization in the direction of extension.

Nonfibrous molecular crystals of polymers are encountered in nature with certain proteins; they can also be prepared with synthetic polymers in single-crystal form. The mechanics and failure properties of these anisotropic crystalline materials have been only recently studied; they tend to experience crystal-plane slip, similar to that for nonpolymeric crystalline materials, and hence they add further varieties of failure modes to those discussed above.

Fibrous molecular crystals of polymers can be prepared by the cold-drawing process or are found in nature as cellulosic and proteinaceous fibers. These exist in various degrees of chemical purity and crystal perfection; their rupture modes, microfracture and submicrofracture habits, and tensile strength also vary widely. These fibrous crystalline polymers fall in categories that encompass *the strongest organic polymers known to man*, as discussed at the beginning of this chapter.

Table 4 presents data on the mechanical properties of several important commercial fibers that fall into the class of oriented semicrystalline systems.

BIBLIOGRAPHY

The following comprehensive treatises are recommended for more detailed information:

T. Alfrey, *The Mechanical Behavior of Polymers*, Interscience, New York, 1946.

R. N. Haward, *Strength of Plastics and Glass*, Interscience, New York, 1949.

L. E. Nielsen, *Mechanical Properties of Polymers*, Academic Press, New York, 1960.

P. H. Geil, *Polymer Single Crystals*, Wiley, New York, 1963.

B. Rosen, *Fracture Processes in Polymeric Solids*, Wiley, New York, 1964.

Diffusion through Polymers 12

A. F. Stancell

A pressure or concentration difference, or, more accurately, a fugacity difference, serves as the driving force for the isothermal diffusion of molecules through polymers. Molecules at the higher fugacity sorb into the solid polymer film and move through the matrix of polymer chains, with subsequent desorption at a lower fugacity from the downstream film face. If the molecules are initially in solution and if a polymer membrane permeable to them separates two solutions in which the molecules have different concentrations, the diffusion of the molecules from the high-concentration side to the low-concentration side is called osmosis or dialysis (Fig. 1).

If, however, the molecules are made to diffuse from the low-concentration side of the polymer membrane to the high-concentration side, the process is called reverse osmosis. This is achieved by applying to the low-concentration side a pressure that is sufficiently high to overcome the opposing force (osmotic pressure) that would normally drive molecules from a high concentration to a low one (Fig. 2). Reverse osmosis is an important application of diffusion through polymers. Currently a great deal of effort is

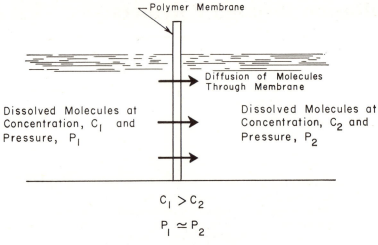

Fig. 1 Osmosis or dialysis.

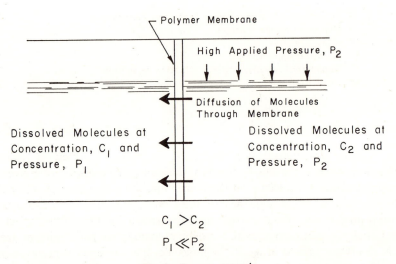

Fig. 2 Reverse osmosis.

being made to find an economical way of using it to make fresh water from seawater or brackish water. In the process high pressure is applied to the salt-water so that water diffusion can take place, and the polymer membrane is specially designed so as to exclude and largely eliminate the diffusion of salt ions.

Another application of diffusion through polymers, and one that is quite evident, is film packaging. Polymer films are used extensively for packaging meat, produce, and baked goods. These applications require definite diffusion properties in the polymer films; for example, polyethylene films used for meat wraps must allow at least a certain minimum diffusion of oxygen from the air in order that the meat may retain an attractive red-pink color.

In the medical field the application of polymer films to artificial-kidney devices is quite important. By selective diffusion, the polymer films can remove urea from the blood when a damaged kidney can no longer do so. Blood is circulated from the patient to a diffusion cell in which the blood contacts a polymer membrane, and urea and uric acid from the blood diffuse across the polymer membrane into a solution. The solution is maintained at a low concentration of urea and uric acid so that the driving force can be maximized for the desired diffusion. The process is sometimes called hemodialysis.

These applications will be discussed further after a presentation of the theory of diffusion through polymers.

1. FUNDAMENTAL CONCEPTS

a. Fick's First Law

There are two general ways of looking at how a molecule diffuses through a solid polymer. Barrer (1, 2) some time ago viewed it as the result of fluctuating thermal energy in the polymer. When the local energy is sufficiently high, the diffusing molecule can move or jump between the polymer chains by cooperative motion with them. Barrer imagined that the polymer-chain segments were conformed to make the opening and that the larger the diffusing molecule, the greater the number of polymer segments involved in the diffusion step.

The other view is the free-volume theory of diffusion, an approach taken by several workers and thoroughly discussed by Kumins and Kwei, (3). It is thought that fluctuating local density in the polymer causes free volume, or holes, to be formed. When a hole of sufficiently large size forms near the penetrant molecule, the molecule can move or jump into it. The diffusion step is assumed to be proportional to the probability of forming such holes of the right size, and the probability can be calculated. Also, the effect of the diffusing molecule on the free volume of the polymer can be taken into account. The theory implies, as does Barrer's theory, that the number of polymer-chain segments involved in the diffusion step increases with the size of the diffusing molecule.

Calculations from the theory give a basis for correlating diffusion data and for determining the physical significance of the diffusion coefficient. The diffusion coefficient is specified by the relationship found to represent the diffusion rate through solids—that is, Fick's first law:

$$J = -D\frac{dC}{dx} \tag{1}$$

where D is the diffusion coefficient (in square centimeters per second) and J is the diffusion rate, in cubic centimeters (STP)/per second per square centimeter of cross-section at a distance x (in centimeters) in the solid; C is concentration of the diffusing molecule, in cubic centimeters (STP) per cubic centimeter of solid.† The units for J and C could also be based on grams or moles rather than cubic centimeters (STP).

The variation of D with temperature follows the form

$$\frac{d(\ln D)}{d(1/T)} = -\frac{E_D}{R} \tag{2}$$

where E_D is the activation energy for diffusion (in calories per gram-mole), T is the temperature (in degrees Kelvin), and R is the gas constant, 1.99 cal/g-mole-°K. The activation energy is thought to be associated with the polymer-chain-segment motion necessary for the passage of a diffusing molecule.

When the diffusion rate J achieves a steady value—that is, it no longer varies with time—a material balance requires that it also be independent of x, in which case Eq. 1 can be integrated to give

$$\int_0^L J\,dx = -\int_{C=C_1}^{C=C_2} D(dC) \tag{3}$$

$$J = -\frac{1}{L}\int_{C=C_1}^{C=C_2} D(dC) \tag{4}$$

Equation 4 indicates that the steady-state diffusion rate is always inversely proportional to the overall membrane thickness L for a given set of boundary conditions. To complete the integration the variation of D with C must be known. In some systems the diffusion coefficient does vary with concentration because the van der Waals forces are high between the diffusing molecules themselves and between the diffusing molecules and the polymer chains. An example is xylene liquid diffusing through polyethylene. In these cases substantial amounts of the diffusing molecules are dissolved in the polymer, thereby altering (loosening) the polymer structure, which continues to change as concentration increases. It is not surprising under these conditions that the

---

† The abbreviation "STP" refers to standard temperature and pressure for gas-volume measurement.

activation energy for diffusion, E_D, also changes (decreases) with increasing concentration of the diffusing molecule. The variation of D and E_D with C seems generally to follow the form

$$D = Ae^{aC} \tag{5}$$

$$E_D = B - bC \tag{6}$$

where A, B are constants for a given polymer and a, b are constants for a specific molelule diffusing through the polymer (4).

When there are relatively weak intermolecular force attractions between the diffusing molecules and the polymer chains, the solubility is low and the polymer structure is not significantly altered. Under these circumstances C can vary without affecting D, and Eq. 4 then becomes

$$J = -\frac{D}{L}(C_2 - C_1) \tag{7}$$

where C_2 and C_1 are respectively the concentrations of the diffusing molecules dissolved at the downstream and upstream polymer-membrane faces. An example of this type of system is oxygen diffusing through polyethylene. Generally such behavior is shown when a gas diffuses through a polymer. The gas–polymer systems also frequently exhibit a simple equilibrium relationship between gas pressure and gas concentration in the polymer. The expression has the form of the familiar Henry's law:

$$C = Sp \tag{8}$$

where p is gas pressure (in atmospheres) and S is the solubility constant for a given gas–polymer system. Various units are used for S, but typically it is expressed in cubic centimeters (STP) per atmosphere per cubic centimeter of polymer.

If for the steady-state gas-diffusion case the polymer-membrane faces are in equilibrium with the external gas pressure, Eq. 8 can be substituted into Eq. 7 and

$$J = -\frac{DS}{L}(p_2 - p_1) \tag{9}$$

Equation 9 expresses the diffusion rate in terms of the external-pressure driving force, a quantity that is easy to measure experimentally. A somewhat simpler form of Eq. 9 that is widely used is

$$J = -\frac{P}{L}(p_2 - p_1) \tag{10}$$

where the parameter P, called the gas permeability, is defined by Eq. 10 and is constant for a given gas–polymer system. Typical units for permeability are cc(STP)-cm/cm^2-sec-atm.

It is clear from Eqs. 9 and 10 that

$$P = DS \qquad (11)$$

and this relationship is frequently used in gas-diffusion studies.

The case often arises in polymer-membrane work of steady-state gas permeation through layers of different membranes. Such a system can be readily analyzed by using Eq. 10 for each membrane and recognizing that J is equal for each membrane. In addition, the overall pressure driving force is equal to the sum of the individual pressure driving forces across each membrane. To illustrate, for a three-layered membrane

$$J_1 = J_2 = J_3 \qquad (12)$$

$$\Delta p_0 = \Delta p_1 + \Delta p_2 + \Delta p_3 \qquad (13)$$

where Δp refers to the pressure driving force. Equations 10, 12, and 13 give

$$\frac{L_0}{P_0} = \frac{L_1}{P_1} + \frac{L_2}{P_2} + \frac{L_3}{P_3} \qquad (14)$$

where P_0 and L_0 are the overall permeability and overall thickness of the layered membrane. Equation 14 indicates that the membrane of greater thickness and lower permeability tends to control the overall diffusion rate.

Considering the various systems of diffusion through polymers, the case of gas diffusion just discussed is the more readily amenable to mathematical analysis. In contrast, liquid systems (the diffusing molecules come from a liquid phase) have relationships that are generally complicated by the variation of D with C. Such a variation, as already noted, is the result of relatively high van der Waals forces between the diffusing molecules themselves and between the diffusing molecules and the polymer chains. Generally speaking, and reverse osmosis is an exception, the overall diffusion driving force to use when liquid contacts both membrane faces would be the difference in liquid concentrations or fugacities. Also, it may be possible to relate the liquid concentrations to the concentrations of diffusing molecules at the membrane faces (C_1 and C_2 in the above equations) in order to calculate the diffusion coefficient. This approach can yield relatively simple relationships under certain conditions—for example, in the special case where liquid contacts the upstream face of the polymer membrane and gas or vapor on the downstream side. Here the concentration at the downstream membrane face can often be assumed to be negligible in comparison with the concentration at the upstream face, and Eq. 4 becomes

$$J = -\frac{1}{L}\int_{C_1}^{0} D \, dC = \frac{1}{L}\int_{0}^{C_1} D \, dC \qquad (15)$$

This allows an integral diffusion coefficient, $\bar{D}$, to be defined:

$$\bar{D} = \frac{1}{C_1} \int_0^{C_1} D \, dC \qquad (16)$$

Using Eq. 16 with Eq. 15 gives

$$J = \frac{1}{L}(\bar{D}C_1) \qquad (17)$$

Thus if C_1 can be measured or approximated from equilibrium sorption of the liquid in the polymer, an integral diffusion coefficient for concentrations up to C_1 can be calculated from experimentally determined diffusion rates J.

b. Fick's Second Law

When the diffusion rate of a component varies through a polymer with time, a mathematical analysis of the system can directly yield information on the diffusion coefficient. The approach is to make a material balance, using Fick's first law, around an element of differential length and unit cross-section:

$$-D\left(\frac{\partial c}{\partial x}\right) - \left[-D\left(\frac{\partial c}{\partial x}\right) + \frac{\partial}{\partial x}\left(-D\frac{\partial c}{\partial x}\right)dx \right] = \left(\frac{\partial c}{\partial t}\right)dx$$

$$\frac{\partial}{\partial x}\left(D\frac{\partial c}{\partial x}\right) = \frac{\partial c}{\partial t} \qquad (18)$$

where t is time. If D is constant,

$$D\frac{\partial^2 C}{\partial x^2} = \frac{\partial C}{\partial t} \qquad (19)$$

The result is shown for diffusion in one dimension, but generalization to diffusion in three dimensions can readily be made in the usual fashion.

Barrer (1) presents the solutions of Eq. 18 for a variety of geometries and boundary conditions. One case that is very useful in experimental studies is the unsteady-state diffusion of molecules through a polymer membrane with the following boundary conditions:

1. $C = C_0$ at $t = 0$ and $0 < x < L$
2. $C = C_2$ at $t \geq 0$ and $x = L$
3. $C = C_1$ at $t \geq 0$ and $x = 0$

where C_0 is the initial concentration of the diffusing species in the membrane and $x = 0$ and $x = L$ refer to the upstream and downstream faces of the membrane. If $C_2 = C_0$ or if C_2 and C_0 are negligible in comparison with C_1, the analysis indicates that

$$D = \frac{L^2}{6\theta} \tag{20}$$

Equation 20 is very useful in measuring the diffusion coefficient of gases in polymers. All that needs to be known is the membrane thickness L and the parameter θ, often called the time lag. An illustration of the time lag is shown in Fig. 3 for typical experimental gas-diffusion data taken at a constant pressure difference across the polymer membrane. At early times the diffusion rate varies with time because the concentration of diffusing species is building up

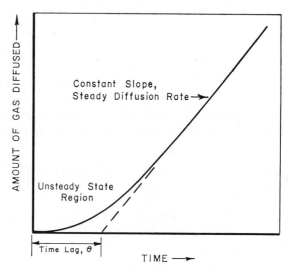

Fig. 3   Illustration of the time lag for gas diffusion through a polymer membrane.

in the membrane. Eventually the concentrations throughout the membrane reach constant values, and when this occurs, the diffusion rate is steady (Fig. 3). From the one set of data then the diffusion coefficient D can be calculated from the measured time lag θ and permeability P can be calculated from the slope of the curve at the steady state. With P and D known, the solubility constant S can be calculated from the relationship $P = DS$. Daynes (5) was the first to note the usefulness of this approach.

Another unsteady-state diffusion case that is very useful in experimental studies is the progressive weight gain or loss of a polymer film as molecules sorb or desorb to equilibrium. The boundary condition is that during the gas sorption or desorption the polymer-film surfaces are at constant gas concentration. The total equilibrium weight gain, for the gas-sorption case, can be used to calculate the gas solubility. In addition, the kinetics of the weight gain or loss can be analyzed so that the diffusion coefficient D can be calculated. Crank (6) shows the result in a convenient form:

$$D = \frac{0.04919}{t_{1/2}/L^2} \tag{21}$$

where D is the diffusion coefficient, assumed to be constant, and $t_{1/2}$ is the time required to complete one-half the total sorption or desorption.

2. GAS-PERMEABILITY MEASUREMENT

There are two general ways of measuring gas permeability. Both of them involve supporting a polymer film in a cell and sealing the film at its edges. A constant gas pressure is applied to one face of the film, and the other side of the film is maintained at a lower pressure. The amount of gas that diffuses through the polymer film is measured as a function of time. One way of determining the volume of gas diffused is to measure the volume change it causes at constant pressure. Another way is to measure the pressure change it causes in a constant-volume reservoir.

A schematic diagram of the variable-volume system is shown in Fig. 4.

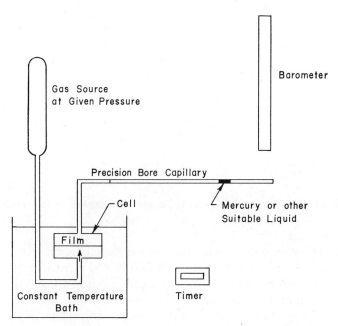

Fig. 4 Primary elements of variable-volume system for gas-permeability measurements.

The diffused gas moves mercury or another suitable liquid along a precision-bore capillary tube, and the position of the liquid indicates the volume of diffused gas at any time. One end of the capillary tube is open to the atmosphere, so that pressure is maintained constant. The local atmospheric pressure over the time of the volume measurements is determined with a barometer. For accurate diffused-gas measurements the system volume downstream of the membrane and not within the constant-temperature bath should be minimized in order to eliminate serious errors that might be caused by slight variations in room temperature. Measurements in the unsteady-state period for time-lag determination can be made if the system is properly designed. An example of such data is shown in Fig. 5.

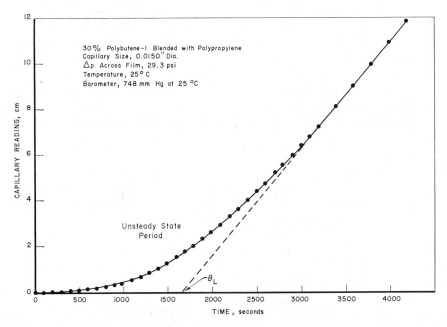

Fig. 5 Unsteady-state and steady-state O_2 diffusion as measured by the variable-volume method.

In the variable-pressure system the gas that has diffused through the polymer film collects in a constant-volume reservoir. The increase in gas pressure with time is noted by using a sensitive vacuum pressure gauge. The measurements are made at very low reservoir pressures, so that little error is introduced in the calculations by assuming that the pressure downstream of the polymer membrane is negligible. The details of this system have been described by Barrer and Skirrow (7).

3. MEASUREMENTS OF GAS-DIFFUSION COEFFICIENT AND GAS SOLUBILITY

The general approach in gas-solubility measurement is to determine the weight change in a polymer-film sample after it has achieved equilibrium with a gas at a known pressure. The rate at which it gains weight when sorbing to equilibrium or the weight-loss rate as gas is desorbed from it can be used to calculate the diffusion coefficient. This has been discussed in Section 1-b.

A versatile instrument that can be used to make the necessary measurements is the quartz-spring balance. The polymer-film specimen is suspended in a chamber (Fig. 6), which is initially evacuated, and then gas at a known pressure is rapidly admitted to it. The weight gain in the polymer film as the gas sorbs can be followed by the extension of the quartz spring. A movable telescope mounted on a vertical graduated scale is used to sight and measure the extension of the quartz spring. After equilibrium has been attained, and if desorption-rate measurements are desired, the chamber can be rapidly evacuated; the contraction of the quartz spring is then followed with time.

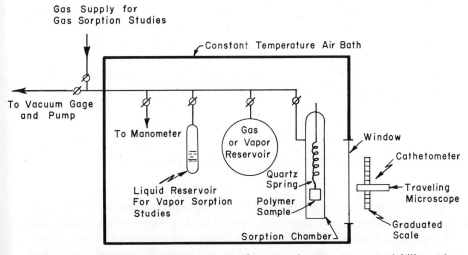

Fig. 6 Schematic of quartz-spring apparatus for measuring gas or vapor solubility and gas or vapor sorption kinetics in polymer films.

4. RELATION TO POLYMER MORPHOLOGY

Several investigators (8–10) have found the solubility constant for relatively-small-size diffusing molecules to be directly proportional to the amorphous content of the polymer:

$$S = \alpha S_a \tag{22}$$

where α is the volume fraction of amorphous polymer and S_a is the gas-solubility constant for a completely amorphous polymer. It follows then that sorption and diffusion occur almost entirely in the amorphous regions of the polymer. This is not surprising since chain mobility is greater in the amorphous regions than it is in the crystalline regions. It is also imagined that as the size of the penetrant molecule increases, less of the total amorphous region of the polymer may be accessible to it (11).

The diffusion coefficient D for a given molecule diffusing through a polymer is a complex function of such properties of the polymer as crosslinking, if any, chain stiffness, crystallinity level, and crystallite size and distribution. When the crystallites are smaller and more uniformly distributed in the polymer, a more tortuous amorphous channel is presented to the diffusing molecule as it moves. The greater path length so imposed acts to lower the diffusion coefficient. Also, the presence of crystallites restricts the motion of polymer-chain segments and thereby makes diffusion more difficult. These effects have been quantitatively characterized by a geometric impedance factor and a chain-immobilization factor (9, 10).

The diffusion coefficient is also affected by orientation or stretching of the polymer (12) and by the glassy or rubbery state of the polymer. In the glassy polymer isolated microvoids are thought to exist, but little effect has been found by them on the steady-state diffusion rate (13).

The myriad of factors in the case of partially crystalline polymers make correlation of diffusion data very difficult. It is only for the completely amorphous rubbery polymers that the diffusion theories we have discussed have had some success.

5. ESTIMATION OF GAS SOLUBILITY IN AMORPHOUS POLYMERS

A correlation has been developed for estimating the gas solubility S_a in completely amorphous nonpolar polymers (4, 10). The method is successful for gases that have low van der Waals force interactions with the polymer. The approach is the same as that used by Jolley and Hildebrand (14) in their correlation of gas solubility in organic liquids. In a crude sense an amorphous solid polymer is considered to be analogous to a low-molecular-weight organic liquid. The correlation is derived by equating the free energy of the gas in equilibrium with the polymer to the partial molar free energy of the gas dissolved in the polymer. The Flory–Huggins equation (15) is used to estimate the free energy of the dissolved gas. The result of the analysis is an equation of the form:

$$\ln S_a = G\frac{\varepsilon}{k} + H \qquad (23)$$

where S_a is the gas solubility at a given temperature in a completely amorphous polymer, G is a parameter that is temperature dependent, H is a constant for a given polymer, and k is the Boltzmann constant. The parameter ε is specific for a given gas and is from the Lennard–Jones potential, which is assumed to describe the van der Waals forces between the gas molecules.

Given the solubility correlation, thermodynamics defines the correlation for the heat of solution:

$$\Delta H_s = I\frac{\varepsilon}{k} + J \qquad (24)$$

where ΔH is the heat of gas dissolution in the amorphous polymer and I and J are constants for a given polymer.

6. PERMEABILITY, DIFFUSION, AND SOLUBILITY PROPERTIES OF POLYMERS

Table 1 lists the permeabilities, solubilities, diffusion coefficients, activation energies, and heats of solution for the diffusion of several gases through polyethylene and glassy polyethylene terephthalate. These polymers were chosen to illustrate polymer types and to give some idea of the magnitudes of the various parameters.

It is useful to note that the diffusion properties of polymers can vary over a broad range. This is often used to advantage in film-packaging applications; for example, in meat-wrap applications for supermarket display the polymer films chosen are those that will allow at least a certain minimum oxygen diffusion from air in order to maintain the meat at a salable red-pink color. On the other hand, polymer films used in cheese-wrap applications are relatively nonpermeable, so as to minimize flavor loss from the cheese. In other packaging applications the diffusion rates of carbon dioxide or water might be important. The selection of a polymer for the various film-packaging applications of course depends also on other factors. However, given certain diffusion requirements, a number of polymers could be selected and then the choices narrowed by considering other properties. A listing that is useful for polymer selection on a diffusion basis is shown in Table 2. Here the permeability P for the more frequently important diffusing gases—oxygen, carbon dioxide, water vapor, and nitrogen—is listed for a host of polymers. The permeability units in this table are different from those in Table 1 (this frustration is often encountered in permeability data). Conversion factors for the various units of permeability are listed in Table 3.

Table 1 Diffusion Parameters for Various Gases in Polyethylene and Glassy Polyethylene Terephthalate at 25°C[a]

Gas	Permeability, $P \times 10^7$ [cc(STP)-cm/cm²-sec-atm][b]		Diffusion Coefficient, $D \times 10^7$ (cm²/sec)		Activation Energy E_D (kcal/mole)		Solubility S_a, 100% Amorphous [cc(STP)/cc-atm]		Heat of Solution ΔH_s (kcal/mole)	
	PE[c]	PET[d]	PE	PET	PE	PET	PE	PET	PE	PET
He	0.38	0.12	68	20	5.9	4.8	0.012	0.008	2.5	0.22
N_2	0.074	0.0005	3.2	0.014	9.9	10.5	0.041	0.058	1.9	−4.4
A	0.21	—	3.6	—	10.1	—	0.103	0.082	0.8	—
O_2	0.22	0.003	4.6	0.037	9.6	11.0	0.077	0.100	0.6	−3.1
CH_4	0.22	0.0002	1.9	0.002	10.9	12.5	0.203	0.240	0.5	−5.3
CO_2	0.96	0.012	3.7	0.006	9.2	12.0	0.451	2.71	0.1	−7.5
C_2H_6	0.52	—	0.68	—	12.8	—	1.28	0.99	−1.5	—

[a] Data for polyethylene from Michaels and Bixler (10). Data for polyethylene terephthalate from charts and graphs presented by Michaels, Vieth, and Barrie (13, 16).

[b] See Table 3 for conversion factors to other permeability units.

[c] Low-density polyethylene, 58 vol % amorphous.

[d] Glassy polyethylene terephthalate, 58 vol % amorphous.

Table 2 Permeability of Polymers[a]

Polymer		Gas or Vapor Permeability[b] {[cc(STP)-mm/cm^2-sec-cm Hg] $\times 10^{10}$}[c]			
Chemical Type	Example	N$_2$	O$_2$	CO$_2$	H$_2$O
Cellulose acetate		1.6–5	4.0–7.8	24–180	15,000–106,000
Chlorosulfonated polyethylene	Hypalon	11.6	28	208	12,000
Epoxy	Epon-1001	—	0.49–16	0.86–14	—
Ethyl cellulose		84	265	410	14,000–130,000
Fluorinated ethylene–propylene copolymer	Teflon FEP	21.5	59	17	500
Natural rubber		84	230	1330	30,000
Phenol–formaldehyde	Bakelite	0.95	—	—	—
Polyamide	Nylon	0.1–0.2	0.38	1.6	700–17,000
Polybutadiene		64.5	191	1380	49,000
Poly(butadiene–acrylonitrile)	Hycar	2.4–25	9.6–82	75–636	10,000
Poly(butadiene–styrene)	Buna S	63.5	172	1240	24,000
Polycarbonate		3	20	85	7,000
Polychloroprene	Neoprene	11.8	40	250	18,000
Polychlorotrifluoroethylene	Kel-F	0.09–1.3	0.25–5.4	0.48–12.5	3–360
Polydimethylbutadiene	Methyl rubber	4.8	21	75	—

[a] Tabulation from Lebovits (17).
[b] Permeability measured at 20 to 30°C. Ranges shown represent values for different crystallinity levels, plasticizer contents, comonomer ratios, etc.
[c] See Table 3 for conversion factors to other permeability units.

Table 2—*Continued* [a]

Polymer		Gas or Vapor Permeability [b] $\{[cc(STP)\text{-}mm/cm^2\text{-}sec\text{-}cm\ Hg] \times 10^{10}\}$ [c]			
Chemical Type	Example	N_2	O_2	CO_2	H_2O
Polyethylene		3.3–20	11–59	43–280	120–2,100
Polyethylene terephthalate	Mylar	0.05	0.3	1.0	1,300–2,300
Polyformaldehyde (acetal)	Felrin	0.22	0.38	1.9	5,000–10,000
Poly(isobutylene–isoprene) (98 : 2)	Butyl rubber	3.2	13	52	400–2,000
Polypropylene		4.4	23	92	700
Polystyrene		3–80	15–250	75–370	10,000
Poly(styrene–acrylonitrile)		0.46	3.4	10.8	9,000
Poly(styrene–methacrylonitrile)		0.21	1.6	–	–
Polytetrafluoroethylene	Teflon	–	–	–	360
Polyurethane	Adiprene	4.9	15.2–48	140–400	3,500–125,000
Polyvinyl alcohol		–	–	–	29,000–140,000
Polyvinyl chloride		0.4–1.7	1.2–6	10.2–37	2,600–6,300
Polyvinyl fluoride	Tedlar	0.04	0.2	0.9	3,300
Polyvinylidene chloride	Saran	0.01	0.05	0.29	14–1,000
Poly(vinylidene fluoride–hexafluoropropylene)	Viton A	4.4	15	78	520
Rubber hydrochloride	Pliofilm	0.08–6.2	0.25–5.4	1.7–18.2	250–19,000
Silicone rubber		–	1,000–6,000	6,000–30,000	106,000

[a] Tabulation from Lebovits (17).

[b] Permeability measured at 20 to 30°C. Ranges shown represent values for different crystallinity levels, plasticizer contents, comonomer ratios, etc.

[c] See Table 3 for conversion factors to other permeability units.

Table 3   Conversion Factors for Permeability Data

To Obtain Values Expressed in	From Values Expressed in cc(STP)-mm/cm²-sec-cm Hg, Multiply by
cc(STP)-cm/cm²-sec-atm	7.6
cc(STP)-mil/m²-day-atm	2.58×10^{12}
cc(STP)-mil/100 in.²-day-atm	1.67×10^{11}
g H_2O-mil/100 in.²-day-mm Hg	1.76×10^{5}
ft³(STP)-mil/ft²-day-psi	5.75×10^{5}

It must be noted that care should be taken in considering permeability values from various sources. Sample preparation may be different, and the morphology may vary for the same polymer type. Differences in polymer crystallinity level and plasticizer content will drastically affect the permeability value.

7. PERMSELECTIVITY

The preferred permeation of one molecule through a polymer with respect to other diffusing molecules is termed membrane permselectivity. It is defined as the concentration ratio of the diffusing molecules leaving the membrane, compared with their concentration ratio prior to diffusion; for example, if components A, B, C have concentrations c_{A_1}, c_{B_1}, c_{C_1} prior to diffusion but concentrations c_{A_2}, c_{B_2}, c_{C_2} after diffusion through the polymer membrane, the permselectivity σ_{AB} is

$$\sigma_{AB} = \frac{c_{A_2}/c_{B_2}}{c_{A_1}/c_{B_1}} \qquad (25)$$

where σ_{AB} is the permselectivity of the polymer membrane to component A compared with that to component B. In the same fashion a quantity σ_{AC} and a quantity σ_{BC} could be defined. The permselectivity is similar to the familiar relative volatility used in separation by distillation.

At the steady state the ratio of components leaving the membrane is set by the individual diffusion rates, that is,

$$\frac{c_{A_2}}{c_{B_2}} = \frac{J_A}{J_B} \qquad (26)$$

The permselectivity can be specified further to include polymer-membrane properties in the case where gas mixtures are permeated. In this case the permselectivity can be expressed in pressure ratios:

$$\sigma_{AB} = \frac{p_{A_2}/p_{B_2}}{p_{A_1}/p_{B_1}} \tag{27}$$

and if the diffusion of the gases through the polymer membrane follows Eq. 9, then

$$\sigma_{AB} = \frac{D_A S_A}{D_B S_B} \left[\frac{1 - (p_{A_2}/p_{A_1})}{1 - (p_{B_2}/p_{B_1})} \right] \tag{28}$$

where the diffusion-coefficient and solubility values are for the particular pure component diffusing through the polymer membrane. It is assumed that the diffusion of one gas does not affect the diffusion of others. Equation 28 is further simplified if the downstream pressures are negligibly small; that is, p_{A_2}/p_{A_1} and p_{B_2}/p_{B_1} are near zero. Then

$$\sigma_{AB} = \frac{D_A S_A}{D_B S_B} \tag{29}$$

For permeation involving liquid mixtures there is no general way of relating the diffusion-coefficient and solubility values of the individual components to the permselectivity of the polymer membrane. This is especially so if the diffusing molecules interact strongly with the polymer chains and swell and alter the polymer structure.

Considering the applications of polymer-membrane separation, it obviously would be advantageous to have a high permselectivity and a high diffusion rate. If the permselectivity is very high, it may be possible to achieve the desired separation in one stage of diffusion; if the diffusion rate is high, a large polymer-membrane surface would not be required. However, it is generally found that polymer membranes with high permselectivity almost invariably have unsatisfactorily low diffusion rates, and vice versa. Attempts to solve the problem by trying to fabricate highly permselective polymer membranes that are very thin in gauge have met with little success. A highly permselective membrane would have to be about 0.5 micron in gauge for desirably high diffusion rate (J is inversely proportional to L) at a reasonable diffusion driving force. It is difficult to fabricate a polymer film that would be pin-hole free at gauges less than 10 microns. The challenge in membrane research is to devise stable polymer membranes capable of high diffusion rates and high permselectivity. Future commercial membrane-separation applications depend on the success of this research.

8. POLYMER MEMBRANES FOR HIGH SEPARATION PERFORMANCE

Studies are under way to form polymer membranes comprising a highly permselective, ultrathin (0.5 micron or less) layer supported on a less selective, much thicker (10 microns or greater), and highly permeable substrate. Such systems should be capable of high diffusion rates and high permselectivity, as can be illustrated by Eq. 14, derived in Section 1-a for gas diffusion through layered polymer membranes:

$$\frac{L_o}{P_o} = \frac{L_u}{P_u} + \frac{L_s}{P_s} \tag{30}$$

where the subscripts o, u, and s refer respectively to the overall membrane, the ultrathin barrier, and the permeable substrate. The highly selective barrier would be expected to have a very low permeability P_u, but since it has a very thin gauge L_u, Eq. 30 indicates that the overall permeability P_o can still be high. Also, the permselectivity for the composite membrane would be expected to be high, since this would be controlled by the ultrathin selective layer.

The most successful of these systems is the Loeb–Sourirajan membrane (18) for water desalination by reverse osmosis. The membrane can be produced in several ways. One way frequently used is to cast a solution of cellulose acetate in dimethyl formamide and acetone, evaporate the acetone for 1.5 min at room temperature, then immerse the film in ice water for 1 hour, and finally heat treat the film for 5 min in hot water, 65–85°C. The mechanism of

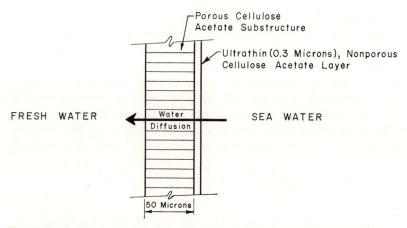

Fig. 7 Loeb–Sourirajan membrane for water desalination by reverse osmosis. In practice the membrane is placed on a rigid structure, for example, a porous metal plate, which provides support against the high applied pressures used.

the membrane formation is not fully understood but in effect the method produces a membrane, shown in Fig. 7, that on one side has an ultrathin (0.3 micron), dense, selective surface layer backed by a highly permeable porous substrate. The dense or "tight" surface layer faces the salt water in desalination by reverse osmosis. If the other side, the porous side, were to be in contact with the brine instead, the salt would enter the membrane and block the pores.

Other means of achieving a gradation in membrane properties have been tried. Rogers (19) has demonstrated that, by radiation initiation of a monomer that has been sorbed into a polymer film, a membrane can be made with a linear gradient in grafted polymer.

9. DIFFUSION CASCADE OF STAGES

When the desired purity cannot be achieved in a single membrane-separation stage, it is necessary to use a number of stages operating in series. As an example, consider the gaseous-diffusion cascade shown in Fig. 8. The principles discussed for this case are applicable to diffusion cascades in general. Feed enters the high-pressure side of a diffusion stage, and enrichment to the more permeable component takes place as some of the feed gas diffuses across the polymer membrane to the low-pressure side. The diffused gas is then compressed so as to enter as feed to another diffusion stage for further enrichment. The gas that did not diffuse at the first stage is also sent to another stage, but for further depletion or stripping of the more permeable component. There is recycling between stages in order to achieve efficient separation without material loss. A heat exchanger may be needed in the cascade to remove the heat of gas compression. Considering the material flows in Fig. 8, it becomes clear that the amounts of gas processed taper off as the top and bottom products of the cascade are approached.

A mathematical analysis of the cascade has been developed by Cohen (20), and a summary of the theory has been presented by Benedict and Pigford (21). The equations are derived on the basis that for maximum separation efficiency, streams if mixed, should have the same composition. The relationship for the number of stages required for a given binary separation is

$$n = 2 \frac{\ln[C_{AP}(1 - C_{AW})/(1 - C_{AP})C_{AW}]}{\ln \sigma_{AB}} \tag{31}$$

where n is the number of stages, C_{AP} and C_{AW} are the concentrations of the more permeable component in the top and bottom product streams, and σ_{AB} is the permselectivity, which is assumed to be constant throughout the

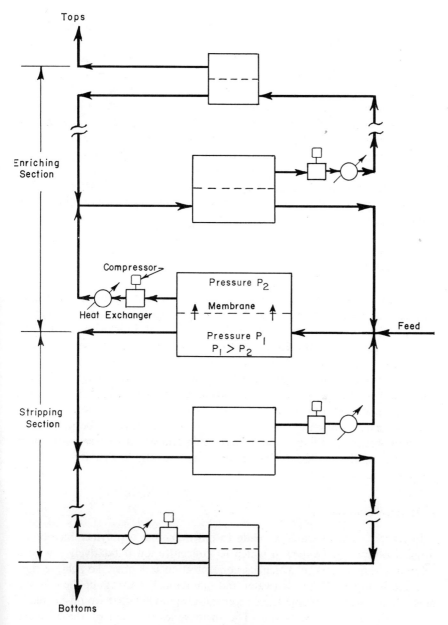

Fig. 8 Gas-diffusion cascade.

cascade of stages. As can be noted, the total number of stages required depends on the permselectivity and the desired separation. The total membrane area required increases with the number of stages and membrane thickness, but it decreases with increased membrane permeability and increased pressure differential across the membrane. The total power required for the separation increases with the number of stages and depends on the absolute pressure level as well as the applied pressure differential across each membrane.

The only commerical example of a diffusion cascade is the enrichment of natural uranium to a higher uranium-235 content. The process was developed during World War II and involves the flow of uranium hexafluoride vapor through totally microporous, rather than solid, membranes. The operation is run under vacuum to obtain such a flow condition in the micropores that separation can be achieved on the basis of molecular weight differences (molecular effusion).

10. WATER DESALINATION

In many areas of the world the supply of fresh water, about 0.05 wt % salt, cannot meet the demand. For coastal regions however, seawater (3.5% salt) is of course readily available, and inland, brackish water (0.2 to 0.8 wt % salt) is plentiful. In the past the desalination of both seawater and brackish water was carried out primarily by distillation—a costly process unless the installation is large, 10^8 gal./day, and unless heat energy is cheaply obtained from a neighboring nuclear power-generating plant (22).

Another way of purifying water, and one that has just recently been receiving a lot of attention, is reverse osmosis. It appears to be potentially competitive with distillation on a large scale and could be less costly than distillation at the modest scales required to meet the fresh-water needs of small, isolated communities.

a. Reverse Osmosis

In reverse osmosis water is made to diffuse across a polymer membrane from a solution (saltwater) in which its concentration is relatively low to a solution (fresh water) in which its concentration is higher. To achieve this, pressure is applied to the saltwater and maintained in excess of the osmotic pressure so as to overcome the concentration gradient that would normally drive fresh water into saltwater. The membranes used are of the cellulose acetate type and are highly permeable to water diffusion as compared with salt diffusion. A schematic diagram of the reverse-osmosis process is shown in Fig. 9 for the purification of seawater. In this case the osmotic pressure

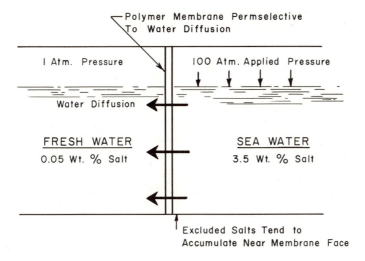

Fig. 9 Schematic of seawater desalination by reverse osmosis.

to be overcome is about 25 atm, and the applied pressure, in practice, is generously higher at 100 atm. This allows high water-diffusion rates to be attained. The salt removal is about 98.5%, and the water throughput is near 15 gal./ft²-day. Accumulation of rejected salt at the upstream membrane face can occur and can detrimentally cause decreased water-diffusion rate and increased salt-diffusion rate. This is because the local relatively high salt concentration at the membrane face in effect causes an increase in the osmotic pressure. Turbulent flow aids rapid dispersion of the rejected salt from the upstream membrane face.

The rate of water diffusion through the polymer membrane, J_w, and the salt-diffusion rate J_s can be expressed by equations of the form

$$J_w = \frac{K_w}{L}(\Delta p - \pi) \tag{32}$$

$$J_s = \frac{K_s}{L}\Delta C_s \tag{33}$$

where J_w is much greater than J_s and Δp is the applied pressure differential across the polymer membrane of thickness L. The parameter π is the osmotic pressure of the saltwater in reference to fresh water. The constants K_w and K_s include diffusion and solubility parameters specific for a given membrane, and ΔC_s is the difference in salt concentrations between saltwater and freshwater.

These expressions are readily derived when the more general form of Fick's first law is used with a chemical potential gradient replacing the concentration gradient. Clearly, from thermodynamics, the difference in water chemical potential across the membrane is proportional to $(\Delta p - \pi)$. In contrast, there is a negligible effect of applied pressure on the salt chemical potential. The ratio of salt concentration in seawater to salt concentration in fresh water is relatively high, and this largely determines the difference in salt chemical potential.

By far the best type of cellulose acetate membrane for water desalination is the one developed by Loeb and Sourirajan (18). As was discussed in Section 8, the membrane has a ultrathin (0.3 micron), nonporous, tightly structured layer backed by a porous, highly permeable substrate. The ultra-thin layer gives excellent salt rejection, and because it is so thin, water-diffusion rates are relatively high.

There are various designs for the reverse-osmosis cell, and this is an area of current engineering study. In one design the cellulose acetate membrane is supported by the surface of a porous tube and saltwater flows inside the tubular structure. Reverse-osmosis pilot units have been operated at fresh-water yields of 1000 and 10,000 gal./day, and units of up to 50,000-gal./day capacity are under construction (22). The main improvement now required in order that the reverse-osmosis process may achieve its full potential is increased membrane life at the 15-gal./ft²-day water-diffusion rate.

b. Electrodialysis

Another method of desalinating water is electrodialysis. In this process purification is achieved by driving salt ions from water by using an applied electric-potential difference. The salt ions are transported through membranes selective to them but not selective to water diffusion. This is in contrast to purification by reverse osmosis, which depends on water diffusion through a polymer membrane.

The membranes used in electrodialysis are ion-exchange membranes. The salt-ion transport through them is complex (6) and is the sum of Fickian diffusion plus ion motion due to the applied electric potential. The ion-exchange membranes can be tailored to favor cation or anion transport and in general are crosslinked polymers with ionizable groups spaced along the polymer chain.

If the ionizable group is a sulfonic acid, an anionic sulfonate group would be bound to the polymer chain with a relatively mobile hydrogen ion in the vicinity of each anion. Such a membrane would allow cation transport but would exclude anions. This is because a large repulsive potential (Donnan potential) builds up between entering anions and the fixed anions anchored to the polymer chains. Similarly, if the polymer contains fixed cationic groups along the chain, anions can be transported through the membrane but cations can not. In the electrodialysis cell, anionic and cationic exchange membranes are alternately stacked and salt water flows in the compartments between the membranes. When an electric potential difference is applied across such an arrangement of membranes, fresh water is produced from alternate compartments as is shown in Fig. 10.

The electrodialysis process is extensively used for the desalination of brackish water. Seawater desalination by this method would be too costly. There are 200 plants with a total fresh-water production of 3×10^6 gal/day (23). A plant at Buckeye, Arizona, has a capacity of 650,000 gal/day.

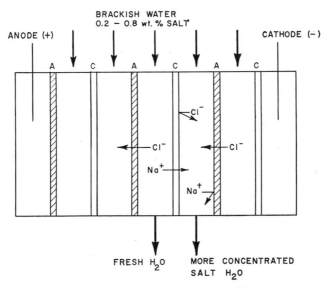

Fig. 10   Schematic of Water Desalination by Electrodialysis.

11. BIOLOGICAL APPLICATIONS

In the medical field diffusion through polymer membranes has been applied to develop the artificial kidney. Polymer films serve by selective diffusion to remove certain components from the blood when a damaged kidney can no longer do so. The process is often called hemodialysis. The components to be removed are urea, uric acid, water, creatine, phosphates, and excess chlorides (24), and patients with damaged kidneys must enter a hospital once or twice a week for treatment. The treatment involves circulating blood from the patient to a diffusion cell where the blood flows inside tubes of cellophane.

Urea, uric acid, etc., selectively diffuse across the cellophane tube wall into a solution dialyzate containing various salts and glucose.

The solution is circulated around the tubes to improve mass transfer. The concentration of diffused species can be maintained at desirably low levels py purging some of the solution while feeding in fresh solution. Heparin, a bolysaccharide that is an anticlotting agent, has to be added to the blood to prevent blood clotting on the membrane surfaces.

Biological applications based on silicone-rubber membranes are just now under study. The silicone membranes have a relatively high permeability to carbon dioxide and to oxygen and are permselective to carbon dioxide. One possible application is carbon dioxide removal from a space cabin during extended journeys into space. Another intriguing possibility was demonstrated by Robb when he showed that an animal enclosed in a very thin silicone-rubber membrane could live under water (25). Life was sustained by oxygen from the water diffusing through the thin silicon-rubber membrane. The necessary removal of carbon dioxide from the enclosure took place by diffusion through the silicone membrane to the water.

A medical application of silicone rubber membranes is the blood oxygenator. The silicone rubber is in the form of very small diameter tubes, about 15 mils, and blood flows through the tubes while oxygen is circulated outside the tubes. Oxygen diffuses into the blood and carbon dioxide diffuses from the blood.

12. COMMERCIAL SCALE HYDROGEN PURIFICATION BY DIFFUSION

Hydrogen can be purified by selective diffusion through a palladium membrane. Although not a polymer diffusion process, it is discussed here because it is the only known fully commercial example of separation by diffusion through solids (25).

The impure hydrogen containing methane and ethylene contacts a palladium membrane through which hydrogen selectively diffuses. The methane and ethylene do not diffuse significantly and are largely excluded from the membrane because of their relatively large molecular size in comparison with hydrogen. The net result is that hydrogen can be purified to 99.99% in a single stage of diffusion starting with as little as 25% hydrogen. The diffusion mechanism involves hydrogen dissociation into atoms on contacting the palladium and then diffusion of the hydrogen atoms through the palladium lattice. A pressure difference across the membrane is the driving force, and for high hydrogen-diffusion rates the applied pressure difference is about 400 to 500 psi and the temperature is 300 to 400°C. Eight plants are reported to be in operation and to produce 25×10^6 ft^3 of pure hydrogen per day (25).

13. ULTRAFILTRATION

The terms ultrafiltration and reverse osmosis have frequently been used synonymously in the past. However, more and more the term ultrafiltration is applied to separations where membranes of a relatively open, porous structure are used. In these separations, large molecular weight components can be filtered out at the membrane surface while low molecular weight components, for example, water, move by what is probably viscous flow through the membrane pores. There is much current activity to apply ultra-filtration to industrial processing. One application of potential importance is ultrafiltration of cheese whey to recover proteins. Whey is the watery by-product of cheese manufacture and its volume is staggering. There are some 20 billion lbs produced annually containing 150 million lbs of protein and 1 billion lbs of lactose (26). About half of the whey produced is discharged to the environment and this is a source of water pollution since bacterial growth is fostered by whey. The first step in whey purification is to remove the relatively high molecular weight proteins by pressuring the whey through cellulose acetate membranes fabricated to have a microporous structure throughout. Since there is a negligible increase in water molar concentration on removing the proteins, especially given their high molecular weight, there is little back diffusion of water to overcome.

REFERENCES

1. R. M. Barrer, *Diffusion in and through Solids*, Cambridge University Press, Cambridge, 1941.
2. R. M. Barrer, *J. Phys. Chem.* **61**, 178 (1957).
3. C. A. Kumins and T. K. Kwei, *Diffusion in Polymers*, Academic Press, New York, 1968.
4. A. S. Michaels and H. J. Bixler, in E. S. Perry, Ed.; *Progress in Separation and Purification*, Interscience, New York, 1968.
5. H. Daynes, *Proc. Roy. Soc. (London)* **97A**, 286 (1920).
6. J. Crank and G. S. Park, *Diffusion in Polymers*, Academic Press, New York, 1968.
7. R. M. Barrer and G. J. Skirrow, *J. Polymer Sci.* **3**, 549 (1948).

8. A. W. Myers, C. E. Rogers, V. Stannett, and M. Szwarc, *Mod. Plastics* **34**, 157 (1957).

9. A. S. Michaels and R. B. Parker, *J. Polymer Sci.* **41**, 53 (1959).

10. A. S. Michaels and H. J. Bixler, *J. Polymer Sci.* **50**, 393 (1961).

11. W. F. Wuerth, Sc.D. Thesis, Massachusetts Institute of Technology, 1967.

12. H. J. Bixler and A. S. Michaels, *Effects of Uniaxial Orientation on the Liquid Permeability and Permselectivities of Polyolefins*, 53rd National Meeting, American Institute of Chemical Engineers, Pittsburgh, 1964, Preprint 32d.

13. A. S. Michaels, W. R. Vieth, and J. A. Barrie, *J. Appl. Phys.* **34**, 13 (1963).

14. J. E. Jolly and J. H. Hildebrand, *J. Am. Chem. Soc.* **80**, 1050 (1958).

15. P. J. Flory, *Principles of Polymer Chemistry*, Cornell University Press, Ithaca, New York, 1953, Chapter XII.

16. A. S. Michaels, W. R. Vieth, and J. A. Barrie, *J. Appl. Phys.* **34**, 1 (1963).

17. A. Lebovits, *Mod. Plastics* **43**, No. 7, 139 (1966).

18. S. Loeb and S. Sourirajan, U. S. Patents 3,133,132 and 3,133,137.

19. S. Sternberg and C. E. Rogers, *J. Appl. Polymer Sci.* **12**, 1017 (1968).

20. K. Cohen, *The Theory of Isotope Separation*, National Nuclear Series, Div. III, Vol. 1B, McGraw-Hill, New York, 1951.

21. M. Benedict and T. H. Pigford, *Nuclear Chemical Engineering*, McGraw-Hill, New York, 1957.

22. A. Sharples, *Science Journal* August 1966.

23. H. Z. Friedlander and R. N. Rickles *Chem. Eng.* **73**, 153 (May 1966).

24. H. Z. Friedlander and R. N. Rickles, *Chem. Eng.* **73**, 111 (Feb. 1966).

25. K. Kammermeyer, in E. S. Perry, Ed., *Progress in Separation and Purification*, Interscience, New York, 1968.

26. R. L. Goldsmith, R. P. de Filippi, S. Hossain, and R. S. Timmins, "Industrial Ultrafiltration," presented at 160th Meeting of Amer. Chem. Soc., September 1970.

Science of Rubbers

13

A. N. Gent

One of the largest and most demanding uses of rubber is in the pneumatic tire, especially in the tread section. This material is required to be flexible, strong, and durable. It must resist wearing away, tearing, cutting, and mechanical fatigue. Furthermore, it must have a high frictional coefficient on a variety of surfaces. It must exhibit these properties over a wide range of temperature and resist deterioration by heat, light, atmospheric oxygen, ozone, water, etc., for long periods. To meet these severe demands materials, recipes, and designs have been developed over the years, largely by empirical means, with extraordinary success. Scientific understanding, on the other hand, has tended to lag somewhat behind these technical advances. [This is is not an unusual situation (1, 2).]

Many of the remarkable physical and chemical properties of rubbers have recently been analyzed in a qualitative, and sometimes in a semiquantitative, way in a number of pioneering investigations. Some of these results have been described in preceding chapters. Others, more directly related to current industrial practice, are reviewed here. We consider first the general manufacture of rubber products and then the stiffness, strength, and durability of model components under certain idealized conditions.

1. PROCESSING AND VULCANIZATION

A simplified, but otherwise representative, tire-tread formulation is given in Table 1. It illustrates the complexity of current recipes. Note, for example, that two, probably incompatible, elastomers are used. It would also be common practice to use two or even more types of "filler" and vulcanization

Table 1   A Typical Tire-Tread Recipe[a]

Component	Parts by Weight
Elastomers:	
1. A high-molecular-weight random copolymer of	
butadiene (75%) and styrene (25%)	75
2. Polybutadiene, predominantly *cis*-1,4	25
	100
Extenders, softeners, and processing aids:	
3. A viscous, highly aromatic petroleum-oil fraction, mixed	
with component 1 during manufacture	28
4. An oil-soluble high-molecular-weight sulfonic acid,	
dissolved in a paraffin oil (acid number ca. 10)	2
Fillers:	
5. A finely powdered carbon black (e.g., ISAF black)	70
Vulcanization agents:	
6. Sulfur	1.8
7. Stearic acid	2
8. Zinc oxide	4
9. A benzothiazole accelerator (e.g., *N*-oxydiethylene	
benzothiazole-2-sulfenamide)	1.5
Antioxidants and antiozonants:	
10. Phenyl-2-naphthylamine	0.5
11. A *p*-phenylenediamine protective agent (e.g.,	
N-(1,3-dimethylbutyl)-*N'*-phenyl-*p*-phenylenediamine)	2
12. A mixture of sparingly soluble microcrystalline waxes	1

[a] Similar to one given in the *Vanderbilt Rubber Handbook* (4). Satisfactory vulcanization of this material is effected by heating for about 30 min at 150°C.

"accelerator." The number of ingredients is thus in the range of 10 to 20. Many of them, moreover, are of indeterminate chemical structure and complex reactivity. Indeed it has been justly remarked (3) that "a more intractable system for study could hardly be imagined." Nevertheless some useful, if rather oversimplified, generalizations can be made regarding the functions of these diverse ingredients and the nature of the product.

a. Elastomers

Almost all polymeric materials can in principle be rubbery. The practical requirement of flexibility at normal temperatures excludes those whose glass-transition temperature T_g is higher than about $-20°C$ and those whose crystal-melting temperature T_m (in the crosslinked state, see below) is higher than about 0°C. It is also necessary to be able to link the polymer molecules together ("crosslinking," or "vulcanization") to form a loose three-dimensional network after the product has been shaped, so that the material becomes a coherent solid at this point. Thus chemical reactivity that allows the introduction of permanent chemical bonds (crosslinks) between the molecules is required.†

Economic considerations limit the range of practical rubbery polymers still further. The principal ones used at present are therefore relatively few (see Table 2).

The methods by which these polymers are prepared are dealt with in other chapters. We note here only that the molecular-weight distribution should be somewhat broad, because this gives a desirable non–Newtonian character to the flow properties; that is, the viscosity is then relatively high under low shear stresses compared with that obtaining under high ones (9). As a result the material will resist flow under gravity stresses but can still be shaped and molded in processing machinery at reasonable stress levels. Also the average molecular weight M should be as high as processing conditions allow in order to yield a strong product. In practice the viscosity, and hence processing difficulty, depends mainly on M, on the number n of physical entanglements per chain [$n = M/cM_e$, where c is the volume fraction of polymer in the mixture of polymer and oil and M_e is a characteristic molecular weight between entanglements for that particular polymeric structure (10)], and in a decreasing way on the temperature difference $T - T_g$ between the processing temperature and the glass-transition temperature of the mixture (10, 11). These are the three main factors that govern flow resistance and allow the effects of temperature, concentration, polymer type, and molecular weight to be treated in a semiquantitative way (10). Values of T_g and M_e, where known, are given in Table 2.

b. Extenders, Softeners, and Processing Aids

Quite large amounts (20 to 50 parts) of a compatible viscous oil can be incorporated into high-molecular-weight elastomers without substantial loss in the strength of the product. This surprising effect is explained later,

---

† Physical crosslinks can also be employed. For example, the end blocks of a three-block copolymer of polystyrene, polybutadiene, and polystyrene will associate to form glassy polystyrene domains linked by rubbery polybutadiene sequences. Such a material resembles vulcanized rubber, although the "crosslinks" are not chemical bonds (5, 6). Similarly, long crystallizing sequences in a block copolymer will associate to form microcrystallites, which act as crosslinks (7, 8). Materials like these are liquid at high temperatures and rubbery at lower ones, where the intermolecular associations form. They can thus be reversibly "crosslinked" on cooling and have been termed thermoplastic elastomers for this reason. Although such developments are likely to yield important rubbers in the future, the present systems are unsatisfactory for the most demanding applications because their physical crosslinks are not sufficiently stable.

Table 2   Some Common Elastomers

Elastomer[a]	T_g (°C)	$T_m{}^c$ (°C)	M_e ($\times 10^{-3}$)
cis-1,4-Polyisoprene (natural and synthetic rubber)	−70	≤15	14
cis-1,4-Polybutadiene and polybutadienes of mixed structure	−100 to −75	≤0	7
Random copolymer of butadiene and styrene (75 : 25) (SBR)	−60	≤−60	—
Random copolymer of butadiene and acrylonitrile (70 : 30) (NBR)	−50	≤−50	—
Random copolymer of ethylene, propylene, and an unsaturated (reactive) termonomer[b] (ca. 64 : 34 : 2) (EPT or EPDM)	∼ −40	—	5
Random copolymer of isobutylene and isoprene (98 : 2) (butyl rubber)	−70	≤−10	17
Polychloroprenes of mixed structure (Neoprene)	−45	≤30	—

[a] Of these only the first three are widely used in tires at present.
[b] Such as ethylidene norbornene.
[c] Crystallization is slow even well below T_m.

when the relationship between molecular strength and macroscopic strength is discussed. We point out here only that a high resistance to segmental motion is desirable—that is, the polymer–oil mixture should have a relatively high viscosity, quite apart from the mixing and storage advantages of viscous materials—and that a high molecular weight of the polymer also enhances strength. The present technical procedure for achieving these results is to polymerize to extremely high molecular weight in an emulsion system and then to coprecipitate the polymer with an emulsion of a relatively-high-molecular-weight oil whose "glass" temperature is close to that of the polymer.

Other oily materials are added as softeners and processing aids. They usually have a surface-active character and probably help to achieve a uniform dispersion of fillers and insoluble vulcanization agents. They may also take part directly in the complex vulcanization reaction in some cases.

c. Fillers (12)

Fillers are finely powdered solids mixed into the elastomer in large amounts (20 to 100 parts by weight) to stiffen it. When the particles are extremely fine, about 1000 Å in diameter or smaller, fillers also cause a striking increase in the strength of relatively weak, noncrystallizing elastomers (13, 14). This "reinforcing" ability seems to involve chemical atatchment of the polymer to the particle surface, at least to some degree, as it is affected by chemical treatment of the surface. By far the most common reinforcing filler is carbon black, consisting of roughly spherical carbon particles about 200 to 300 Å in diameter, prepared by incomplete combustion of hydrocarbon gases in a furnace. The surfaces have a complex chemical structure, with a variety of reactive organic groups.

The stiffening power of solid additives is readily accounted for in terms of the restraints they impose on the deformation of the elastic matrix bonded to them. It can be described in a quantitative way for large nonreinforcing particles (14). Two difficulties arise with reinforcing fillers, however. The particles associate together in long chains and have a disproportionately large stiffening action because of these "structures." As the interparticle associations are rather weak, the filled material is softened by an extension of only about 1 % or less; it then stiffens again slowly on standing.

A second stress-softening process is observed at large extensions, of the order of 100 % (Fig. 1). It is attributed to detachment of highly stressed rubber molecules from the carbon particles (10). Again, the original stiffness is recovered slowly on standing, and rapidly when the material is heated to about 100°C or higher. Both softening processes give rise to mechanical hysteresis, or energy absorption, as reflected in the very different stress-versus-strain relations for increasing and decreasing strain shown in Fig. 1.

The phenomenon of strength reinforcement by fillers is still one of the least understood aspects of rubber technology. However, a general treatment of strength outlined in Section 3 makes it clear that both mechanical hysteresis and stiffness are important factors in determining the macroscopic strength of solids, and both are greatly enhanced by reinforcing fillers. Moreover, the course of tearing is found to deviate from a linear path in filled rubbers, a feature that gives rise to enhanced tearing resistance. In fact the combined effects of hysteresis, stiffness, and tear deviation account for most, if not all, of the strength reinforcement contributed by fillers, so that future attention should be focused on these phenomena.

d. Vulcanization Agents

One way of linking polymer molecules together to form a network is by the combination of radical entities R· on different molecules to yield a covalent bond between them:

$$R\cdot + R\cdot \longrightarrow R{-}R$$

Polymer radicals are readily formed by high-energy irradiation or chemical

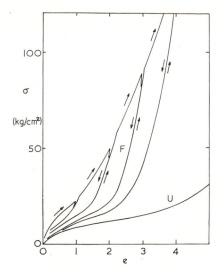

Fig. 1 Tensile stress–strain relations for typical filled (F) and unfilled (U) rubber vulcanizates.

reaction with free-radical sources, for example, di-*tert*-butyl peroxide and dicumyl peroxide (15). These materials decompose at high temperatures to give *tert*-butoxy or cumyloxy radicals, which abstract hydrogen atoms from the polymer molecules. The resulting polymer radicals can in principle undergo several possible reactions besides the coupling reaction considered above. Some of them are the following:

1. Intermolecular addition to a double bond, yielding a crosslink and an adjacent radical, which may then add to another molecule, and so on, giving rise to a train of crosslinks from each initial radical.
2. Cyclization.
3. Bimolecular disproportionation.
4. Combination with initiator radicals.
5. Reaction with oxygen to form polymer peroxy radicals. (These may in turn enter into combination, addition, disproportionation, or scission reactions.)
6. Decomposition, resulting in chain scission.

The fate of the polymer radicals depends on their chemical structure as well as their environment. For example, in polyisobutylene scission predominates, so that crosslinking by radical sources is not possible. In 1,4-polyisoprenes the most common tertiary allylic radical is resonance-stabilized to a high degree and appears to be capable only of coupling (at least when present at high concentration relative to dissolved oxygen), the other reactions being of negligible importance. As a result a quantitative relationship (ca. 1 : 1) is found to hold between the number of molecules of initiator and the number of crosslinks formed in this case (15). The most readily formed radical by hydrogen abstraction in 1,4-polybutadiene is a secondary allylic one, and it is not as strongly resonance-stabilized. It is correspondingly more reactive and appears to take part in the addition reaction extensively, so that many more crosslinks are obtained for a given amount of initiator (16).

The products obtained by free-radical crosslinking are relatively weak in comparison with those obtained by crosslinking with sulfur, and the process is commonly used only with saturated polymers (e.g., polyethylene and polydimethylsiloxane), for which sulfur vulcanization is not feasible. For most other materials sulfur is the normal crosslinking agent. We now turn to the mechanism of crosslinking in this case, as revealed by recent studies (17–20).

A zinc mercaptide coordination compound (**1**) is first formed from the accelerator and soluble zinc. The mercaptide, normally almost insoluble, is

1

solubilized by coordination with nitrogen bases. Free molecular sulfur then reacts (sparingly) with compound **1** to yield a zinc perthiomercaptide complex (**2**) of the general structural form

$$\overset{\text{ligand}}{\underset{\text{ligand}}{XS-S_x-\overset{\downarrow}{\underset{\uparrow}{Zn}}-S_x-SX}}$$

2

where X denotes the benzothiazole moiety, for example, and the value of x is controlled by equilibrium considerations, namely, the relative concentration of sulfur and soluble mercaptide. Compound **2** is believed to be the active sulfurating agent. It reacts with the rubber hydrocarbon RH to give the direct precursor to crosslinking (**3**), probably by allylic substitution.

$$R-S_x-SX$$

3

Crosslinking is finally achieved either by the coupling of two molecules of **3** or by reaction of **3** with another rubber molecule to yield polysulfidic cross-linked structures

$$R-S_x-R$$

and the zinc mercaptide complex again. Alternatively the polymer polysulfide **3** may form a complex with zinc mercaptide and regenerate compound **2**, a "quenching" reaction (20).

These reactions yield long polysulfidic crosslinks. However, two important "maturing" reactions go on during vulcanization as well:

1. The crosslinks shorten by expulsion of sulfur, reaching eventually the thermally stable monosulfidic form. This reaction is promoted by the zinc mercaptide complex **1** acting as a desulfurating agent and leads to more efficient use of sulfur.

2. The crosslinks undergo decomposition, and sulfur becomes combined in other ways, notably as cyclic monosulfides and disulfides. In addition, various main-chain modifications occur, including molecular scission. These reactions lead to less efficient use of sulfur, with as many as 20 sulfur atoms combined per crosslink.

Thus high concentrations of soluble mercaptide yield short crosslinks with few main-chain modifications or scissions (i.e., efficiently vulcanized rubber), whereas in the absence of vulcanization activators the reverse is true. Furthermore, the detailed structure of complex **1** and its sulfurating product **2** will play an important role in vulcanization because their reactivities will be affected by steric effects from substituents and complexing ligands, and in **2** by the electrophilicity of the penultimate sulfur atoms. The general principles of activated-sulfur vulcanization and delayed-action accelerators (using the quenching reaction already mentioned) are now fairly clear, however.

Polysulfides readily undergo exchange reactions (21). Similarly, polysulfidic crosslinks are labile at moderate temperatures and interchange, leading to rapid stress relaxation in a stretched vulcanizate containing them (21–23). This process has been suggested as the main reason why sulfur-crosslinked rubbers, particularly those with a high proportion of polysulfidic crosslinks, are stronger than rubbers crosslinked by radical sources and having carbon–carbon crosslinks. A rearrangement of polysulfide crosslinks is proposed to take place under stress, relieving the tension in disproportionately stressed network chains and so allowing the entire network to bear stress equitably (24). It has recently been pointed out that a similar rearrangement of crosslinks during vulcanization will minimize chain tensions set up in the undeformed state as a result of the continued crosslinking of an already partially crosslinked material (25). This annealing process will also lead to greater strength. A critical experimental test is needed to decide which of these two proposed strengthening mechanisms is more important.

e. Antioxidants and Antiozonants

Before dealing with the functions of antioxidants and antiozonants it is necessary to describe the general mechanisms by which vulcanized rubbers undergo chemical deterioration. The three principal causes of deterioration are heat, oxygen, and ozone. Most hydrocarbon polymers undergo purely thermal decomposition at about 200°C or higher; this places an upper limit on their long-term service temperature range, as such a process is not easily prevented. They also react with oxygen at all temperatures, with two deleterious consequences: further crosslinking, causing an increase in hardness and a decrease in extensibility; and molecular scission, causing a decrease in hardness and strength. These reactions are treated later. Finally, unsaturated hydrocarbon polymers react extremely rapidly with ozone, and this causes molecular scission.

i. Ozone Cracking

The scission is so rapid in comparison with the rate of diffusion of ozone into the interior that it is normally confined to a microscopically thin surface layer, the reaction products themselves acting as a protective shield against the further penetration of ozone. However, in stretched samples the severed molecules are pulled apart and the interior is left unprotected. As a result large cracks develop at points of stress concentration, where sufficient strain energy is available to form new surfaces.

The overall tensile stress σ_c necessary for a crack to grow thus depends on the stress-raising effect of small flaws present in the surface. It is given approximately by Griffith's fracture criterion (26):

$$\sigma_c = \left(\frac{2E\theta}{\pi c}\right)^{1/2} \tag{1}$$

where E is Young's modulus for the rubber, θ is the energy required for the formation of a unit area of new surface (about 25 ergs/cm^2 for a simple hydrocarbon liquid), and c is the effective depth of a surface flaw. For a typical die-cut or molded surface c is about 10^{-3} cm (13). Thus the critical stress σ_c is predicted to be about 0.6 kg/cm^2 for a soft rubber vulcanizate, with $E = 20$ kg/cm^2, corresponding to a critical tensile strain of about 3%. Experimentally observed values are in reasonably good agreement with these predictions (27).

As the stress level is raised above the critical value, numerous weaker stress raisers become effective, and more cracks form. The presence of a large number of fine, mutually interfering cracks is actually less detrimental than a few widely separated cracks that develop into deep cuts, so that the most harmful condition is just above the critical stress level.

When the critical energy condition is satisfied, the rate at which a crack grows depends on both the rate of molecular motion in the neighborhood of the crack tip and the rate of incidence of ozone at the crack tip. When either of these processes is sufficiently slow, it becomes rate-controlling. If both of them are taken into account in an approximate way (28), the overall rate R (in centimeters per second) of crack growth is given by

$$R^{-1} = 8 \times 10^{11} \phi^{-1} + 1.2 \times 10^3 c^{-1}$$

where ϕ (sec^{-1}) is the natural frequency of Brownian motion of molecular segments, discussed in the following section, and c (in milligrams per liter) is the concentration of ozone in the surrounding air. For a typical external atmosphere c is of the order of 10^{-4} mg/l. The second term is then dominant for values of ϕ greater than about 10^4 sec^{-1}, that is, at temperatures greater than about $T_g + 25°$C (Eq. 5).

ii. Antiozonants

Three classes of protective agents are employed. The first causes a reduced rate of crack growth, either by preferential reaction with ozone or by interference with the scission reaction. These materials are usually antioxidants as well; typical ones are the diaryl-*p*-phenylenediamines. The second class probably reacts with the products of ozonolysis to yield a coherent surface layer that requires a higher strain energy to crack. These materials may thus prevent cracking altogether if used in sufficient amount, whereas the others reduce the speed of cracking, but not its general character. Only certain dialkyl-*p*-phenylenediamines are known to act in the second manner, and the detailed mode of their action is obscure.†

The third class of antiozonants consists of sparingly soluble waxes that slowly migrate to the surface to form a protective wax film. If the film is broken by abrasion or changes in deformation, however, it becomes ineffective.

iii. Oxidation

The deleterious effects of oxidation—that is, scission and further cross-linking—are only minor consequences of oxidation; for example, many atoms of oxygen are usually incorporated for one scission. However, only minor

† It is important to note that if too little of this type of antizonant is added, the service stresses may lead to growth of a few deep cracks only, rather than the fine cracking that would occur in the absence of protection; the additive would then be harmful rather than beneficial.

extents of reaction are highly detrimental because of the high molecular weight per network chain; less than 1 % of combined oxygen renders a normal vulcanizate unserviceable. We are thus concerned with side reactions in the early stages of oxidation of a highly complex system. Nevertheless some progress has been made in establishing the principal reactions (29–33).

Although the precise initiating step is still not clear, once a free-radical species R· is formed from the polymer RH, oxidation proceeds by a series of successive radical reactions in which the main product is the unstable hydroperoxide RO_2H:

$$R\cdot + O_2 \longrightarrow RO_2\cdot \tag{I}$$

$$RO_2\cdot + RH \longrightarrow RO_2H + R\cdot \tag{II}$$

The second reaction is rate-controlling, being only about 10^{-6} times the speed of the first, so that the process is not much dependent on the concentration of oxygen unless this is quite small.

When the α-methylenic hydrogen atom is not easily abstracted by a peroxy radical (i.e., less labile), an alternative reaction that may occur is one in which the peroxy radical adds directly to a double bond in another molecule:

$$RO_2\cdot + RC{=}CR \longrightarrow RO_2RC{-}\overset{\bullet}{C}R \quad (\text{i.e., } R\cdot) \tag{III}$$

This reaction gives an increase in molecular weight by branching and crosslinking. It is clearly facilitated by accessible or reactive unsaturated groups—for example, the pendant $-CH{=}CH_2$ groups in polymers containing 1,2-butadiene repeat units, as in SBR. Such polymers harden as oxidation proceeds. In *cis*-1,4-polyisoprene, on the other hand, this third reaction is unimportant.

Decomposition of the unstable hydroperoxide yields more radicals and hence renders the oxidation process autocatalytic. Also, trace amounts of certain metal ions (cobalt, copper, manganese, and iron) probably catalyze the decomposition and thus promote oxidative deterioration.

The key reaction leading to scission in polyisoprene is believed to involve tertiary alkoxy radicals

$$-CH_2-\underset{\underset{\bullet}{\overset{|}{O}}}{\overset{\overset{CH_3}{|}}{C}}-CH{=}CH-$$

formed either by hydroperoxide decomposition or more probably by a bimolecular interaction of tertiary peroxy radicals (31, 32), an alternative to reaction II or III. These alkoxy radicals are then proposed to undergo disproportionation and scission, yielding a ketone and a primary alkyl radical. A somewhat different scission mechanism that has also been proposed involves a single tertiary alkoxy radical adjacent to cyclic peroxide structures (33). Thus the detailed reaction is still uncertain, although a tertiary alkoxy radical is generally accepted to be the precursor to scission.

Oxidation of polyolefins in the presence of sulfur compounds is still more complex. In addition to main-chain scission in sulfur vulcanizates, the crosslinks undergo oxidation as well. The mechanism of crosslink scission appears to involve first oxidation of a dialkenyl monosulfidic crosslink to the sulfoxide, which then decomposes with cleavage at the C—S bond. Eventually for every two monosulfidic crosslinks broken one new crosslink is formed, and various conjugated structures are introduced into the main chain (33).

iv. *Antioxidants*

Inhibitors can function in a variety of ways:

1. They can terminate the free-radical chain reactions I and II by providing labile hydrogen sources and yielding unreactive stable radicals.
2. They can react with organic hydroperoxides in various ways.
3. They can deactivate other oxidation catalysts.

Secondary aromatic amines and hindered phenols appear to act in the first way, probably reducing the peroxy radical and yielding an unreactive antioxidant radical. Metal dialkyldithiocarbamates are particularly effective because they also decompose hydroperoxides. Thiosulfinates and sulfoxides, notably di-*tert*-butyl sulfoxide, appear to act in the second way also. Some successful examples of sequestering agents for metal ions have been noted, notably dithiocarbamates, which render copper inactive.

2. DEFORMATION

a. Small Elastic Deformations

Under small deformations rubbers are linearly elastic solids, like metals. As they have a high modulus of bulk compression, about 2×10^4 kg/cm^2, compared with their shear modulus G, about 2 to 50 kg/cm^2, they can be regarded as relatively incompressible materials. Their elastic behavior under small strains can thus be described by a single elastic constant G. Poisson's ratio is effectively 0.5, and Young's modulus E is given by $3G$, to a good approximation.

A wide range of values of G can be obtained by varying the degree of crosslinking, oil dilution, and filler content. However, soft materials whose shear moduli are less than about 2 kg/cm^2 prove to be extremely weak and are seldom used. Also, particularly hard materials made by crosslinking to high degrees prove to be brittle and inextensible. In both cases this probably occurs because such materials are relatively good in an elastic sense, dissipating little strain energy internally. They are therefore particularly vulnerable to tearing, as discussed later. The practical range of shear modulus obtained by changes in degree of crosslinking and oil dilution is thus from about 2 to 10 kg/cm^2. Stiffening by fillers increases the upper limit to about 50 kg/cm^2, but fillers that have a particularly pronounced stiffening action also give rise to the stress-softening effects mentioned in Section 1-c, so that the modulus is a somewhat uncertain quantity in these cases.

It is customary to characterize the modulus, stiffness, or hardness of rubbers by measuring the elastic indentation of a rigid die, of prescribed size and shape, under specified loading conditions. Various nonlinear scales are employed to derive a value of hardness from such measurements (34). Corresponding values of shear modulus G for two common hardness scales are shown in Fig. 2.

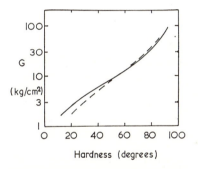

Fig. 2 Relations between shear modulus G and indentation hardness. Solid curve: Shore A scale; broken curve: international rubber hardness scale.

b. Large Deformations

Rubber products are normally subjected to fairly small deformations, rarely exceeding 25% in extension or compression, or 75% in simple shear. A good approximation for the corresponding stresses can therefore usually be obtained by conventional elastic analysis, assuming linear relationships, except for two striking instances when nonlinear large-deformation theory is mandatory. An example of the first is afforded by a large simple shear deformation. Inwardly directed *normal* stresses, proportional to the square of the amount of shear, are required to maintain the deformation in addition to the conventional shear stresses, which are directly proportional to the amount of shear (35). Such *second-order* stresses cannot of course be predicted by linear elasticity theory. Again, when the relations between load and deformation are so nonlinear that a maximum exists, the corresponding deformation state is mechanically unstable. This circumstance can develop at quite moderate strains. Examples are provided by the inflation of a thin-walled spherical balloon and a cylindrical tube, which become unstable at diametral expansions of about 30 and 60%, respectively (36, 37). At these points homogeneous deformation ceases and the balloon or tube undergoes a characteristic bimodal deformation process, with one region becoming highly strained with respect to the other. Indeed the highly strained portion may well be brought to the point of rupture, in the same way that a strut of relatively brittle material may break at the Eulerian buckling load. In these cases where failure occurs as a consequence of elastic instability the maximum load before instability can be regarded as the "strength" of the product, although it can obviously be calculated from purely elastic considerations. A characteristic feature of such processes will thus be a direct dependence of the maximum load on the stiffness of the rubber, rather than on its strength under a purely homogeneous deformation. One example is mentioned below, for rubbers under triaxial tension. It seems likely that there are other instabilities, as yet unrecognized, for materials with strongly nonlinear elastic behavior, like rubbers.

c. Extension or Compression of a Bonded Block

One particularly important deformation is now treated: the extension or compression of a thin rubber block, bonded on its major surfaces to rigid plates (Fig. 3). A general approximate treatment of such deformations has recently been reviewed (38). The deformation is assumed to take place in two parts: a purely homogeneous compression or extension of amount e, requiring a uniform compressive or tensile stress $\sigma_1 = Ee$, and a shear deformation restoring points in the planes of the bonded surfaces to their original positions in these planes. For a cylindrical block of radius a and thickness h the corresponding shear stress t acting at the bonded surfaces at a radial distance r from the cylinder axis is given by

$$t = \frac{Eer}{h} \tag{2}$$

The associated normal stress σ_2 is given by

$$\sigma_2 = Ee\,\frac{a^2}{h^2}\left(1 - \frac{r^2}{a^2}\right) \tag{3}$$

These stress distributions are shown schematically in Fig. 3. Although they must be incorrect right at the edges of the block, because the assumption of a simple shear deformation cannot be valid at these points, they should provide reasonably satisfactory approximations elsewhere.

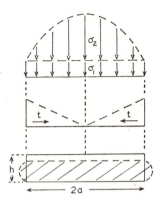

Fig. 3 Sketch of a bonded rubber block under a small compression. The distributions of normal stress σ and shear stress t acting at the bonded surfaces are represented by the upper two diagrams.

By integrating the sum of the normal stresses $\sigma_1 + \sigma_2$ over one bonded surface, the total compressive force F is obtained in the form

$$\frac{F}{\pi a^2 e} = E\left(1 + \frac{a^2}{2h^2}\right) \tag{4}$$

Clearly for thin blocks of large radius the effective value E' of Young's modulus (given by the right-hand side of Eq. 4) is much larger than the real value E, due to the restraints imposed by the bonded surfaces. Indeed when the ratio a/h exceeds a value of about 10, a significant contribution to the observed displacement comes from volume compression or dilation because the value of E' is now comparable to the modulus of bulk compression (39).

The stiffening effect of the restraints imposed by the bonded surfaces is shown in Fig. 4, where the ratio E'/E is plotted against the ratio a/h, with logarithmic scales being used for both axes. Similar stiffening effects would be expected for blocks restrained by friction, like the elements of a tire tread compressed against the road.

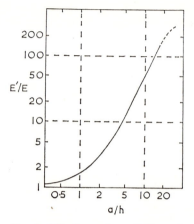

Fig. 4 Effective value of Young's modulus, E', for bonded blocks as a function of a/h, the ratio of radius to thickness.

d. Internal Rupture

In the center of a bonded block in tension a set of approximately equal triaxial tensions is set up, the magnitude of the stress in each direction being given by the stress σ_2 at $r = 0$ [i.e., $Ee(a^2/h^2)$, Eq. 3].

It has been shown from the theory of large elastic deformations that a small hole in the center of a block of rubber under outwardly directed tensions acting at infinity will undergo an unbounded expansion (a form of elastic instability, of the same form as that observed in the inflation of a balloon or tube) at a critical value of the tension, of about $5E/6$ (40). If a small cavity is present in the central region of a bonded block it will therefore develop into a large cavity at a critical tensile strain e_c of the block, sufficient to generate this critical level of triaxial tension in the center. The critical strain e_c is given approximately by

$$e_c = \frac{5h^2}{6a^2}$$

The corresponding critical tensile load is obtained by substituting this value of e in Eq. 4.

Internal cracks and voids are found to develop suddenly in bonded rubber blocks, at well-defined tensile loads that agree with those predicted by the treatment outlined above (40). In particular the loads are found to increase in proportion to Young's modulus E for rubbers of different hardness, in support of the proposed mechanism of fracture due to elastic instability. To avoid internal fractures of this kind it is necessary to limit the mean tensile stress applied to thin bonded blocks to less than about $E/3$. In compression, on the other hand, quite large mean stresses can be supported because the maximum tensile deformation is restricted to points right at the bonded edges (and even at these points the shear stress is only a fraction of the applied compressive stress, assuming the approximate treatment outlined above is applicable everywhere). A compressive stress limit can be obtained by assuming that the maximum shear stress should not exceed G; that is, that the maximum shear deformation should not exceed about 100%. This yields a value for e, the allowable overall compressive strain, of $h/3a$, corresponding to a mean compressive stress of the order of E for disks with values of the ratio a/h between about 3 and 10.

e. Effect of Rate of Deformation and Temperature

Rubbers are stiffer at higher rates of deformation or at lower temperatures, approaching the glass-transition temperature T_g, because their molecular segments move in a viscous environment, composed of other segments, and do so with greater difficulty under these conditions. The natural frequency of segmental motion ϕ for the migration of chain segments to new positions under Brownian motion is a strong function of temperature and follows a characteristic law, due to Williams, Landel, and Ferry (11):

$$\log_{10} \frac{\phi}{\phi_g} = \frac{17.4(T - T_g)}{52 + (T - T_g)} \tag{5}$$

where ϕ is the segmental jump frequency at temperature T and ϕ_g is that at a reference temperature T_g. A number of measurements indicate that ϕ_g is about 0.1 sec^{-1} (10). Indeed this feature reveals the nature of the glass transition; when segmental motion requires a time comparable to the experimental time scale, generally about 10 sec, the material fails to respond by segmental motion and becomes glasslike. When the experimental time scale is shorter (e.g., in impact tests), the effective value of T_g is correspondingly higher, by an amount that can be calculated from Eq. 5.

The Williams–Landel–Ferry relation reflects changes in free volume (i.e., the volume available for segmental motion) with temperature as a result of thermal expansion. As nearly all rubbery polymers have similar thermal expansions, their segmental frequencies ϕ are found to be in reasonable agreement with Eq. 5 for temperatures between about T_g and $T_g + 100°C$.

Many physical properties, like elastic modulus, are functions of the rate of deformation $\dot{e}$ as well as temperature. As the fundamental variable is obviously the ratio of the rate $\dot{e}$ to the rate of segmental motion ϕ, the effects of both rate of deformation and temperature can be described simultaneously by means of a reduced rate $\dot{e}(\phi_g/\phi)$, or $\dot{e}a_T$, where $a_T = \phi_g/\phi$ is obtained from Eq. 5.

The elastic modulus E for oscillatory loads or deformations has a complex character for viscoelastic materials like rubbers. A component $E'e$ of the stress is in phase with the deformation, and a component $E''e$ is 90° out of phase. For simple glass-forming polymers, both components vary with the angular frequency ω in terms of a reduced frequency ωa_T, when changes in temperature are taken into account. Similarly, time-dependent stresses and deformations depend on an appropriate reduced time t/a_T. Velocity-dependent properties (e.g., sliding friction) depend on a reduced velocity Va_T. Indeed many physical properties, including many aspects of strength as described in the following section, depend on the rate of deformation and temperature in this simple way. It follows that all these properties are governed by segmental motion (i.e., viscous factors) rather than thermodynamic equilibria.

Of course, the dependence on temperature is more complex for materials of a semicrystalline nature or with other temperature-dependent reinforcing mechanisms. However, the simple and characteristic dependence for glass-forming polymers serves as a useful diagnostic test when the mechanism of time dependence is obscure, and it becomes dominant at low temperatures in all cases.

At high temperatures and low rates of deformation the elastic modulus of pure vulcanized rubbers does not vary much with the rate of deformation; such materials are almost ideally elastic. The modulus increases with temperature, as predicted by the thermodynamic treatment of elastic networks (Chapter 9), roughly in proportion to the absolute temperature.

3. STRENGTH

a. Tensile Rupture

Figure 5 shows several relations for the tensile stress at break, σ_b, for an unfilled vulcanizate of SBR, as a function of the rate of elongation $\dot{e}$ at various temperatures (41). A small correction factor (T_s/T) has been applied to the measured values to allow for changes in the elastic modulus with temperature. The corrected values are denoted σ_b'.

Logarithmic scales are used for both axes in view of the wide ranges. It is seen that the experimental relations form parallel curves, superimposable by horizontal displacements. The strength at a given temperature is thus equal to that at another temperature, provided the rate is adjusted appropriately, by a factor depending on the temperature difference. (With logarithmic scales

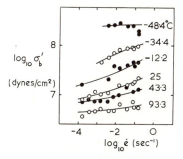

Fig. 5 Tensile strength of an SBR vulcanizate as a function of the rate of extension at various temperatures [After Smith (41)].

a constant multiplying factor is equivalent to a constant horizontal displacement.) The factor is found to be the ratio ϕ_{T_2}/ϕ_{T_1}, or a_{T_1}/a_{T_2}, where T_1 and T_2 are the temperatures concerned. This factor is readily calculated from the Williams–Landel–Ferry relation (Eq. 5).

A master curve can thus be constructed for a reference temperature T_s, often chosen for convenience as $T_g + 50°C$, by applying the appropriate shift factors to relations determined at other temperatures. The master curve for tensile strength, obtained from the relations shown in Fig. 5, is presented in Fig. 6. Good superposition is found to obtain.

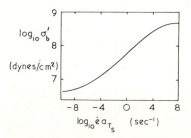

Fig. 6 Master curve for the tensile strength of an SBR vulcanizate as a function of the rate of extension, reduced to a reference temperature T_s of $-10°C$, that is, $T_g + 50°C$ [After Smith (41).]

The variation of tensile strength with temperature is thus primarily due to a change in segmental viscosity. Moreover, the master curve under isoviscous conditions has the form expected of a viscosity-controlled quantity: it rises sharply with increased rate of elongation to a maximum value at high rates when the segments do not move and the material breaks as a brittle glass (41, 42).

In fact a general parallel is observed between the strain energy dissipated internally when rubbers are deformed and the energy required to bring about rupture. This is demonstrated in a striking way by the observation of Grosch, Harwood, and Payne (43) that there is a direct relationship between the rupture energy W_b per unit volume and the energy W_d dissipated during stretching up to the breaking elongation. Their relation, largely independent of the mechanism of energy loss—that is, for reinforced or unreinforced, strain-crystallizing, or amorphous elastomers—is

$$W_b = 4.1 W_d^{2/3} \tag{6}$$

W_b and W_d being measured in ergs per cubic centimeter (Fig. 7). Materials that require the most energy to bring about rupture (i.e., the strongest elastomers) are precisely those in which the major part of the energy is dissipated before rupture.

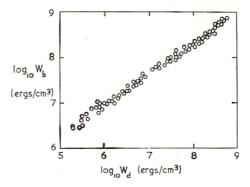

Fig. 7   Relation between work-to-break W_b and energy dissipated W_d for a wide variety of elastomer systems (43).

There are many reasons for attributing the rupture of rubber components in tension to the catastrophic growth of a small nick or flaw, generally from one edge, by tearing (13). The criterion for rupture consequently resembles Griffith's criterion for the rupture of brittle solids (Eq. 1), namely, that sufficient elastic energy be available to meet the energy requirements for tearing at the flaw. The generalized rupture criterion for highly deformable, imperfectly elastic materials becomes (13)

$$4cW \geq 2\theta \tag{7}$$

where c is the depth of the initial flaw (generally about 10^{-3} cm for die-cut surfaces), W is the stored energy per unit volume at break (given by $W_b - W_d$), and 2θ is the energy required to form two units of new surface, on growth of the flaw by unit area. Thus θ is related to the tearing strength of the material and can be determined independently by simple tearing experiments; the values obtained are discussed below. However, it is already clear from the tensile-strength results that θ depends strongly on the temperature and the rate of tear—that is, the rate at which material is deformed to rupture at the tear tip.

b. Tearing Strength

Several critical values of θ can be distinguished. The smallest possible one is given by the surface energy, about 25 ergs/cm^2 for nonpolar hydrocarbons. This value is indeed found to govern the growth of surface cracks due to chemical scission of the elastomer molecules (by atmospheric ozone), when the function of the applied forces is merely to separate molecules already broken (27). Another critical value of θ is that necessary to break all the molecules crossing a random plane. This has been estimated to be of the order of 10^4 ergs/cm^2 for typical hydrocarbon elastomers (44). Measurements of the minimum value of θ necessary to cause any cut growth by mechanical rupture are in reasonable agreement with this value, as described in the following section.

In simple tearing measurements the observed values of θ are considerably larger, ranging from 10^5 to 10^8 ergs/cm^2. The reason for the enhanced strength is made clear by considering the process by which the energy θ is dissipated at the tear tip. Thomas has shown that θ can be expressed in terms of the effective radius r of the tip of the tear and the "intrinsic" breaking energy $\overline{W}_b$ of the material by the approximate relation (13)

$$\theta = r\overline{W}_b \qquad (8)$$

The "intrinsic" breaking energy can be defined as the energy required to break unit volume of the material in the absence of a significant nick or flaw. It will be generally similar to, but larger than, the value of W_b determined for carefully prepared tensile test pieces in which chance edge flaws are minimized. Both $\overline{W}_b$ and r depend on the conditions of tear. However, for unfilled non-crystallizing elastomers r remains small (of the order of 0.01 cm) and relatively constant. In these cases θ is proportional to W_b (45) and changes in a parallel fashion with temperature and rate of tearing (rate of extension). Indeed θ is found to be the same for elastomers of widely different chemical composition under conditions of equal segmental mobility (46), as shown in Fig. 8. Thus

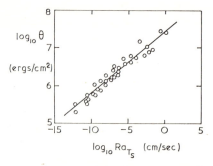

Fig. 8 Master relation for tear energy θ as a function of the rate of tearing R, reduced to $T_g + 20°C$, for vulcanizates of butadiene–styrene and butadiene–acrylonitrile copolymers (46, 47).

the internal viscosity determines the intrinsic breaking energy and the tearing resistance of viscoelastic materials.

In strain-crystallizing elastomers (e.g., natural rubber) the tearing resistance and tensile strength are greatly enhanced. Such materials also show enhanced energy dissipation because of their crystallization, and their strength has been accounted for in this way (48). Adding reinforcing particulate fillers to non-crystallizing elastomers brings about a similar strengthening. This effect is principally due to a pronounced change in the character of the tearing process, from relatively smooth tearing in unfilled materials to a discontinuous stick-slip process, in which the tear develops laterally or even circles around under increasing force until a new tear breaks ahead and the tear force drops abruptly. The process then repeats itself. This form of tearing has been termed "knotty" tearing (49).

Although its mechanism is still obscure, there are some indications that tear deviation is also associated with the viscoelastic response of the polymer. Pronounced knotty tearing is restricted to a limited range of tear rates and temperatures (47, 49), depending on the particular filler and elastomer combination employed.

c. Strength of Adhesive Bonds

The energy necessary to separate an elastomer from a different material to which it is adhering can be treated in a way analogous to the tear energy θ. Denoted by θ', it will have a much smaller lower limit because detachment can in principle take place without the breaking of primary chemical bonds. (Indeed θ' may be negative if the elastomer does not wet the substrate.) The normally expected value would be of the same order as typical surface energies, about 25 ergs/cm^2, instead of about 10^4 ergs/cm^2 for breaking network chains (44). In practice the adhesion energy, like the tear energy, is found to be much larger than this and to vary with the rate R of separation of two adhering materials and with temperature in accordance with the Williams–Landel–Ferry rate–temperature equivalence (Eq. 5) when a model viscoelastic adhesive is used (50). This is shown in Fig. 9. Thus adhesion also is greatly enhanced by energy dissipation within the adhesive, probably for the same reason as in tearing: the stresses set up at a crack tip or tip of a separation band are reduced by viscous or inelastic effects within the material. When the adhesive becomes highly inelastic in response, the adhesive strength becomes astonishingly high.

Two special features should be noted. When the adhesive is a ductile solid or fluid, yielding and drawing under stresses somewhat below those the bond will withstand (at the particular rate and temperature employed), the resulting large energy of deformation is reflected in a correspondingly large adhesion energy (Fig. 9, region A). When the material fails to draw out at a sufficiently low stress, this advantage is lost; for example, the full curve in Fig. 9 shows the adhesion of a crosslinked sample. Second, at high rates of deformation, when the adhesive becomes glasslike, the observed strength depends to a great degree on the mode of separation. Only when the adhesive is compelled to

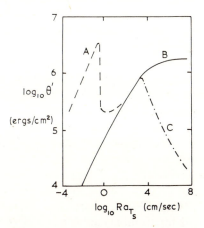

Fig. 9 Master relation for adhesion energy θ' as a function of the rate of peeling R, reduced to a reference temperature T_s of 23°C (i.e., $T_g + 63$°C), for a butadiene–styrene copolymer (60:40) in contact with a Mylar substrate (50).

deform by bending it away from the substrate in order to bring about separation is the high strength shown in Fig. 9, region *B* observed. When the substrate is bent away from the adhesive, the observed strength is much smaller (Fig. 9, region *C*).

4. DURABILITY

Deterioration due to heat, oxygen, and ozone has been discussed already in connection with protective agents. We turn now to two mechanical processes of deterioration: fatigue due to repeated deformation and abrasive wear. Both are strongly affected by chemical factors, of course, but a basic tearing process can be recognized in each case, as a result of recent work.

a. Mechanical Fatigue (51–53)

As described in the preceding section, when a test piece of an amorphous rubber like SBR is subjected to a constant stress, a crack or cut in one edge advances by tearing, at a rate dependent on the applied stress and temperature (Fig. 8). An appropriate measure of the tearing tendency of an applied tensile stress is given by $2cW$ (Eq. 7), denoted θ_t here to indicate that its magnitude may not be large enough to effect steady tearing at a significant rate. Indeed for strain-crystallizing rubbers like natural rubber the crack advances only when θ_t exceeds a critical value. For values less than this the tear does not grow continuously.

For both types of material some growth Δc of a crack is brought about by the actual process of imposing a stress. When a given stress or deformation is applied and removed repeatedly, these growth steps are cumulative. They are thus the main cause of failure when small loads or deformations are applied many times, and constitute mechanical fatigue.

The growth laws have been found experimentally to take the approximate form

$$\Delta c = B\theta_t^{\,n} \qquad (9)$$

for $\theta_t \geq \theta_0$, and $\Delta c = 0$ for $\theta_t < \theta_0$, where B is a cut-growth constant, n is a numerical index (1 to 2 for strain-crystallizing rubbers and 3 to 6 for amorphous rubbers) and θ_0 is a minimum value below which no growth is observed at all. Some experimentally determined growth steps Δc are shown in Fig. 10 as a function of θ_t; the cutoff point is evident. Moreover, the value of θ_0 determined in this way is in good agreement with that calculated theoretically as necessary to cause any molecular rupture (44). For a typical soft rubber vulcanizate, with a small cut about 10^{-3} cm deep in one edge, it corresponds

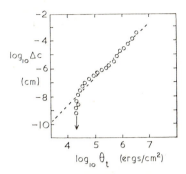

Fig. 10 Relation between cut growth Δc per stress application and the energy θ_t available for tearing, for a natural rubber vulcanizate (53).

to a simple extension of 50 to 75%. No mechanically induced tearing would therefore be expected for strains smaller than this. It is noteworthy that this value is of the same order as the deformation limits adopted empirically for rubber spring design.

On substituting for θ_t from Eq. 7 in Eq. 9 and integrating for the number N of stress applications necessary to cause the growth of a cut from an initial depth c_0 to an infinite depth, we obtain

$$\frac{1}{N} = (n-1)2^n BW^n c_0^{n-1} \tag{10}$$

This provides a quantitative prediction for the fatigue life N in terms of only two material constants: the cut-growth constants B and n, which can be determined experimentally. Measured fatigue lives for specimens with initial cuts of different depth and for imposed deformations of different magnitude have been found to be in good agreement with this relation. Moreover, when no deliberately introduced cuts are present, the observed fatigue lives still depend on the imposed deformation in satisfactory agreement with this relationship. They now correspond numerically to initial flaws equivalent to cuts about 10^{-3} cm deep, in agreement with other estimates of the severity of chance flaws (13).

For noncrystallizing rubbers the index n is generally larger than that for strain-crystallizing rubbers, and as a result their fatigue lives are much more strongly dependent on the size of accidental flaws and the magnitude of the deformation. Such elastomers are therefore generally longer lived at small deformations and with no accidental cuts but much shorter lived under more severe conditions.

Finally mention should be made of the remarkably beneficial effect on strain-crystallizing rubbers of not relaxing the stress to zero on each cycle but retaining a significant stress on the test piece at all times. Under these circumstances the highly stressed region at the tip of a crack remains crystalline throughout the stress cycle, and virtually no crack growth occurs due to mechanical tearing.

b. Friction and Wear (54)

Before treating the process of abrasive wear it is necessary to consider the origin of sliding friction, which is responsible for wear.

Friction is naturally associated with energy dissipation, and the principal mechanism of dissipation again turns out to be energy losses within the elastomer. The connection between the coefficient of friction μ and the energy losses in a simple deformation cycle is a complex one, however, two distinct modes of deformation having been distinguished (55).

On a lubricated rough track the value of μ increases with increasing sliding velocity and then passes through a maximum (Fig. 11). The relation closely resembles the dependence of the energy absorption per deformation cycle on the frequency of deformation. Indeed the speed of sliding at which μ has a maximum value, divided by the spacing between asperities, corresponds accurately to the frequency of deformation at which the energy dissipation

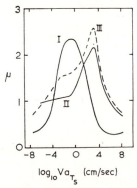

Fig. 11 Coefficient of sliding friction μ for a butadiene–acrylonitrile vulcanizate as a function of the speed of sliding V, reduced to 20°C (55). Curve I, dry smooth track; curve II, lubricated rough track; curve III, dry rough track.

is maximal at the same temperature. The dominant role of energy dissipation in lubricated sliding friction is thus established. For sliding over a clean, smooth surface the relation for μ is found to be similar, but displaced toward lower velocities. It corresponds, therefore, to "asperities" of much closer spacing than those in the rough surface. By comparing the velocity for maximum friction with the frequency for maximum energy absorption the spacings are calculated to be only about 60 Å. Thus friction between dry, smooth surfaces is associated with deformations on a molecular scale. It has therefore been attributed to transitory molecular adhesions between elastomer and track. The high frictional coefficient is, however, again due to dissipation of energy in the rubber as it undergoes local shearing deformations around the temporary bonds, rather than to the strength of the bonds themselves. This is shown by the characteristic dependence on speed and temperature. It is exactly analogous to the strength of adhesion and the tearing strength in these respects.

On dry, rough surfaces the effects of both surface asperities and molecular adhesions are evident in the master relation for μ as a function of the speed of sliding (Fig. 11, curve III). On lubricated, smooth surfaces both deformation processes are minimized and the coefficient of friction is correspondingly small.

As both the tearing resistance and the tearing (frictional) force depend on speed and temperature in accord with viscosity-controlled processes, it is not surprising that abrasive wear as a function of the speed of sliding and temperature should do so as well. A suitable measure of the rate of wear is provided by the ratio A/μ, where A is the volume abraded away per unit normal load and unit sliding distance and μ is the coefficient of friction. This ratio, termed the abradability, represents the abraded volume per unit energy dissipated in sliding.

It is found to decrease with increasing speed, pass through a minimum, and then rise again at high speeds as the material becomes glasslike in response (Fig. 12). This behavior resembles the variation of the reciprocal of the breaking energy W_b with rate of deformation (a reciprocal relationship because high abradability corresponds to low strength.) Indeed Grosch and Schallamach (56) found a general parallel between A/μ and $1/W_b$. Moreover, the coefficient of proportionality was found to be similar, about 10^{-3}, for all the unfilled elastomers examined. This coefficient represents the volume of rubber abraded away by unit energy applied frictionally to a material for which unit energy per unit volume is necessary to cause tensile rupture. It can be regarded as a measure of the inefficiency of rupture by tangential surface tractions; large volumes are deformed, but only small volumes are removed.

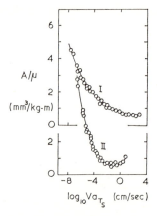

Fig. 12 Abradability A/μ versus the speed of sliding V, reduced to 20°C, for a butadiene–styrene vulcanizate (Curve I) and a butadiene–acrylonitrile vulcanizate (Curve II) (56).

5. CONCLUSIONS

This chapter has dealt with current rubber formulations and their behavior under simple, somewhat idealized, conditions in an attempt to explain the basic principles of rubber technology. The materials and processes used are so complex that only an outline of a proper analysis can be drawn at this stage. However, the fact that this is now possible in a general way for most of the important aspects of rubber usage is a notable landmark in the transition from an art to a science.

For brevity, and in view of the wide range of subjects touched on, reference has been made primarily to other more specialized review articles rather than to original sources. As an unfortunate consequence of this procedure, some major contributors to the subject have received little or no recognition. Also, none of the topics can be dealt with thoroughly in the space available. To counteract these shortcomings the reader is urged to consult the references cited in the text and a number of other recent monographs (57–66).

Acknowledgments

This review was prepared at Queen Mary College (University of London) while the author was a visiting professor in the department of materials under the Science Research Council senior research fellowship program. Thanks are due to Prof. E. H. Andrews for making facilities available there as well as to Dr. D. Barnard and Dr. M. Porter of the Natural Rubber Producers' Research Association for helpful comments.

REFERENCES

1. S. C. Gilfillan, *The Sociology of Invention*, Follet, Chicago, 1935.
2. J. Langrish, *Science Journal* **5A**, 81 (1969).
3. D. Craig, in E. M. Fettes, Ed., *Chemical Reactions of Polymers*, Interscience, New York, 1964, Chapter IX, Part C.
4. G. G. Winspear, Ed., *The Vanderbilt Rubber Handbook*, R. T. Vanderbilt Company, Inc., New York, 1968.
5. J. T. Bailey, E. T. Bishop, W. R. Hendricks, G. Holden, and H. R. Legge, *Rubber Age* **98**, 69 (1966).
6. R. Zelinski and C. W. Childers, *Rubber Chem. Technol.* **41**, 161 (1968).
7. C. S. Schollenberger, H. Scott, and G. R. Moore, *Rubber Chem. Technol.* **35**, 742 (1962).
8. S. L. Cooper and A. V. Tobolsky, *Textile Res. J.* **36**, 800 (1966).
9. G. Kraus and J. T. Gruver, *J. Appl. Polymer Sci.* **9**, 739 (1965).
10. F. Bueche, *Physical Properties of Polymers*, Interscience, New York, 1962.
11. J. D. Ferry, *Viscoelastic Properties of Polymers*, Wiley, New York, 1961.
12. G. Kraus, Ed., *Reinforcement of Elastomers*, Interscience, New York, 1965.
13. H. W. Greensmith, L. Mullins, and A. G. Thomas, in L. Bateman, Ed., *The Chemistry and Physics of Rubberlike Substances*, Maclaren, London, 1963, Chapter 10.
14. L. Mullins, in L. Bateman, Ed., *The Chemistry and Physics of Rubberlike Substances*, Maclaren, London, 1963, Chapter 11.
15. L. D. Loan, *Rubber Chem. Technol.* **40**, 149 (1967).
16. B. M. E. Van der Hoff, in H. Keskkula, Ed., *Polymer Modification of Rubbers and Plastics*, Applied Polymer Symposium No. 7, Interscience, New York, 1968, p. 21.
17. L. Bateman, C. G. Moore, M. Porter, and B. Saville, in L. Bateman, Ed., *The Chemistry and Physics of Rubberlike Substances*, Maclaren, London, 1963, Chapter 15.
18. C. G. Moore, in L. Mullins, Ed., *Proceedings of the Natural Rubber Producers' Research Association Jubilee Conference, Cambridge, 1964*, Maclaren, London, 1965, p. 188.
19. M. Porter, in A. V. Tobolsky, Ed., *The Chemistry of Sulfides*, Interscience, New York, 1968, p. 165.

20. A. Y. Coran, *Rubber Chem. Technol.* **37**, 679 (1964); *ibid.*, **38**, 1 (1965).

21. M. D. Stern and A. V. Tobolsky, *J. Chem. Phys.* **14**, 93 (1946).

22. B. A. Dogadkin and Z. N. Tarasova, *Kolloid Zh.* **15**, 347 (1953).

23. B. A. Dogadkin and Z. N. Tarasova, *Rubber Chem. Technol.* **27**, 883 (1954).

24. L. Bateman, J. I. Cunneen, C. G. Moore, L. Mullins, and A. G. Thomas, in L. Bateman, Ed., *The Chemistry and Physics of Rubberlike Substances*, Maclaren, London, 1963, Chapter 19.

25. A. V. Tobolsky and P. F. Lyons, *J. Polymer Sci.* **A2**, **6**, 1561 (1968).

26. A. A. Griffith, *Phil. Trans. Roy. Soc.* **A221**, 163 (1921).

27. M. Braden and A. N. Gent, *J. Appl. Polymer Sci.* **3**, 100 (1960).

28. A. N. Gent and J. E. McGrath, *J. Polymer Sci.* **A**, **3**, 1473 (1965).

29. A. Mercurio and A. V. Tobolsky, *J. Polymer Sci.* **36**, 467 (1959).

30. F. R. Mayo, *Ind. Eng. Chem.* **52**, 614 (1960).

31. D. Barnard, L. Bateman, J. I. Cunneen, and J. F. Smith, in L. Bateman, Ed., *The Chemistry and Physics of Rubberlike Substances*, Maclaren, London, 1963, Chapter 17.

32. P. M. Norlin, T. C. P. Lee, and A. V. Tobolsky, *Rubber Chem. Technol.* **38**, 1198 (1965).

33. J. I. Cunneen, *Rubber Chem. Technol.* **41**, 182 (1968).

34. A. L. Soden, *A Practical Manual of Rubber Hardness Testing*, Maclaren, London, 1952

35. R. S. Rivlin, in F. R. Eirich, Ed., *Rheology, Theory and Applications*, Vol. 1, Academic Press, New York, 1956, Chapter 10.

36. W. Johnson and P. D. Soden, *Intern. J. Mech. Sci.* **8**, 213 (1966).

37. J. M. Charrier, A. N. Gent, and E. A. Meinecke, unpublished work.

38. A. N. Gent and E. A. Meinecke, *Polymer Eng. Sci.* **10**, 48 (1970).

39. A. N. Gent and P. B. Lindley, *Proc. Inst. Mech. Engrs.* (*London*) **173**, 111 (1959).

40. A. N. Gent and P. B. Lindley, *Proc. Roy. Soc.* (*London*) **A249**, 195 (1958).

41. T. L. Smith, *J. Polymer Sci.* **32**, 99 (1958).

42. F. Bueche, *J. Appl. Phys.* **26**, 1133 (1955).

43. K. A. Grosch, J. A. C. Harwood, and A. R. Payne, *Nature* **212**, 497 (1966).

44. G. J. Lake and A. G. Thomas, *Proc. Roy. Soc.* (*London*) **A300**, 108 (1967).

45. H. W. Greensmith, *J. Appl. Polymer Sci.* **3**, 183 (1960).

46. L. Mullins, *Trans. Inst. Rubber Ind.* **35**, 213 (1959).

47. A. N. Gent and A. W. Henry, *Proc. Int. Rubber Conf., Brighton, 1967*, Maclaren, London, 1968, p. 193.

48. E. H. Andrews, *Rubber Chem. Technol.* **36**, 325 (1963).

49. H. W. Greensmith, *J. Polymer Sci.* **21**, 175 (1956).

50. A. N. Gent and R. P. Petrich, *Proc. Roy. Soc.* (*London*) **A310**, 433 (1969).

51. A. N. Gent, P. B. Lindley, and A. G. Thomas, *J. Appl. Polymer Sci.* **8**, 455 (1964).

52. G. J. Lake and P. B. Lindley, *Rubber J.*, **146**, (10 and 11), 24, 79; 30, 39 (1964).

53. G. J. Lake and A. G. Thomas, *Kautschuk und Gummi* **20**, 211 (1967).

54. A. Schallamach, *Rubber Chem. Technol.* **41**, 209 (1968).

55. K. A. Grosch, *Proc. Roy. Soc.* (*London*) A**274**, 21 (1963).

56. K. A. Grosch and A. Schallamach, *Trans. Inst. Rubber Ind.* **41**, 80 (1965).

57. P. J. Flory, *Principles of Polymer Chemistry*, Cornell University Press, Ithaca, N.Y., 1953.

58. G. S. Whitby, Ed., *Synthetic Rubber*, Wiley, New York, 1954.

59. L. R. G. Treloar, *The Physics of Rubber Elasticity*, 2nd ed., Oxford University Press, Oxford, 1958.

60. A. V. Tobolsky, *Properties and Structure of Polymers*, Wiley, New York, 1960.

61. W. J. S. Naunton, Ed., *The Applied Science of Rubber*, Arnold, London, 1961.

62. G. Alliger and I. J. Sjothun, Eds., *Vulcanization of Elastomers*, Reinhold, New York, 1964.

63. P. W. Allen, P. B. Lindley, and A. R. Payne, Eds., *The Use of Rubber in Engineering*, Maclaren, London, 1967.

64. W. Hofmann, *Vulcanization and Vulcanizing Agents*, Maclaren, London, 1967.

65. E. H. Andrews, *Fracture in Polymers*, American Elsevier, New York, 1968.

66. J. P. Kennedy and E. G. M. Toernqvist, Eds., *Polymer Chemistry of Synthetic Elastomers*, Part 1, Interscience, New York, 1968.

The Science of Plastics

14

R. D. Deanin

The largest commercial use of polymeric materials is in the manufacture of plastics. The name "plastics" comes from the fact that most of them are converted into useful end products by heating to make them soft and pliable, and then molding into the desired final form. The majority of plastics are stable high-molecular-weight polymers, either linear or branched, and retain their chemical identity throughout processing and use; these are called *thermoplastics*. The other large class, and actually the older commercially, are called *thermosets* and are usually polymerized into two stages: in the first stage they are converted to reactive low-molecular-weight viscous liquids or fusible solids; in the second stage, during molding, they are further converted into highly crosslinked three-dimensional structures that are then stable, insoluble, and infusible. Both thermoplastic and thermosetting polymers are also used in a variety of more complex forms, such as coatings, adhesives, foams, reinforced plastics, and high-temperature materials, which may broadly be described as *composites*. In end-use *applications* choice of the proper material is based first on design considerations, second on material properties, and third on process techniques for converting the material into the desired product. These aspects of plastics science are summarized in this chapter.

1. THERMOPLASTICS

Thermoplastics are stable high-molecular-weight polymers, either linear or branched. Their processability, mechanical, and thermal-mechanical properties depend on their molecular weight, molecular flexibility, crystallinity, and

307

polarity. Their stability, solubility, and permeability depend on their chemical composition and crystallinity. The interaction of these structural factors and properties can be seen in each of the major thermoplastics.

a. Materials

Polyethylene is the most fundamental plastic, scientifically because of its simple structure and commercially because of its low cost and useful balance of properties. Linear polyethylene (**1**) is a flexible regular molecule that crystallizes readily to rigid, strong products of good electrical and chemical resistance.

$$\sim CH_2-CH_2-CH_2-CH_2\sim$$
1

The older and more common branched polyethylene (**2**) contains an ethyl (*a*) or butyl (*b*) side group about every 50 carbon atoms along the chain; these side groups break up the regularity, reduce the crystallinity, and produce lower density, lower melting, more flexible products. These molecules also contain an occasional long branch (*c*), which somewhat lowers their viscosity (at constant molecular weight).

$$\sim CH_2CH_2(CH_2CH)CH_2CH_2(CH_2CH)CH_2CH_2(CH_2CH)CH_2CH_2\sim$$

	CH_2		CH_2		CH_2
(*a*)		(*b*)		(*c*)	{
	CH_3		CH_2		
			CH_2		
			CH_2		
			CH_3		

2

Polypropylene (**3**) has a regular succession of methyl side groups on alternate carbon atoms.

$$\sim CH_2-CH-CH_2-CH-CH_2-CH\sim$$
$$\qquad CH_3 \qquad\quad CH_3 \qquad\quad CH_3$$
3

The methyl groups stiffen the main chain and make it brittle at low temperatures. When polymerized into a regular isotactic configuration, the molecule is forced by these methyl groups into a preferred helical-coil conformation (Fig. 1), whose regularity produces high crystallinity, which in turn produces rigid strong products of higher melting point. As in polyethylene, the hydrocarbon structure produces good electrical and chemical resistance.

Polystyrene (**4**) has a succession of bulky phenyl groups on alternate carbon atoms.

~CH₂—CH—CH₂—CH—CH₂—CH~

4

The phenyl side groups stiffen the main chain severely, producing hard, rigid products; their normal random configuration prevents crystallization, producing a clear, amorphous, glassy solid that is brittle at room temperature and softens gradually at higher temperatures. The hydrocarbon composition produces good electrical properties, but the amorphous structure produces little resistance to organic solvents.

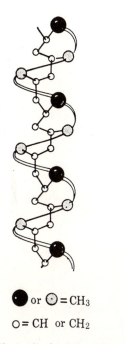

● or ◉ = CH₃

○ = CH or CH₂

Fig. 1 Helical coil of polypropylene (1).

Copolymerization with about 30% of acrylonitrile produces *SAN* copolymers (**5**); these have intermolecular polarity and hydrogen bonding between adjacent molecules, raising the softening temperature and chemical resistance.

$$\sim CH_2-CH-CH_2-CH-CH_2-CH\sim$$

5

Graft copolymerization of polystyrene onto a butadiene elastomer backbone disperses submicroscopic rubbery particles throughout the continuous polystyrene matrix phase, retaining most of polystyrene's rigidity and strength but providing resilient spots within this rigid matrix, so that the shock energy of sudden impact can be absorbed and dispersed harmlessly, thus preventing brittle failure. These materials (**6**), containing up to 20% rubber, are therefore called *impact styrenes*.

6

Finally, joint graft copolymerization of styrene and acrylonitrile onto 20 to 30% of a butadiene elastomer backbone combines both benefits, giving impact-resistance plastics, commonly called *ABS*, of good rigidity, strength, hot strength, and chemical resistance.

Polyvinyl chloride (**7**) has a succession of large negative chlorine atoms on alternate carbon atoms, which stiffen the main chain by both steric hindrance and electronic repulsion and also provide intermolecular attraction between adjacent chains. The configuration is also fairly regular, producing 5 to 10% crystallinity.

$$\sim CH_2-CH-CH_2-CH-CH_2-CH\sim$$
$$\quad\quad Cl \quad\quad\quad Cl \quad\quad\quad Cl$$

7

Altogether these forces produce fairly rigid, strong products that have good chemical resistance and soften gradually at higher temperatures. The high crystalline melting point, however, makes for difficult processability, and the aliphatic chlorine atoms permit dehydrochlorination, which produces poor thermal stability. Polyvinyl chloride is more important in plasticized form, which is discussed later in this chapter.

Polytetrafluoroethylene (**8**) is a completely perfluorinated carbon chain in which the negative repulsion of adjacent fluorine atoms stiffens the molecule into an almost rodlike structure of extremely high (unprocessable) melt viscosity.

$$
\begin{array}{ccccccc}
 & F & F & F & F & F & F \\
 & | & | & | & | & | & | \\
\sim C & - C & - C & - C & - C & - C & \sim \\
 & | & | & | & | & | & | \\
 & F & F & F & F & F & F
\end{array}
$$

8

Packing of these rodlike molecules produces high crystallinity, which in turn produces a high melting point and chemical resistance. The strength of the C−F bond produces high thermal and chemical stability, and the symmetrical nonpolar structure has high electrical and chemical resistance. On the other hand, low intermolecular attraction produces low rigidity and strength, and compact electronic structure produces low surface energy, surface tension, adhesion, and friction—and high lubricity.

Polymethyl methacrylate (**9**) is stiffened by the disubstitution of successive quaternary carbon atoms along the main chain, which yields hard, rigid products. Their normal random configuration prevents crystallization, producing amorphous glassy solids that have high clarity but soften gradually at higher temperatures and have poor resistance to organic solvents.

$$
\begin{array}{ccccc}
CH_3 & & CH_3 & & CH_3 \\
| & & | & & | \\
\sim CH_2 - C - CH_2 - C - CH_2 - C \sim \\
| & & | & & | \\
CO_2 & & CO_2 & & CO_2 \\
| & & | & & | \\
CH_3 & & CH_3 & & CH_3
\end{array}
$$

9

Polyoxymethylene (**10**) is a flexible linear molecule that crystallizes readily, producing rigid, strong, resilient products with good resistance to organic chemicals. The acetal structure is sensitive to thermal oxidation, and particularly to inorganic acids and bases, limiting use in these media.

$$\sim CH_2 - O - CH_2 - O - CH_2 - O \sim$$

10

Cellulose acetate (**11**) has ester groups in place of most of the hydroxyls in the cellulose molecule, preventing crystallization and minimizing intermolecular attraction.

$$
\begin{array}{c}
O \\
\parallel \\
CH_2OCCH_3 \\
| \\
HC - O \\
\diagup \qquad \diagdown \\
\sim HC \qquad\qquad CH - O \sim \\
\diagdown \qquad \diagup \\
HC - CH \\
| \quad | \\
OH\ OCCH_3 \\
\parallel \\
O
\end{array}
$$

11

With the addition of high-boiling-point esters (plasticizers), such as dibutyl phthalate, this yields stiffly flexible products of good processability, toughness, and clarity, but low hot strength and solvent resistance.

Poly-2,6-dimethylphenylene oxide (**12**) is an inflexible chain of bulky rings, which produce high melt viscosity and difficult processing but contribute good rigidity, strength, dimensional stability, and hot strength.

12

Commercially it is generally modified by the addition of impact styrene to improve processability, with some sacrifice in hot strength.

Polysulfone (**13**) is an inflexible chain of bulky rings, which produce high melt viscosity but good rigidity, strength, dimensional stability, and heat resistance. Here, however, processability appears to be adequate without any major additives.

13

Bisphenol polycarbonate (**14**) is an unusual molecule because it contains a beneficial balance of opposing forces.

14

On the one hand the bulky rings and sterically hindered quaternary carbon atom in the main chain produce molecular stiffness, giving rigid, strong products of high hot strength and melting point. On the other hand, the oxygen links in the main chain provide some molecular flexibility, producing resilient resistance to brittle failure. The stiffly kinked chain also minimizes crystallizability, giving an amorphous structure of high clarity. But low intermolecular attraction produces low resistance to organic solvents, and the ester groups are also sensitive to alkaline hydrolysis.

Nylons (e.g., nylon 6, **15**, and nylon 66, **16**) are composed of flexible hydrocarbon chains alternating with regularly repeating polar amide groups.

$$
\begin{array}{cc}
\text{O} & \text{H} \\
\| & | \\
\end{array}
$$
$$\sim CH_2CH_2CH_2CH_2CH_2C{-}N\sim$$
15

$$
\begin{array}{ccccc}
\text{H} & \text{O} & & \text{O} & \text{H} \\
| & \| & & \| & | \\
\end{array}
$$
$$\sim CH_2CH_2CH_2CH_2CH_2CH_2N{-}CCH_2CH_2CH_2CH_2C{-}N\sim$$
16

The regular structure and the intermolecular attraction due to polarity and hydrogen bonding combine to cause high crystallinity, which produces rigid, strong products of good hot strength and solvent resistance, whereas the flexible hydrocarbon segments produce resilience and processability. Moisture absorption and acid hydrolysis are the two major weaknesses.

b. Properties

i. Mechanical Properties

(*1*). *Stress and Strain.* When a force (stress σ) is applied to a mass of solid polymer, the entire mass tends to deform (strain ε) in the direction of that force, and the ratio between the force and the deformation (modulus $E = \sigma/\varepsilon$) is a measure of the rigidity of the material. If the rate of deformation is constant, the generalized relation between stress and strain is like that shown in Fig. 2a.

In actual materials different molecular structures respond in different ways, which are usually portions or modifications of this general stress-versus-strain curve (Fig. 2). For example, in very stiff molecular structures, such as polystyrene and polymethyl methacrylate, the segments of the polymer molecules cannot flex and uncoil to relieve the stress. They respond slightly by bending and stretching interatomic bonds, which takes considerable force to produce even slight deformation, giving them high modulus (rigidity) (Fig. 2c). If any greater force is applied, they cannot respond any further; the force concentrates at points of imperfection, such as microscopic nicks or strains or impurities, from which breakage propagates rapidly across the mass, producing ultimate failure.

On the other hand, in very flexible molecular structures, such as rubber and plasticized polyvinyl chloride, medium to large segments of the polymer molecules can rotate past each other easily and displace in the direction that will decrease the stress, giving low modulus (flexibility) (Fig. 2d). Beyond a certain extent of deformation (yield point), the normally coiled molecules have disentangled sufficiently to flow past each other more easily; but with continued flow they become so nearly parallel that they all resist the stress

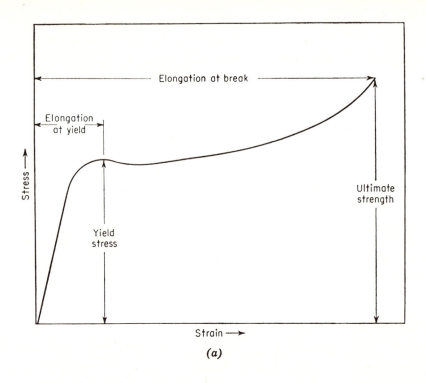

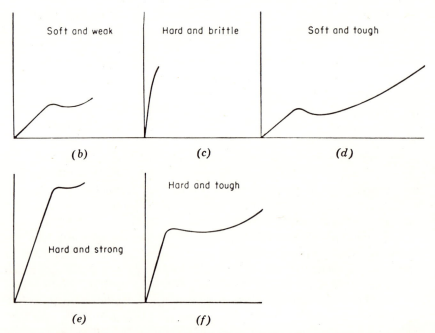

Fig. 2 Types of stress-versus-strain curves: (a) typical curve for thermoplastic materials; (b) soft and weak; (c) hard and brittle; (d) soft and tough; (e) hard and strong; (f) hard and tough (2).

simultaneously, and the force must increase again until it concentrates at imperfections and leads to breakage and ultimate failure.

A third type of stress-and-strain behavior is observed in flexible regular molecules, which crystallize during processing. Crystallinity immobilizes the molecular segments, resisting deformation and producing high modulus (rigidity) (Figs. 2e and f). Beyond the yield point, the crystallites realign or even melt temporarily, permitting easier deformation. At still higher strain molecules and crystallites align parallel to the direction of stress, immobilizing the structure again and producing higher modulus and strength.

(2). *Time Dependence.* When a rigid plastic is subjected to a moderate stress for a long time, larger molecular segments gradually disentangle, and eventually whole molecules begin to slip past each other, producing a gradual, continual, permanent increase in deformation, called *creep* strain or compliance (Fig. 3). Conversely, when the material is subjected to a fixed

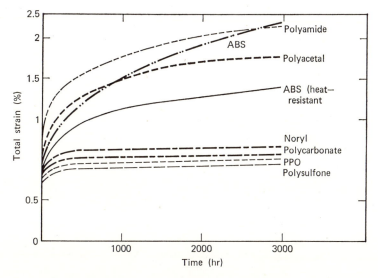

Fig. 3   Creep strain or compliance (from reference 3, p. 44).

deformation, the stress required to maintain this strain gradually decreases, producing a decrease in apparent (creep) modulus; this is called stress relaxation (Fig. 4). Creep rates increase with the initial stress or strain and can eventually lead to rupture and total failure (Fig. 5). These phenomena are of vital importance whenever plastic products are designed for use under continual load.

At the other extreme, when a rigid plastic is stressed very rapidly, even

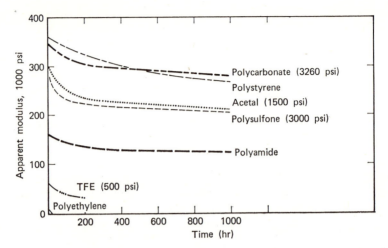

Fig. 4 Stress relaxation (from reference 3, p. 44). Measurements made at 73°F, except that for polysulfone, which was made at 210°F.

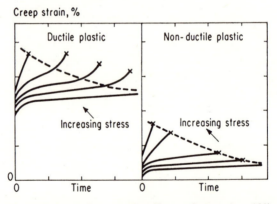

Fig. 5 Creep strain to rupture (from reference 4, p. 939).

short molecular segments do not have sufficient time to disentangle and move past each other to relieve the stress; the stress concentrates at any points of imperfection, propagates rapidly through the material, and produces brittle breakage. Resistance to such brittle failure is called *impact strength.* It is most commonly measured by making a pendulum break a notched bar and observing the transfer of energy required for this. Since some plastics are particularly brittle at a notch or sharp corner, the test is sometimes run on unnotched bars for comparison. In a drop-weight test a standard weight is dropped vertically onto a plastic sample, the height of drop being increased until breakage occurs; this requires more time and a larger number of samples but correlates better with actual use. For theoretical study tensile stress can be applied at increasing speeds, up to and beyond the normal speed of impact, and the energy absorbed in breakage can be measured from the area under the stress-versus-strain curve (Fig. 6).

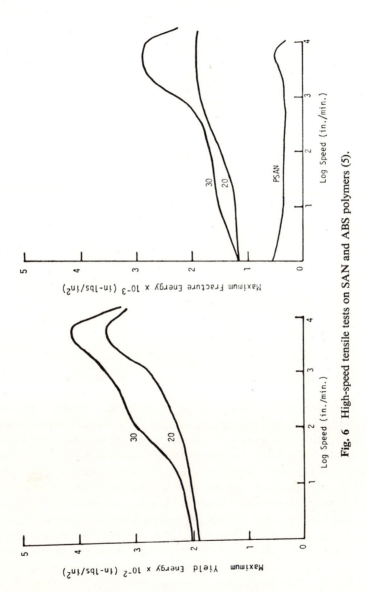

Fig. 6 High-speed tensile tests on SAN and ABS polymers (5).

(3). *Temperature Dependence.* Temperature is simply the mobile energy of atoms and molecules. With increasing temperature, the atoms in the polymer molecule vibrate more vigorously, increasing the empty space, or "free volume," between them, and larger molecular segments can move past each other and disentangle with greater frequency, responding to a stress more rapidly. Beyond about 2.5% free volume, their segmental mobility is sufficient to change this response from rigid to stiffly flexible or leathery, and this change is generally called the *glass transition* (Fig. 7); this is usually close to the *heat-deflection*, or maximum use, temperature of rigid amorphous thermoplastics. With further increase in temperature and segmental mobility the material actually becomes flexible or rubbery in nature; if the polymer molecules are very long and intertangled or joined by an occasional crosslink,

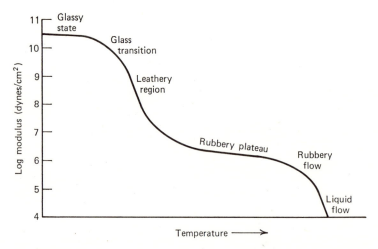

Fig. 7   Effect of temperature upon the modulus of a linear amorphous polymer.

as in vulcanized rubber, this *rubbery plateau* can extend over a considerable temperature range. In linear polymers the mobility of large segments eventually increases until whole molecules can move past each other under stress, first fairly readily in *rubbery flow* and eventually quite easily in *liquid flow*. This final region is most commonly used in plastics processing.

The above description applies primarily to linear amorphous polymers. When crystallization occurs, the crystalline regions are much more compact and immobile, up to the melting temperature, which is frequently 1.5 to 2.0 times the glass-transition temperature in degrees Kelvin. In this region between the glass transition of the amorphous phase and the melting of the crystalline phase the degree of crystallinity determines the degree of immobilization and stiffening of the otherwise rubbery plateau (Fig. 8). At the

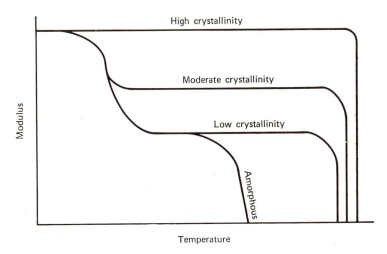

Fig. 8   Effect of crystallinity on the modulus-versus-temperature curve.

melting point, providing that the molecules are inherently flexible and not excessively long, there is a sharp transition from a rigid plastic to a fluid liquid, which alters processing technology according.

ii. *Other Thermal Properties*

(*1*). *Thermal Stability*. As organic materials, plastics are often unstable at higher temperatures, particularly in the presence of oxygen. For practical use they are conventionally stabilized by small amounts of additives. Halogen-containing polymers like polyvinyl chloride tend to lose HCl autocatalytically,

$$\sim CH_2-CH-CH_2-CH-CH_2-CH-CH_2-CH\sim$$
$$\quad\ \ \ \ |\qquad\quad\ |\qquad\quad\ |\qquad\quad\ |$$
$$\quad\ \ \ Cl\qquad\ \ Cl\qquad\ \ Cl\qquad\ \ Cl$$

$$\downarrow$$

$$\sim CH_2-CH-CH_2-CH-CH_2-CH-CH=CH\sim\ +HCl\uparrow$$
$$\quad\ \ \ \ |\qquad\quad\ |\qquad\quad\ |$$
$$\quad\ \ \ Cl\qquad\ \ Cl\qquad\ \ .Cl$$

$$\downarrow$$

$$\sim CH=CH-CH=CH-CH=CH-CH=CH\sim\ +\,n HCl\uparrow$$

and this is generally retarded by additives that react with HCl and/or interrupt the reaction at one of its steps (Table 1). Polymers containing tertiary or allylic hydrogen, such as polyolefins and impact grafts on rubber, tend to peroxidize and then crosslink (reaction V) or depolymerize (reaction VI),

Table 1   Stabilizers for Polyvinyl Chloride

Type	Compound
Lead salts	Lead carbonate
	Lead maleate
	Lead pthalate
	Lead stearate
	Lead sulfate
Barium–cadmium synergistic systems	Barium Cadmium Zinc $\left\{\begin{array}{l}\text{laurate}\\\text{naphthenate}\\\text{octoate}\\\text{phenate}\\\text{stearate}\end{array}\right.$
	+
	Epoxidized soybean oil
	+
	Organic phosphite esters
Organotins	Dibutyl tin dilaurate
	Dibutyl tin dimaleate
	Dibutyl tin mercaptoates
	Dioctyl tin analogs

$$R{:}H \longrightarrow R{\cdot} + {\cdot}H \tag{I}$$

$$R{\cdot} + O_2 \longrightarrow RO_2{\cdot} \tag{II}$$

$$RO_2{\cdot} + R'H \longrightarrow ROOH + R'{\cdot} \tag{III}$$

$$ROOH \longrightarrow RO{\cdot} + {\cdot}OH \tag{IV}$$

$$R{\cdot} + {\cdot}R' \longrightarrow R{:}R' \tag{V}$$

$$RCH_2CHX{\cdot} \longrightarrow R{\cdot} + CH_2{=}CHX \tag{VI}$$

and this is generally delayed by additives that inhibit free-radical oxidation or help to destroy peroxides after they are formed (Table 2).

(2). *Flammability.* At still higher temperatures most polymers are flammable, causing fire hazards, particularly in such applications as building and transportation. Polymers containing aromatic rings in the main chain are generally fairly flame retardant, presumably because they cannot liberate volatile combustible monomers by an unzipping depolymerization chain reaction. Halogenated polymers like polyvinyl chloride are inherently flame retardant, presumably because the volatilized halogen gases interrupt the free-radical oxidation chain reaction, and this action is greatly synergized by

Table 2   Antioxidants

OH
|
(CH₃)₃C—⟨ring⟩—C(CH₃)₃
|
CH₃

$(CH_3)_3C$—⟨ring-OH⟩—CH_2—⟨ring-OH⟩—$C(CH_3)$
with CH₃ substituents

$(CH_3)_3C$—⟨ring-OH⟩—S—⟨ring-OH⟩—$C(CH_3)_3$
with CH₃ substituents

$$CH_3(CH_2)_{11}OCCH_2CH_2SCH_2CH_2CO(CH_2)_{11}CH_3$$
(with O double-bonded to each carbonyl carbon)

the addition of antimony oxide. Other polymers can be made flame retardant by the addition of highly halogenated liquids or waxes, particularly with conjoint use of antimony oxide again. Organophosphorus compounds are even more effective, presumably through the formation of a phosphorus oxide glass on the surface of the burning plastic, which keeps oxygen from reaching the reaction. Their use, however, has been limited mainly to liquid additives, such as tricresyl phosphate.

iii. Other Properties

(*1*). *Electrical.* Being organic materials with low electronic mobility and little or no ionic mobility, plastics are generally good electrical insulators. Nonpolar hydrocarbons and electronically balanced structures, such as polytetrafluoroethylene, are least likely to be influenced by electrical fields and are most desirable for insulation applications, particularly at high frequencies. On the other hand, flexible bonds with unbalanced polarity, as in polyvinyl chloride, try to orient in an alternating electric field, converting electrical energy into thermal energy. Compounds that contain them are therefore particularly adapted for processing by dielectric heating methods.

(*2*). *Optical.* Rigid amorphous polymers are isotropic to visible light and have high clarity, approaching that of glass, particularly when they are cast with care to avoid strain orientation and flow marks; polymethyl metha-crylate is particularly noteworthy. Grafted and blended impact plastics and crystalline moldings, on the other hand, contain dispersed areas of different refractive indices, making them translucent to opaque.

Longer infrared wavelengths are absorbed selectively by the functional groups in the polymer molecule, providing useful means for the analysis of structure (Fig. 9). At the other extreme, short ultraviolet wavelengths from the sun are apt to be absorbed and activate unstable bonds in the polymer molecule, causing poor resistance to weathering.

(3). Chemical. Organic polymers are generally resistant to water and inorganic reagents, although nylons do swell in moist atmospheres and polyesters and polyamides may hydrolyze in acid or alkali. Resistance to organic solvents is more problematic and depends largely on differences in polarity and hydrogen bonding.

Adhesive bonding between a polymer and another surface may depend on mechanical factors, surface energy, similarity in polarity, or actual interfacial reaction. Conversely, the lubricity required in self-lubricating plastic gears, cams, and bearings is generally obtained through their low surface energy. This surface lubricity is often involved in resistance to abrasion as well.

c. Processing

Stable high-molecular-weight linear thermoplastic polymers are converted into finished end products primarily by heating them to the liquid state and applying pressure to make them flow into a mold or through a die to produce the desired shape and then cooling below the melting or the glass-transition temperature to make them retain that shape. The small molecules in low-molecular-weight liquids flow easily past each other, and the rate of flow is proportional to the applied force, according to conventional Newtonian principles. In molten polymers, however, the long, intertwined molecules disentangle gradually with increasing pressure (shear stress τ) and rate of flow (shear rate $d\gamma/dt$). Hence the apparent viscosity μ_a decreases and is called *pseudoplastic* (Fig. 10). When the temperature is too low or the rate of flow through a narrow orifice is too high, processability retrogresses from the liquid into the rubbery flow region, producing ripples or discontinuities, known as melt fracture (Fig. 11).

Rapid, economical processing requires more than the easy flow of the molten polymer: it is also important to heat the polymer rapidly before flow and to cool it rapidly after flow. Heating is most often accomplished by forcing the granular or powdered solid through a heated tube, generally by using a tapered screw to apply pressure and provide mixing and forward

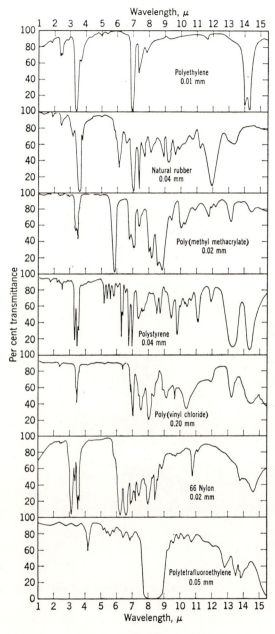

Fig. 9 Typical infrared spectra of polymers (from reference 6, p. 93).

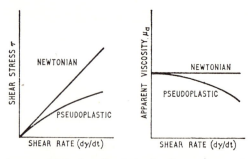

Fig. 10   Melt viscosity of plastics (from reference 7, p. 113).

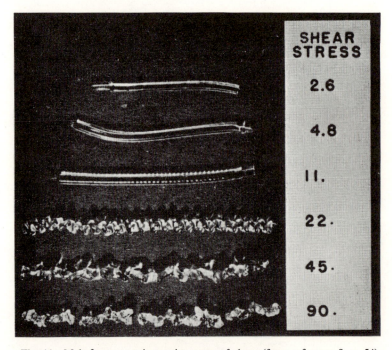

Fig. 11   Melt fracture at increasing rates of shear (from reference 8, p. 54).

flow (Fig. 12). Cooling is generally accomplished by circulating cold water against the mold or by circulating water or air directly against the hot polymer emerging from the die. Other processing techniques occasionally use hot air or infrared heating and refrigerated or even liquid carbon dioxide cooling, all for maximum speed and economy.

The most common type of plastic processing is *injection molding*, in which

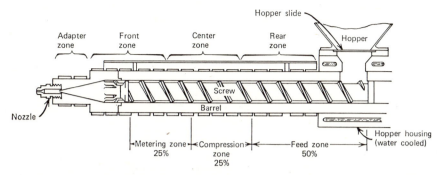

Fig. 12   Screw for melting and conveying plastic powder and granules in injection molding and extrusion (from reference 4, p. 426).

the molten polymer is forced into a mold (Fig. 13) that is kept below the melting or the glass-transition temperature of the polymer. The mold is then opened to eject the plastic product in finished form. Continuous lengths of standard profile—such as film, sheet, tubing, pipe, wire coating, and gasketing—are *extruded* through a die of the desired shape and then cooled rapidly

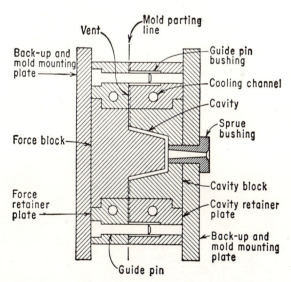

Cross-section of simple center-gated injection mold showing principal parts. Piece being molded is a cup-shaped object

Fig. 13   Cross section of typical injection mold (from reference 9, p. 715).

by air or water bath (Fig. 14). Continuous film and sheet are also made by *calendering* (Fig. 15), adjusting the shape, spacing, and temperature of the rolls very precisely to control the quality of the final product. Simple hollow shapes are made rapidly in low-cost molds by a process called *thermoforming*,

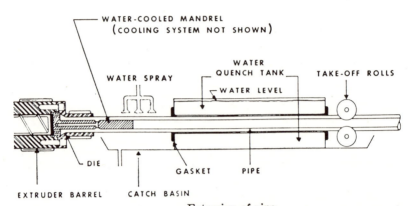

Extrusion of pipe.

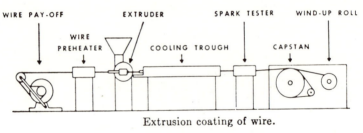

Extrusion coating of wire.

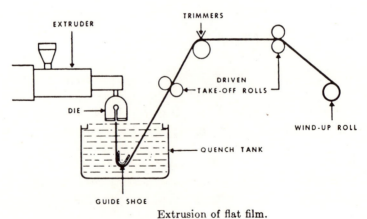

Extrusion of flat film.

Fig. 14   Typical extrusion trains (from reference 8, p. 283).

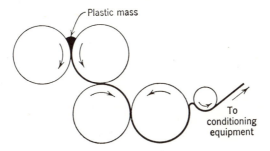

Fig. 15 Typical Z-style calender (from reference 6, p. 493).

in which continuous extruded sheet is heated rapidly by infrared and then drawn into the mold by vacuum or blown in by air pressure to produce the desired shape (Fig. 16).

More difficult hollow shapes, such as bottles, are made by *blow molding*, in which a molten tube of polymer is inserted into the mold, air is blown into the tube to expand it against the walls of the mold, and the mold is finally opened to eject the bottle (Fig. 17). Small runs of large hollow shapes are made by *rotational molding*, in which the powdered polymer is tumbled in a heated, low-cost mold and fuses to a solid, continuous layer on the surface of the mold

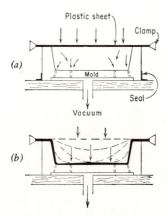

Fig. 16 Simple thermoforming (from reference 3, p. 777). The plastic sheet is clamped and heated. Vacuum beneath sheet (*a*) causes atmospheric pressure to push sheet down into mold (*b*).

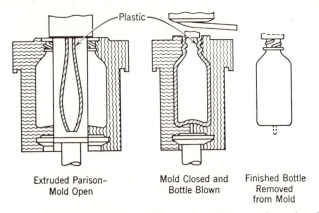

Extruded Parison– Mold Open

Mold Closed and Bottle Blown

Finished Bottle Removed from Mold

Fig. 17   Blow molding (from reference 6, p. 492). A molten tube of polymer is inserted into the mold (*a*), the mold is closed and air is blown to expand the polymer against its walls (*b*), and the finished bottle (*c*) is removed from the mold.

(Fig. 18). One of the newest and most exciting process techniques is *cold forming* or *warm forging*, adapted from metallurgy, in which plastic sheet is stamped into finished form without ever heating it to the molten state.

In molding and extrusion the polymer undergoes true liquid flow. In calendering and rotational molding its behavior is more likely rubbery liquid flow. In thermoforming and blow molding it is generally believed to be rubbery, and in cold forming and warm forging it may be approaching leathery behavior. In all cases, however, it must be recooled rapidly below the melting or the glass-transition temperature to retain its final shape, and this cooling step is often the rate-controlling one.

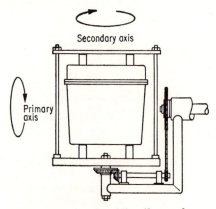

Fig 18   Rotational molding (from reference 4, p. 26).

d. Films

Extrusion and calendering can convert plastic materials into continuous films rapidly and economically. These not only are competitive with textiles and paper but also offer a number of advantages: three-dimensional design, strength, flexibility, clarity, decorative effects, impermeability, and impact, flame, and chemical resistance. They therefore find wide use in packaging, clothing, drapery, and electrical applications.

i. *High-Modulus Films*

Polymers of stiff molecular structure and/or high crystallinity can form films of high strength and low extensibility. Often the film as extruded tends to be brittle or opaque (or both). It is then improved by stretching, either directly after extrusion or in a separate later step. Stretching makes the film thinner and more flexible and increases the area yield. More important, however, it orients the molecules and crystallites in the direction of stretch, thus increasing strength and toughness, and it reduces the size of the crystallites below the wavelength of visible light, thus making the film transparent. Polypropylene, polystyrene, polyvinylidene chloride (**17**), and polyethylene terephthalate (**18**) films are all made in this way.

$$\begin{array}{ccc} & Cl & Cl & Cl \\ & | & | & | \\ \sim CH_2-C-CH_2-C-CH_2-C\sim & & \\ & | & | & | \\ & Cl & Cl & Cl \\ & & \mathbf{17} \end{array}$$

$$\sim CH_2CH_2O\overset{O}{\overset{\|}{C}}-\text{\Large$\bigcirc$}-\overset{O}{\overset{\|}{C}}O\sim$$

18

Historically cellophane was the first polymeric film—and the hardest to produce. The high polarity, hydrogen bonding, and crystallinity of cellulose make it very difficult to process. It is first swollen in aqueous alkali, treated with carbon disulfide to produce the soluble xanthate (**19**), and then extruded through a slit die into an acid bath that decomposes the xanthate and reprecipitates the cellulose in film form.

$$\begin{array}{c} S \\ \| \\ CH_2O\overset{}{C}S^-\ Na^+ \\ | \\ CH-O \\ \diagup \qquad \diagdown \\ \sim CH \qquad\qquad CH-O\sim \\ \diagdown \qquad \diagup \\ CH-CH \\ | \qquad | \\ O \qquad O \\ | \qquad | \\ Na^+\ {}^-SC \qquad CS^-\ Na^+ \\ \| \qquad \| \\ S \qquad S \end{array}$$

19

Cellulose acetate film is either extruded for maximum economy or solution-cast for maximum clarity. In solution casting a viscous solution of polymer is spread over a polished, continuous, stainless-steel belt, passed through an oven to evaporate the solvent, and rolled up at the other end.

ii. Low-Modulus Films

(*1*). *Polyethylene*. Branched polyethylene film is extruded and stretched in a single continuous process (Fig. 19). The polymer is first extruded into a thin-walled tube, which is pulled away from the extruder so rapidly that it is stretched in the lengthwise direction. At the same time air is blown into the hot tube to expand it radially, thus producing stretching in both directions simultaneously. Quick cooling of the thin film causes rapid formation of many small crystallites, smaller than the wavelength of visible light, and therefore produces good clarity. Since the branched polymer is only partly crystalline, the amorphous portions provide good flexibility and extensibility, whereas the crystalline portions provide strength. Combined with the low cost of material and processing, these make polyethylene the most widely used of all the plastic films.

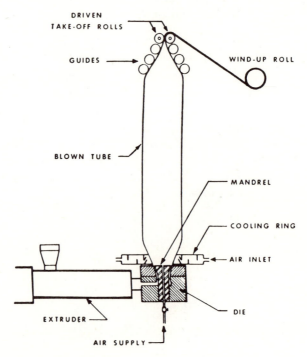

Fig. 19 Extrusion of blown film (from reference 8, p. 285).

(2). Plasticized Polyvinyl Chloride. Polyvinyl chloride alone, as already noted, is a fairly rigid material, stiffened by bulky polar chlorine atoms, intermolecular attractions, and some crystallinity. It can be softened greatly by adding high-boiling-point esters, such as dioctyl phthalate, to serve as plasticizers, providing that their polarity and hydrogen bonding make them compatible with the polymer. Most commonly 100 parts of polymer are blended with 40 to 70 parts of plasticizer and calendered to produce soft, flexible film and sheet. In this form it is widely used for clothing, drapery, and upholstery. For greater strength, particularly in upholstery, it may be bonded to a fabric backing; this can also be done by calendering, but it is often accomplished by spreading a viscous paste of polymer dispersed in plasticizer, called a *plastisol*, over the surface of the fabric and then passing it through an oven to warm it above the fusion point and thus convert it into the final film. In plasticized polyvinyl chloride the small plasticizer molecules, floating freely between the polymer molecules, provide flexibility, and the low crystallinity of the polymer provides anchor points to keep the polymer molecules from sliding completely past each other, thus giving strength and dimensional stability to the entire structure.

iii. *Permeability*

Most of the properties of film-forming polymers have already been discussed. One of particular importance—permeability—requires further consideration. The largest application of plastic films is in packaging, where they are often used to replace metal and glass for the storage of foods and other materials with volatile components. Whereas the dense structures of metals and glass retain these volatile components quite permanently, the relatively open structures of organic polymers offer considerable "porosity" at the atomic level. Small molecules, such as water vapor—and even such larger organic molecules as alcohol, solvents, oils, odors, and flavors—can permeate gradually through the polymer, causing serious losses during storage. Also small molecules of atmospheric gases, such as oxygen, can permeate into the package and cause spoilage of the material stored in it.

In general the larger molecules permeate best through polymers of similar polarity; thus they are best stored in polymers of very different polarity. The small gas molecules appear indifferent to such considerations and are retained best by polymers that are densely crystalline, particularly polymers that are symmetrical in structure (Table 3). In many cases the outstanding impermeability of polyvinylidene chloride is the final resort for plastic packaging, even though it is considerably higher in cost than some of the other plastic films.

Table 3   Permeability of Several Gases through Typical Polymers[a]

Polymer	Permeability[b]				Nature of Polymer
	N_2	O_2	CO_2	H_2O^c	
Polyvinylidene chloride	0.0094	0.053	0.29	14	Crystalline
Polychlorotrifluoroethylene	0.03	0.10	0.72	2.9	Crystalline
Polyethylene terephthalate	0.05	0.22	1.53	1,300	Crystalline
Rubber hydrochloride (Pliofilm ND)	0.08	0.30	1.7	240	Crystalline
Nylon 6	0.10	0.38	1.6	7,000	Crystalline
Polyvinyl chloride (unplasticized)	0.40	1.20	10	1,560	Slightly crystalline
Cellulose acetate	2.8	7.8	68	75,000	Glassy
Polyethylene ($d = 0.954$ to 0.960)	2.7	10.6	35	130	Crystalline
Polyethylene ($d = 0.922$)	19	55	352	800	Some crystalline
Polystyrene	2.9	11	88	1,200	Glassy
Polypropylene	—	23	92	680	Crystalline
Butyl rubber	3.12	13.0	51.8	—	Rubbery
Methyl rubber	4.8	21.1	75	—	Rubbery
Polybutadiene	64.5	191	1380	—	Rubbery
Natural rubber	80.8	233	1310	—	Rubbery

[a] Data from reference 7, p. 82.
[b] At 30°C.
[c] 90% RH at 25°C.

2. THERMOSETS

In general, fluidity during processing is favored by low molecular weight, whereas optimum final properties depend on high molecular weight. Even though this is a gross oversimplification of a complex field, it describes a major paradox for the plastics scientist. In the thermoplastics the paradox is resolved by compromise: molecular weight is chosen to be low enough for economical processing yet high enough for good final properties.

In the thermosetting plastics the dilemma is resolved in another way: the polymerization reaction is carried out in two or more stages. In the first stage the monomers are partially polymerized to viscous liquids or low-melting-point soluble fusible solids, all of which still provide high fluidity for processing. In the final stage the polymerization reaction is carried to completion, producing a highly crosslinked three-dimensional structure.

This ingenious solution carries both advantages and disadvantages. In processing it means that liquids can be poured into low-cost molds, but they must be left there until the final polymerization reaction is completed; fusible solids can be "cured" and ejected from hot molds without cooling, but scrap is not reusable. In properties the tight, rigid structure immobilizes the polymer molecules, providing high hardness, hot strength, and chemical resistance; but it also makes the pure polymers inflexible and brittle, requiring fibrous reinforcement for good mechanical properties, as discussed later.

a. Major Types

There are six major types of thermosetting plastics, each with its own balance of properties and consequent applications.

i. *Allyls*

Two diallyl monomers have found general commercial use. Diallyl phthalate (**20**) is the easier to produce and lower in cost; it is partially polymerized to a low-molecular-weight fusible solid, blended with glass fibers, catalyzed with peroxide, and molded and cured into highly crosslinked thermoset parts for electrical equipment.

$$CH_2=CHCH_2O\overset{\overset{O}{\|}}{C}\quad\overset{\overset{O}{\|}}{C}OCH_2CH=CH_2$$

20

Diethylene glycol bis(allyl carbonate) (**21**) is a more specialized material.

$$CH_2=CHCH_2O\overset{\overset{\displaystyle O}{\|}}{C}OCH_2CH_2OCH_2CH_2O\overset{\overset{\displaystyle O}{\|}}{C}OCH_2CH=CH_2$$

21

It is catalyzed with peroxide, poured into a mold, and polymerized completely in one step, primarily to produce plastic lenses that are harder and more scratch resistant than acrylic ones.

ii. *Epoxy Resins*

When bisphenol A (**22**) is condensed with epichlorohydrin (**23**) in the presence of alkali, the resulting diglycidyl ethers (**24**) are called epoxy resins.

By using excess epichlorohydrin and minimum alkali to control molecular weight, these prepolymers are made to be viscous liquids or soluble low-melting-point solids, which can be poured into molds or spread on surfaces as coatings or adhesives. When mixed with polyamines, they cure rapidly at room temperature.

When mixed with polybasic acid anhydrides, they can be cured at higher temperatures to give harder and more heat-resistant final properties.

When the plastic is cast around delicate electrical and electronic equipment. this is referred to as potting or encapsulation. When used for coatings and adhesives, the reactivity of the epoxy groups toward active hydrogen on the surface of the substrate helps provide high adhesion, and the high crosslinking provides heat and chemical resistance.

iii. *Polyesters*

Propylene glycol is copolymerized with maleic and phthalic anhydrides to produce low-molecular-weight, viscous, liquid or soluble solid polyesters (**25**).

$$\underset{\substack{|\\CH_3}}{\sim CHCH_2O}\overset{\overset{\displaystyle O}{\|}}{C}CH=CH\overset{\overset{\displaystyle O}{\|}}{C}O\underset{\substack{|\\CH_3}}{CHCH_2O}\overset{\overset{\displaystyle O}{\|}}{C} \quad \overset{\overset{\displaystyle O}{\|}}{C}O\sim$$

25

These are mixed with about one-half their weight of styrene monomer to form viscous liquids, catalyzed with peroxide, and often synergized with amines or cobalt soaps, impregnated into glass-fiber mats or woven fabrics, and polymerized in low-cost molds at low pressures and often at room temperatures—although an oven bake is usually used to finish the cure. The peroxide initiates styrene polymerization, and the growing polystyrene molecules copolymerize with the maleic unsaturation in the polyester chain to produce extensive crosslinking. For flame resistance in building and transportation applications the phthalic anhydride may be replaced by a heavily chlorinated monomer; for hydrolytic resistance in chemical plant equipment the propylene glycol may be replaced by the glycol ether of bisphenol; and for high-temperature resistance the styrene may be replaced by triallyl cyanurate.

iv. *Urea-Formaldehyde Resins*

When urea (**26**) and formaldehyde (**27**) are mixed in the presence of water, they condense rapidly to a highly crosslinked material.

$$\underset{\substack{\\ \textbf{26}}}{H_2N-\overset{\overset{\displaystyle O}{\|}}{C}-NH_2} + \underset{\substack{\\ \textbf{27}}}{CH_2O} \longrightarrow \sim N-\overset{\overset{\displaystyle O}{\|}}{C}-\underset{\substack{|\\CH_2\\ \}}}{N}-CH_2\sim$$

Precise control of pH and temperature can limit the initial reaction to the formation of low-molecular-weight soluble and fusible prepolymers. Later catalysis and/or heating is then used to finish the crosslinking and cure. One of the first commercial plastics half a century ago, urea resins suffered severely from atmospheric moisture, until reinforced with cellulosic cotton or wood fibers. They now find their major use as binders in plywood and granulated-wood products.

v. Melamine-Formaldehyde Resins

Condensation of melamine (28) with formaldehyde provides even higher functionality, and the resulting structure has much greater hardness and resistance to heat and moisture.

Reinforced with cellulose from cotton, the fusible intermediate is used for molding dinnerware. Dissolved in water and impregnated into paper, the soluble intermediate is then pressed and cured to form decorative laminates for countertops and furniture.

vi. Phenol–Formaldehyde Resins

Phenol (29) and formaldehyde condense rapidly to highly crosslinked structures.

Known for a century, they were worthless because the exothermic reaction produced gas bubbles, and the cured resins were very brittle. Dr. Leo Baekeland solved both these problems by molding under pressure to prevent bubbling and adding wood-flour fibers to provide strength and resilience toward impact, thus producing the first commercial synthetic plastic, Bakelite, in 1908. Although limited by the dark color of quinoid by-products from phenol, the high rigidity, strength, hot strength, flame retardance, and chemical resistance of these resins have made them the perennial leader among the thermosetting plastics.

b. Processing

The rapid crosslinking reaction of most thermosetting plastics has generally limited process techniques to casting liquids and compression-molding fusible solids. In *compression molding* (Fig. 20) the fusible reactive prepolymer

(B-stage resin) is pressed between two hot mold faces until cured, being then ejected directly without the need for cooling. For improved flow and faster cure the resin is sometimes preheated and fused in an adjacent chamber, and then forced to flow through a short orifice into the final mold (Fig. 21); this is called *transfer molding*. Most recently improvements in both the resins and the equipment have introduced the possibility of *screw-injection molding* (see Figs. 12 and 13) of thermosets, and this new development may well lead to greater vitality and interest in the thermosetting sector of the plastics field.

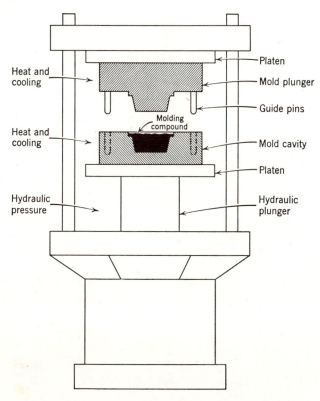

Fig. 20   Compression molding (from reference 6, p. 488).

3. COMPOSITES

The properties of simple homogeneous polymers are sufficient for many applications. On the other hand, materials scientists are noting with increasing frequency and urgency that heterogeneous dispersion of a second phase in the polymer, on a submicroscopic or microscopic scale, can often produce

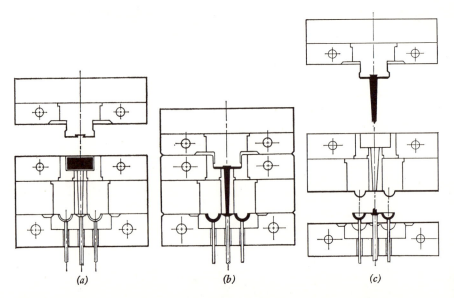

Fig. 21 Transfer molding (from reference 3, p. 716). The material is first placed in the transfer pot (*a*) and then forced through an orifice into the closed mold (*b*). When the mold opens (*c*), the cull and sprue are removed as a unit, and the part is lifted out of the cavity by ejector pins.

tremendous improvement in properties and versatility for a greater variety of more demanding applications. Some examples have already been noted in this discussion: partial crystallinity, impact plastics, polyblends, glass fibers and cellulosic fibers from cotton and wood. The subject can easily be extended to include adhesives, coatings, and larger, macroscopic, combinations of materials.

It has been suggested that the mechanical properties of composites in general depend on the combination of a high-modulus, high-strength phase with a low-modulus, high-elongation phase, in the proper geometrical arrangement (10). Typical examples are the following:

1. Laminated safety glass, containing a soft adhesive layer of organic polymer between two layers of rigid glass.

2. Automobile tires, containing strong, high-modulus cord in a very-low-modulus rubber.

3. Reinforced plastics, containing very strong, higher modulus glass fibers in a lower modulus resin matrix.

4. Reinforced rubber, in which carbon black contributes much greater strength and tear resistance to the rubber.

5. Impact plastics, in which a colloidal dispersion of elastomer particles in a brittle, high-modulus matrix provides an increase in impact strength.

6. Plywood, in which anisotropic layers of wood are crosslaminated by polymeric adhesive to provide two-dimensional isotropy.

7. Wood itself, in which strong cellulose fibrils are cemented together by softer lignin.

8. Plastic foams, in which a gas is dispersed throughout the structure of thin-walled polymeric cells.

In a more general sense materials scientists are beginning to recognize the tremendous versatility that can be achieved by combining two or more materials on a submicroscopic, microscopic, or even macroscopic scale to obtain properties not previously available for a much greater variety of applications. A review of several general types is sufficient to illustrate the tremendous range of possibilities.

a. Coatings

Although the greatest volume of coatings is primarily decorative in nature (house paints and inks), most coating materials and formulations are designed to provide either protection against weathering and chemical corrosion or special effects, such as electrical insulation and lubricity. Once they have solved the problem of bonding the coating to the substrate (often with the aid of an intermediate layer of mutual adhesive called a primer), coatings engineers can combine the mechanical and thermal-mechanical properties of the substrate with the decorative, protective, and functional qualities of the topcoat. For example, an extruded coating of polyethylene on paperboard is far superior to the old-fashioned waxed cardboard for milk cartons and other liquid containers. Coatings of rubber and plasticized polyvinyl chloride on fabrics combine the strength and dimensional stability of the fabric with the impermeability and cleanliness of the polymeric coatings for use in clothing, drapery, and upholstery. Epoxy coatings on concrete and steel provide long-term protection against wear, weather, and chemical attack. All these are typical of the benefits obtained by combining a polymeric coating with a substrate.

b. Adhesives

Such natural materials as solder, mortar, starch, glue, and pitch have long been used for bonding two solid materials to each other or for filling and sealing the gaps between them. More recently synthetic polymers have often proved much more versatile in both application techniques and performance properties. For example, high-melting-point films can be made heat sealable by coating them with a thin layer of lower melting polyethylene. Blends of nitrile rubber with phenolic resin produce useful adhesives for bonding brake linings and even metals. Polyvinyl acetate (**30**) is useful for bonding paper and cardboard, and in formulating a variety of home and workshop adhesives.

$$\sim CH_2-CH-CH_2-CH-CH_2-CH\sim$$

$$\begin{array}{ccc} | & | & | \\ OCCH_3 & OCCH_3 & OCCH_3 \\ \| & \| & \| \\ O & O & O \end{array}$$

30

Epoxy adhesives are two-part systems that are mixed just before use; easily applied, they cure and bond strongly to so many different substrates that their universality has become almost legendary. Urea–formaldehyde and phenol–formaldehyde resins are widely used in bonding plywood and may well copolymerize with the cellulose as they do so. In addition, dozens of other polymer systems are used for special adhesive applications.

The mechanism of adhesive action is the subject of many diverse theories. From a purely mechanical approach it has been pointed out that when stress and strain are applied to the substrate, they are transmitted to the adhesive layer, which must therefore be able to respond similarly to them without failure. Another approach is based on the importance of intimate contact between layers. Related to this is the fact that the adhesive must be reasonably fluid to flow and fill the irregularities of a solid surface. Also related is the need for an adhesive of low surface tension to flow across a substrate of higher surface tension, thus reducing the total surface energy and producing good wetting.

A somewhat different approach is the need for similar (preferably high) polarity and hydrogen bonding to produce attractive forces between adjacent layers. Still further in this direction, some of the best adhesives, such as epoxy, urea, and phenolic resins, may react directly with the surface of the substrate, particularly with active hydrogen and especially hydroxyl groups, to produce actual covalent bonds to support adhesion. In some systems it is actually postulated that both surfaces have enough molecular mobility to migrate and penetrate into each other and produce a transitional phase that thereby creates adhesion.

Obviously some theories may apply best to certain adhesives and substrates, other theories to others. In many cases a balanced combination of mechanisms may ultimately provide the most meaningful explanation of adhesive action.

c. Foams

A foamed plastic is actually a dispersion of gas in a solid polymeric matrix and derives properties from both phases. The polymeric structure contributes most of the mechanical and chemical properties, whereas the gas phase contributes most to thermal and electrical insulation. Their interacting effect on mechanical properties depends primarily on the structure of the cell walls (Fig. 22): if they are closed cells, external stresses, such as compression, are resisted by gas pressure within the cells, and the foam is quite stiff, far more than an equal weight of solid polymer; if, however, the cells are open and interconnected so that the gas can flow in and out freely, the foam is quite soft and flexible. Energy absorption, in furniture, packaging, and crash padding, is also a function of both phases and geometry.

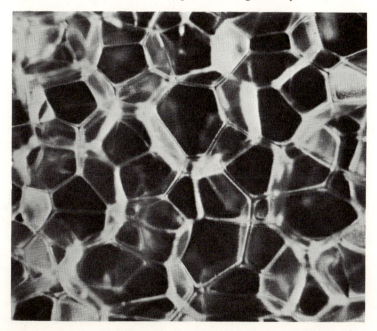

Fig. 22 Photomicrograph of a typical foam structure (11).

Three polymers are used in most of the present commercial plastic foams:

1. *Polystyrene* is swelled by a low-boiling-point liquid like pentane and then molded or extruded, producing a stiff, closed-cell structure that is used mainly in the packaging of delicate products for safety in shipment.

2. *Plasticized polyvinyl chloride* plastisols are formulated to contain an

unstable organic nitrogen compound (blowing agent), such as azobisforma-
mide (31).

$$H_2N-\overset{\overset{\displaystyle O}{\|}}{C}-N=N-\overset{\overset{\displaystyle O}{\|}}{C}-NH_2$$

31

When the plastisol is heated during fusion, the blowing agent decomposes,
releasing nitrogen gas and producing a soft foam that has the rich feel of high-
quality leather—highly desirable in clothing and upholstery.

  3. *Polyurethanes* (32) are made by liquid-phase reaction between low-
molecular-weight polyols and polyisocyanates;

$$HOR'OH + O=C=NR'N=C=O \longrightarrow \sim RO-\overset{\overset{\displaystyle O}{\|}}{C}-\overset{\overset{\displaystyle H}{|}}{N}R'\overset{\overset{\displaystyle H}{|}}{N}-\overset{\overset{\displaystyle O}{\|}}{C}-O\sim$$

32

These can be foamed as they polymerize and can then be crosslinked by
incorporation of some triol

$$HO\frown\curlyvee\frown OH$$
$$\underset{OH}{|}$$

to stabilize the foam structure and make it permanent. Flexible foams are
made by using a small amount of triol, reacting excess isocyanate and water
to produce gas for foaming,

$$RNCO + H_2O \longrightarrow RNH_2 + CO_2\uparrow$$

and crushing the gelled foam temporarily to produce open cells before final
cure. They are used primarily in mattresses and furniture cushions. Rigid
foams are made by using a large amount of high-functionality polyol and
adding low-boiling-point liquids like chlorofluoromethanes and ethanes,
which are volatilized by the exothermic polymerization reaction, to produce
gas for foaming; these foams are used primarily for thermal insulation in
appliances.

  A number of other polymers are converted into specialized plastic foams,
and some of these may find growing use in the coming years. One of the major
current areas of growth is in the molding of high-density rigid foams to
replace wood in furniture.

d. Reinforced Plastics

  When glass fibers are dispersed in a rigid polymeric matrix, any stress
applied to the polymeric matrix is transmitted to the fibers as well. This

distributes the load along its length and adds greatly to the modulus, strength, impact resistance, and dimensional stability of the total plastic product.

Maximum strength-to-weight ratios are obtained in *filament-wound* simple hollow shapes, such as tanks (Fig. 23), by passing continuous glass roving through a bath of liquid thermosetting resin, winding it on a mandrel of the desired shape, and heating to produce final cure; epoxy resins are generally used for maximum final properties.

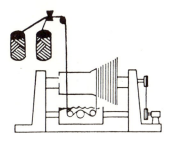

Fig. 23 Filament winding (from reference 3, p. 638).

Simple shapes, such as flat and corrugated sheet and boat hulls, are made by impregnating liquid polyester resin into *woven glass fabric* (Fig. 24), laying the assembly in a low-cost mold, and applying mild pressure during room-temperature gelation and preferably final oven cure.

More complex shapes are made by using a *nonwoven mat* of random glass fibers (Fig. 25).

High-speed production, for mass markets like transportation, can be achieved by using more costly *matched molds* resembling light-weight compression-molding equipment. Major applications are in boats, transportation equipment, construction, and tanks.

The production of such reinforced thermoset plastics is obviously more laborious and costly than most plastics processing, and this has certainly limited its growth. A newer trend is appearing in the use of *short glass fibers* ($\frac{3}{8}$ in. or less) in thermoplastic injection-molding resins like polystyrene and nylon. These short fibers permit normal, high-speed, economical processing, and produce high modulus, strength, and dimensional stability, creating a new class of plastic materials that are competitive with die-cast metals in many applications.

Aside from glass fibers, a great deal of kraft paper is impregnated with partially polymerized phenolic resin, stacked, and compression-molded into laminated boards that find wide use in electrical construction and, especially

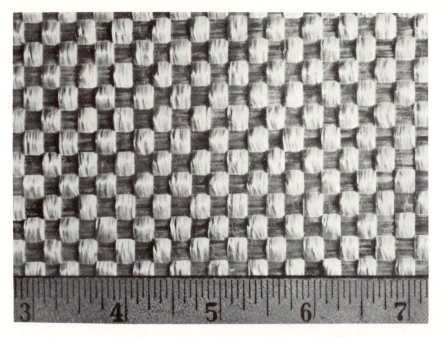

Fig. 24   Woven glass fabric (from reference 12, p. 166).

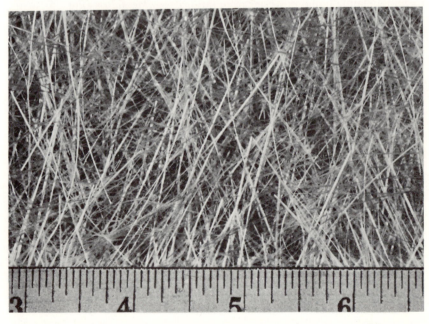

Fig. 25   Nonwoven mat of random glass fibers (from reference 12, p. 159).

344

with a decorative melamine topcoat, in countertops, cabinets, tabletops, and furniture. The essential use of short cellulose fibers from cotton or wood flour in urea, melamine, and phenolic resins has already been discussed in the initial description of these plastics.

Another fibrous material frequently used in plastics is asbestos. The combination of low cost, reinforcement, and flame resistance is particularly useful in polyvinyl chloride flooring and in some of the more recent polyolefin molding resins.

e. High-Temperature Materials

Still mainly in the future is the development of exotic, high-performance composites from new binder resins of high thermal stability (Table 4) and new

Table 4   Typical New High-Temperature Polymers

Polymer	Formula
Polybenzimidazole	
Polyimide	R = variable groups.
Polybenzothiazole	

reinforcing fibers of very high modulus, strength-to-weight ratios, and heat resistance (Tables 5 and 6). They were developed originally for highly specialized and highly demanding instrument and aerospace applications, and their future utility is yet to be explored.

Table 5   Properties of Experimental Epoxy Composites[a]

Composite[b] Fiber/Wire	Density (lb/in.³)	Specific Gravity	Volume % Matrix	Tensile Strength		Modulus of Elasticity	
				Ultimate (× 10³ psi)	Ratio to Density (× 10⁶ psi)	× 10⁶ psi	Ratio to Density (× 10⁶ psi)
S-Glass[c]	0.074[d]	2.05	35	260	3.53	7.6	103
None	0.043	1.20	100	12	0.28	0.5	11
Beryllium	0.058	1.61	35	97	1.67	28.0	482
Boron	0.074	2.05	35	320	4.32	36.0	486
Graphite	0.067	1.87	35	160	2.39	20.0	299
Carbon	0.054	1.50	35	80	1.48	4.3	80

[a] Data from reference 3, p. 598. Epoxy-resin tensile strength, 12,000 psi; E, 0.5 × 10⁶ psi; and elongation, 4.47%; compressive strength, 18,000 psi; E, 0.5 × 10⁶ psi; and deformation, 6%.
[b] Unidirectional orientation.
[c] S-Glass epoxy composite; tensile elongation, 3.2%; compressive strength, 175,000 psi; compressive E, 7.5 × 10⁶ psi; interlaminar shear, 11,000 psi; flexural strength, 190,000 psi; and Poisson's ratio, 0.2.
[d] 20 Wt %.

Table 6   Physical Properties of Whiskers[a]

Material	Density (gm/cm³)	Melting Point (°F)	Tensile Strength (psi ×10⁶)	Young's Modulus (psi × 10⁶)	Strength Density (in. × 10⁶)
Aluminum oxide	3.9	3780	2–4	100–350	14–28
Aluminum nitride	3.3	3990[b]	2–3	50	13–21
Beryllium oxide	1.8	4620	2.0–2.8	100	31–43
Boron carbide	2.5	4440	1	65	11
Graphite	2.25	6500	3	142	37
Magnesium oxide	3.6	5070	3.5	45	27
Silicon carbide (alpha)	3.15	4200[b]	3–5	70	26–44
Silicon carbide (beta)	3.15	4200[b]	1–3	100–150	0.9–26
Silicon nitride	3.2	3450	0.5–1.5	55	4.2–13

[a] Data from reference 3, p. 599.
[b] Disassociates.

4. APPLICATIONS

Current use of plastics materials leans heavily to those of low cost and easy processability (Table 7). Largest applications are in packaging and building (Table 8), with strong growth in a number of smaller fields.

Table 7   U.S. Production of Plastics Materials in 1969[a]

Material	Production (lb × 10⁶)
High-density polyethylene	1478
Low-density polyethylene	3778
Polypropylene	1135
Polystyrene and copolymers	3250
Polyvinyl chloride and copolymers	2750
Acrylic	352[b]
Polyoxymethylene	56[b]
Cellulosics	181
Polycarbonate	31[b]
Nylons	87[b]
Epoxy	172
Polyesters	655
Urea and melamine resins	787
Phenolic resins	1127
Polyurethane foam	758[b]
All others	1933
Total	18,530

[a] Data from *Modern Plastics* (13).
[b] Consumption.

Table 8   Major U.S. Markets for Plastics in 1969[a]

Industry	Quantity (lb × 10⁶)
Agriculture	189
Appliances	537
Building construction	2644
Electrical and electronic	1135
Furniture	655
Housewares	850
Packaging	3458
Toys	538
Transportation	852

[a] Data from *Modern Plastics* (14).

The choice of an optimum material for any desired end product must be based on its balance of important properties, and especially the cost of obtaining these properties. The design of a new product

$$\begin{array}{ccc}
\text{Material} & \xleftarrow{\quad 1 \quad} & \text{Product} \\
\text{selection} & \xrightarrow{\quad 2 \quad} & \text{design} \\
& \updownarrow 3 & \\
& \text{Process} & \\
& \text{technique} &
\end{array}$$

starts with a general sketch to estimate the property requirements, then a search of the available materials to select one or more that may be suitable, then a redesign of the product to make optimal use of each of these materials, and finally exploration of the processing techniques required for each, and the economics and performance that will result.

At present the materials engineer must still select his optimum materials by inspection of tables and data sheets. With more than 50 classes of polymers already in commercial use—each in a variety of copolymers, molecular weights, and compounding additives—there are already too many for rational choice by such inspection techniques. In the near future design engineers may well convert to a computerized searching system. Such a system would start first with the absolute property requirements P_i and select only those materials that passed this first screening test. Second, it would balance the relative importance f_i of other properties p_i, preferably on the basis of cost per unit property. Third, it would present the design engineer with the one or several materials whose balance of properties would be best suited to his needs.

$$P = P_1 P_2 P_3 \ldots P_i \ldots P_n(f_1 p_1 + f_2 p_2 + f_3 p_3 + \ldots + f_i p_i + \ldots + f_n p_n)$$

Such computerized searching may well be one of the major new developments in plastics science during the coming decade.

REFERENCES

1. M. L. Miller, *The Structure of Polymers*, Reinhold, New York, 1966, p. 503.
2. C. C. Winding and G. D. Hiatt, *Polymeric Materials*, McGraw-Hill, New York, 1961, p. 66.
3. *Modern Plastics Encyclopedia*, Vol. 45, McGraw-Hill, New York, 1968.
4. *Modern Plastics Encyclopedia*, Vol. 46, McGraw-Hill, New York, 1969.
5. V. E. Malpass, *SPE ANTEC* 13, 624 (1967).
6. F. W. Billmeyer, Jr., *Textbook of Polymer Science*, Interscience, New York, 1962.

7. J. A. Brydson, *Plastics Materials*, Iliffe, 1966.

8. E. C. Bernhardt, *Processing of Thermoplastic Materials*, Reinhold, New York, 1959.

9. *Modern Plastics Encyclopedia*, Vol. 44, McGraw-Hill, New York, 1966.

10. T. Alfrey and E. F. Gurnee, *Organic Polymers*, Prentice-Hall, Englewood Cliffs, N. J., 1967.

11. E. Baer, *Engineering Design for Plastics*, Reinhold, New York, 1964.

12. G. Lubin, *Handbook of Fiberglass and Advanced Plastics Composites*, Van Nostrand Reinhold, New York, 1969, p. 166.

13. *Modern Plastics* **47** (1), 70–76 (1970).

14. *Modern Plastics* **47** (1), 77–80 (1970).

15. R. D. Deanin, Engineering Thermoplastics from a Commercial Development Viewpoint, *Advances in Chemistry* **96**, 9 (1969).

16. L. E. Nielsen, *Mechanical Properties of Polymers*, Reinhold, 1962.

17. A. V. Tobolsky, *Properties and Structure of Polymers*, Wiley, 1960.

Science of Fibers

15

L. Rebenfeld

1. THE PRODUCTION OF NATURAL AND MAN-MADE FIBERS

One of the most important forms of polymeric materials is the fiber, which can be described as a flexible, macroscopically homogeneous body, with a high length-to-thickness ratio and a small cross-section. Such a description does not limit fibers to those composed of organic polymers, and indeed fibers can be obtained from inorganic and metallic substances, including asbestos, glass, graphite, carbon, boron, boron nitride, quartz, silicon carbide, and many others. Within the realm of organic polymers the fibrous state can be described—in a manner that is more satisfying to molecular scientists—as an irreversibly oriented polymeric system with a high level of three-dimensional regularity, being characterized by highly anisotropic physical properties.

The textile and paper industries are the prime converters of fibers into end products whose characteristics can be directly traced to the unique combination of properties that are typical of fibers. The paper industry uses a naturally occurring cellulosic fiber derived from wood almost exclusively, and the textile industry uses a variety of naturally occurring and man-made fibers in the manufacture of its wide range of products. Textile fibers are one of the principal outlets for synthetic organic polymers, having amounted to approximately 6 billion pounds in 1969 in the United States and nearly 10 billion pounds on a worldwide basis. The world production of natural and man-made textile fibers since 1960 is shown in Table 1. It is quite evident that the man-made fibers based on synthetic polymers represent the most rapidly growing segment of the fiber industry.

Table 1   World Production of Textile Fibers[a]

Year	Man-made Fibers		Natural Fibers		
	Rayon and Acetate	Synthetic	Cotton	Wool	Silk
1960	5749	1548	22,295	3225	68
1961	5930	1831	21,647	3267	69
1962	6315	2381	23,052	3257	73
1963	6744	2942	24,130	3320	68
1964	7245	3728	24,932	3263	72
1965	7360	4521	25,523	3289	73
1966	7370	5473	23,248	3385	72
1967	7312	6239	22,941	3429	74
1968	7776	8290	25,102	3488	75

[a] In millions of pounds. Data from *Textile Organon* (1).

a. Natural Fibers

Textile fibers are conveniently classified according to their origin, as natural and man-made. The natural fibers are derived from the animal, vegetable, and mineral kingdoms and can be used directly in textile-manufacturing operations after preliminary cleaning. With the exception of silk, which is extruded by the silkworm as a continuous filament, natural fibers are of discrete and finite length. The length of these staple fibers, along with such other geometric properties as cross-sectional area and cross-sectional shape, is an important quality factor of natural fibers and determines their industrial utility. Other characteristics of natural fibers that must be considered are mechanical properties, spinnability, dyeability, and presence and ease of impurity removal. The principal naturally occurring fibers used in the textile industry are cotton, linen, jute, hemp, silk, asbestos, wool, and a number of other animal hairs, such as mohair, camel hair, cashmere, and vicuña. With the exception of asbestos, natural textile fibers are based on the two most commonly occurring natural organic polymers: proteins and cellulose. In fact the fibrous state in nature is indeed most common, with the cell walls of most plant life being cellulosic and the muscle tissue of all animals being proteinaceous.

b. Man-Made Fibers

Man-made fibers are manufactured from naturally occurring polymers, synthetic polymers, and inorganic substances. Glass fiber is the only inorganic man-made fiber in common use today, although other inorganic and metallic fibers are being developed, principally for fiber-reinforced composite materials and certain special space and military applications. Table 2 lists the principal man-made fibers according to their generic name classification, as established by the Federal Trade Commission, and describes their chemical structures.

Fibers made from naturally occurring organic polymers can be either regenerated or derivative. In a regenerated fiber the chemical structure of the polymer in the final product is the same as it is in the natural state, whereas in a derivative fiber the chemical structure of the polymer in the fiber is not that of the naturally occurring polymer but an appropriate derivative of that polymer. Man-made fibers manufactured from synthetic polymers, frequently referred to as synthetic fibers, are numerous and constitute the most rapidly growing segment of the fiber industry. Both addition and condensation polymers are used, with polyamide, polyester, acrylic, polyolefin, and polyurethane

Table 2   Chemical Nature of Important Man-Made Fibers

Fiber	Chemical Nature
Acetate	Acetyl derivative of cellulose; triacetate designation can be used when not less than 92% of the cellulose OH groups are acetylated
Acrylic	Composed of not less than 85% by weight of acrylonitrile units
Modacrylic	Composed of less than 85% but at least 35% by weight of acrylonitrile units except when it is a rubber
Nylon	Long-chain synthetic polyamide with recurring amide groups as an integral part of chain; aliphatic and aromatic
Olefin	Composed of at least 85% by weight of ethylene, propylene, or other olefin except when amorphous as rubber; principally polypropylene
Polyester	Composed of at least 85% by weight of an ester of a dihydric alcohol and terephthalic acid
Rayon	Regenerated cellulose; regular rayon and polynosic fibers
Saran	Composed of at least 80% by weight of vinylidene chloride units
Spandex	Composed of at least 85% by weight of a segmented polyurethane
Vinyon	Composed of at least 85% by weight of vinyl chloride units

homopolymers and copolymers being the most extensively used in fiber manufacture.

Man-made fibers are produced by three principal spinning or extrusion
systems: melt, dry, and wet spinning. In melt spinning the polymer is heated
on a grid well above its melting point and after filtration and deaeration is
pumped through a die, known as a spinneret, that contains many small
holes. The diameter and cross-sectional shape of the filament are largely
determined by the geometric configuration of the holes in the spinneret. The
extruded stream of molten polymer passes through a cooling zone, where it
solidifies into a filament. Polyester, polyamide, and polyolefin fibers are
manufactured by this process. In dry spinning the polymer is dissolved in an
appropriate solvent and, after filtration and deaeration, extruded through a
spinneret into a heating zone. Solidification into a filament occurs in this
zone on evaporation of the solvent. Cellulose acetate and triacetate, and some
polyacrylic filaments are manufactured by this process. In wet spinning the
polymer or some chemical derivative of the polymer is dissolved in a suitable
solvent, and the solution is extruded through a spinneret into a liquid bath,
where the filament is formed. Usually at least three processes occur during
filament formation, the relative importance of each depending on the particular system under consideration. These processes are polymer regeneration, if
a chemical derivative of the polymer was formed, precipitation of the polymer,
and coagulation of the newly formed solid phase. Regenerated-cellulose
fibers (rayon) and certain polyacrylic filaments are manufactured by wet-
spinning methods.

The production of fibers from polymeric sheets or film by slitting and splitting processes is of increasing importance, and it may be expected that in the
immediate future many textilelike products will be produced by such techniques.

2. GEOMETRIC CHARACTERISTICS

The traditional methods of converting textile fibers into useful structures,
such as yarns and fabrics, are based on the principal naturally-occurring
fibers cotton and wool. The various systems of yarn manufacture are based on
alignment, parallelization, and attenuation of assemblies of these fibers into
linear, uniform strands, eventually resulting in a spun yarn. In a typical spun
yarn there may be anywhere from 50 to 100 individual fibers in the yarn's
cross section. It thus becomes readily apparent that the geometric and shape
properties of textile fibers are extremely important, and indeed such characteristics as length, linear density, cross-sectional area, cross-sectional shape,
and crimp have been the first fiber properties to be measured and used as a

means of indicating fiber quality. Crimp is a form factor that is used to designate longitudinal waviness and can be measured either in geometric terms, such as crimp frequency and amplitude, or in energistic terms, such as the force or energy required to uncrimp (straighten) the fiber.

The cross-sectional area of a fiber is evaluated most accurately by direct observation and measurement of a cross section under a microscope. Measurement of fiber diameters from observations of longitudinal views under a microscope is somewhat easier, but the ellipticity of the cross section in certain fibers can lead to serious errors. An important nondestructive means of determining the fiber's cross-sectional area is the vibroscope, which consists of a system for applying an oscillatory force of known frequency to a fiber held under tension and also a means of detecting mechanical resonance under the applied oscillatory force. From the values of frequency, tension, length, and bulk density the cross-sectional area can be calculated from the classical vibrating-string formula. Corrections taking into consideration fiber stiffness, cross-sectional shape, and nonuniformity along the length of the fiber have been developed (2).

Although the cross-sectional area of a fiber is occasionally used to designate fiber size or fineness, the more usual method of designating this important geometric property is by the linear density of the fiber in units of mass per unit length. If the bulk density (mass per unit volume) is known, the cross-sectional area can be computed by dividing the linear density by the bulk density. In fiber and textile terminology three special linear density units are used: the denier, the grex, and the tex. A denier, defined as unit of linear density, is the weight in grams of 9000 meters of the fibers. Thus 9000 meters of a 1-denier fiber weigh 1 gram. The grex value is defined as the weight in grams of 10,000 meters of the fiber. The tex value is the weight in grams of 1000 meters of the fiber. The tex system has been adopted by the American Society for Testing and Materials as the standard unit for designating the linear density of textile fibers (3).

There is a great deal of nonuniformity in the properties of naturally occurring fibers, both within one individual fiber and among fibers in a given population. In measuring geometric fiber properties, as well as bulk mechanical properties, it is important to indicate not only mean values but also some measure of the dispersion. Processability of fibers into yarns and the properties of all fiber assemblies, such as yarns and fabrics, are a function of the geometric and mechanical properties of the fibers, as well as the uniformity of these properties (4, 5).

Recent studies have indicated that the fiber's surface characteristics are of great importance in the cohesion of fiber assemblies. The surface roughness of fibers, which determines the effective total area of contact between fibers in an assembly, coupled with fiber friction, determines the mechanical properties of such important fiber assemblies as card webs, slivers, and loosely twisted roving (6, 7).

Values for some of the important geometric properties of the two principal naturally occurring fibers are shown in Table 3. In the case of man-made fibers geometric properties can be varied almost at will. A number of important developments in man-made-fiber technology are based on geometric properties of the fibers rather than on bulk mechanical properties or molecular structure.

Table 3   Typical Geometric Properties of Cotton and Wool Fibers

Property	Cotton	Wool
Mean length (in.)	0.8–1.5	3–6
Linear density (tex)	0.1–0.3	0.5–0.7
Cross-sectional area (cm²)	1×10^{-6}	4×10^{-6}
Cross-sectional shape	Elliptical, ribbonlike	Nearly circular
Crimp frequency	Nonuniform	10–20 per inch

3. MECHANICAL BEHAVIOR

The mechanical properties of fibers describe the response of a fiber to deforming loads under conditions that induce tension, compression, torsion, or bending. Torsional deformations are normally analyzed in terms of shearing stresses, whereas bending can be considered to produce simultaneous tensile and compressive stresses around a neutral fiber axis. Mechanical properties are usually evaluated under standard conditions of temperature and humidity (65% relative humidity, 70°F) and under closely specified conditions of load application. Specific values of various mechanical properties are obtained from stress-versus-strain or load-versus-deformation curves, which are graphical records of tensile, compressive, or shearing stresses as a function of deformation. In view of a fiber's shape and dimensions, these curves are usually evaluated under uniaxial tension. The procedure for obtaining a load-versus-deformation curve is to subject the fiber to increasing loads while recording the extension. Alternatively the fiber can be subjected to controlled extension, the force generated being recorded by some suitable device.

A schematic stress-versus-strain curve of an uncrimped, ideal textile fiber is shown in Fig. 1, and the stress-versus-strain curves of some of the more common textile fibers are shown in Fig. 2. When subjected to a load that places it under a stress higher than its yield value, the fiber will not completely return to its original length on removal of the deforming load. It will in this

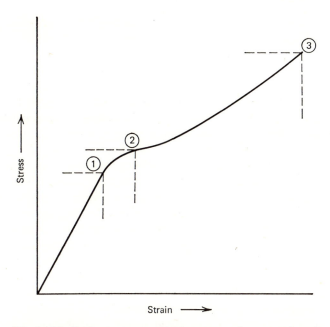

Fig. 1 Idealized stress-versus-strain curve of an uncrimped textile fiber. Point 1 is the proportional limit; point 2 is the yield point; point 3 is the break, or rupture, point.

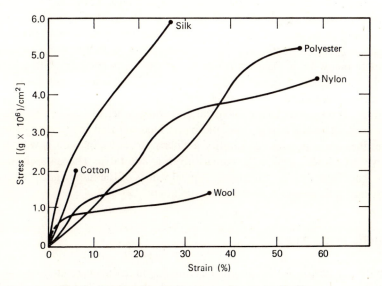

Fig. 2 Stress-versus-strain curves of some textile fibers.

case develop a permanent set. Not all of the original deformation will be retained as permanent set; some will be recoverable. Hamburger (8) divides the process of recovery from deformation into two major components. The first takes place immediately on removal of the deforming load and is called the immediate elastic deflection. The other is the delayed deflection, which can be further subdivided into a time-dependent recoverable component and a nonrecoverable component. The latter is also known as permanent set or secondary creep. Permanent, nonrecoverable deformation is a manifestation of plasticity or internal-stress-induced plastic flow.

Another aspect of plasticity is the time-dependent progressive deformation under constant load, known as creep. This process occurs when a fiber is loaded above the yield value and continues over several logarithmic decades of time. The extension under fixed load, or creep, is analogous to the relaxation of stress under fixed extension. Stress relaxation is the process whereby the stress generated as a result of a deformation is dissipated as a function of time. Both of these time-dependent processes are reflections of plastic flow resulting from various molecular motions in the fiber. As a direct consequence of creep and stress relaxation, the shape of a stress-versus-strain curve is strongly dependent on the rate of deformation.

An important aspect of the mechanical properties of fibers is their response to time-dependent deformations. Fibers are frequently subjected to conditions of loading and unloading at various frequencies, and it is important to know their response to these dynamic conditions. In this connection the fatigue properties of textile fibers are of particular importance. Prevorsek and Lyons (9) have extensively studied the fatiguing of various textile fibers in cyclic tension and have interpreted their results in terms of molecular processes. Fatigue failure of fibers, and presumably other semicrystalline, oriented polymers below their glass-transition temperature, is considered to proceed by a brittle-fracture process resulting from the stress-induced growth of sub-microscopic flaws and molecular imperfections.

Table 4 summarizes the mechanical and some other physical properties of the important textile fibers. It should be pointed out that in the case of man-made fibers it is possible to modify a fiber's structure in such a way as to achieve properties outside the ranges given in the table. The data shown in Table 4 are intended to indicate the typical values of the important properties of various natural and man-made textile fibers.

4. FIBER STRUCTURE

The organization of the polymer chains in three-dimensional space determines to a large extent the chemical, physical, and particularly the mechanical properties of fibers. Characteristic fiber properties are achieved through the

Table 4 Chemical and Physical Properties of Important Textile Fibers

Fiber	Tenacity at Break (g/denier)		Extension At Break (%)		Elastic Modulus (g/denier)		Density	Moisture Regain at 65% RH (%)	Approximate Volume Swelling in Water (%)	Thermal Stability
	Dry	Wet	Dry	Wet	Dry	Wet				
Cotton	3–5	3–6	4–9	5–12	40–90	30–60	1.54	7.0–8.0	40	Decomposes at 150°C
Wool	1–1.7	0.8–1.6	25–35	25–50	25–35	20–30	1.32	15.0–17.0	35	Decomposes at 130°C
Silk	3–5	3–4	20–25	30	80–120	80–120	1.25	11.0	30–40	Decomposes at 130°C
Rayon:										
Regular	1.5–4.0	0.8–2.5	10–30	15–40	40–70	15–35	1.52	11–13	45–85	Decomposes at 150°C
Polynosic	3.0–4.0	2.5–3.5	7–10	9–12	45–70	35–50	1.54	10–12	40	Decomposes at 150°C
Acetate:										
Secondary	1.0–1.5	0.7–1.1	25–40	30–45	25–40	20–35	1.33	6.4	10–30	Softens at 165°C
Triacetate	1.2–2.1	0.8–1.4	20–28	30–40	35–40	25–35	1.32	4.5	5–15	Softens at 235°C
Acrylic	2.5–5.0	2.5–4.5	15–30	20–35	40–70	30–60	1.17	1.6–2.5	2–5	Softens at 235°C
Nylon	4.8–8.0	3.5–7.0	15–35	20–40	30–50	20–40	1.14	4.0–4.5	2–10	Melts at 220–260°C
Polyester	3.5–6.0	3.5–6.0	15–40	15–40	90	90	1.38	0.4	None	Melts at 250°C
Polypropylene	5.0–9.0	5.0–9.0	15–30	15–30	30–45	30–45	0.90	0	None	Softens at 140°C
Spandex	0.9	0.9	600	600	0.05	0.05	1.21	1.3		Softens at 225–270°C

development of an intermolecular chain organization that can generically be described as highly oriented and semicrystalline. The extent or degree of crystallinity can be estimated by X-ray diffraction techniques, but it is important to note that structural regularity or lateral order can exist even in noncrystalline systems. The ability to diffract X-rays is dependent on the size and perfection of the ordered domain, and a lack of X-ray crystallinity should not be taken to mean a total lack of order. In addition to X-ray techniques for the determination of the degree of crystallinity, other methods for estimating the degree of order are equilibrium moisture sorption, accessibility to chemical attack under standard conditions, density, deuterium–hydrogen exchange, calorimetric methods, and nuclear-magnetic-resonance (NMR) spectroscopy.

The configuration of the polymer molecules in the crystalline, or highly ordered, regions of fibers can be linear or helical. In linear configurations—such as those found in polyethylene, polyesters, polyamides, and cellulose—the chains can be either fully extended or folded, depending on the extent of mechanical drawing and thermochemical annealing. The fact that these and other polymer molecules can crystallize from solution or from the melt in a folded configuration both in single-crystal and in polycrystalline form has been established unequivocally (10). As discussed below, the extent of chain folding in commercial fibers remains the subject of some controversy. The crystalline regions of other polymer molecules—for example, polyisobutene, polytetrafluoroethylene, isotactic polypropylene, and α-polypeptides—are not linear but helical. The helix is usually stabilized by both intermolecular and intramolecular interactions. In some cases the linear or helical molecules associate via lateral forces into sheet structures, the polyamides and certain polypeptides providing the best examples.

Although the crystalline, or highly ordered, regions of a fiber have been the primary target of structural investigations, the disordered, or amorphous, regions may indeed play a dominant role in terms of fiber properties. Characterization of the amorphous region is quite difficult, and most of the available physicochemical methods are not suitable. The existence of a glass-transition temperature is to be associated with the amorphous region; this suggests that these regions should not be considered to be totally devoid of structure. Although it is semantically inconsistent to speak of structure in an amorphous region, it is important to recognize that it does indeed exist.

It may be appropriate to discontinue the use of the term "amorphous," as many workers have, and to think in terms of regions with a low level of lateral order. Indeed the concept of a lateral-order distribution, first proposed for cellulosic fibers (11), may be the best means of describing the three-dimensional organization of polymer chains in semicrystalline or polycrystalline materials like fibers. Some studies of the response of nylon fibers to various chemical treatments have been interpreted in these terms (12, 13).

The coexistence of crystalline and amorphous regions, being treated in the extreme as two separate phases, has provided the basis for various theories of fiber structure since the 19th century. The crystalline region, frequently referred to as a micelle, was taken to be a chemically inaccessible domain, which from a mechanics point of view acted as a stiff, rodlike reinforcement for the amorphous region. As molecular dimensions became better established by both chemical and physical methods, it became evident that a polymer molecule could not be accommodated in one discrete crystallite or micelle; this in turn gave rise to the concept of a fringed micelle. In the fringed-micelle theory polymer molecules may pass through a crystallite into an amorphous region and possibly into another crystallite. The most important feature of this concept is the disappearance of a sharp boundary between crystalline and amorphous regions, and the postulation of regions of intermediate levels of order. The crystallite ceased being a totally inaccessible domain with perfect structural regularity, and the amorphous region was no longer considered to be a liquidlike matrix. A schematic visualization of the fringed-micelle theory and other modifications of the polyphase concept are shown in Fig. 3.

With the discovery in 1956 that polymer molecules could fold on themselves (14, 15), the passage of a polymer molecule from a crystalline region to an

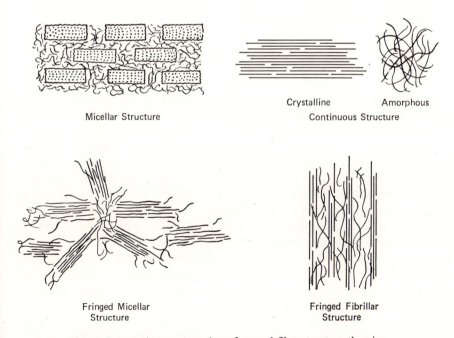

Micellar Structure

Crystalline   Amorphous
Continuous Structure

Fringed Micellar
Structure

Fringed Fibrillar
Structure

Fig. 3 Schematic representation of several fiber-structure theories.

amorphous one was no longer a necessary means of accommodating molecular dimensions. Many workers began considering fibers as fully crystalline materials with varying levels of crystalline imperfections (16). Molecular dislocations, impurities, and chain ends are the principal sources of these imperfections, which from a chemical accessibility as well as from a mechanics point of view provide the means of interpreting the macroscopic properties previously associated with amorphous regions.

At the present time fiber structure is viewed as somewhere intermediate between the fringed polyphase model and the fully crystalline folded model. The crystalline regions certainly do contain polymer-chain folds, and imperfections in these regions may be taken to account for many fiber properties. On the other hand, the accessibility of fibers to large and bulky molecules, such as other polymers and dyes, does suggest the existence of fairly large domains with at least localized regions of low density. Nevertheless, the existence of polymer-chain folds in most synthetic fibers is undisputed, and infrared-spectroscopy evidence of molecular folds in polyesters (17) and polyamides (18) has recently been presented. The role of chain folding in fibers and in oriented, semicrystalline polymer systems in general has recently been reviewed by Peterlin (19).

5. ORIENTATION AND DRAWING

Fibers, natural and man-made, have one particular structural feature in common: a preferential orientation of their elemental units with respect to the fiber axis. The elemental units may be either polymer molecules or various aggregates, such as crystallites and micelles. In the case of natural fibers, as will be discussed later, these elemental or structural units may be of even higher organization, usually referred to as fibrils.

In the case of man-made fibers the preferential orientation is achieved through mechanical drawing operations, during which the filament is extended, immediately after extrusion, several times its initial length. The draw ratio is the extension factor, which can be defined as the length of drawn filament per unit length of undrawn filament. Drawing has the dual purpose of orienting structural elements with respect to the fiber axis in order to bring these elements into optimal stress-bearing positions and to allow the development of a three-dimensional structural regularity. Undrawn filaments have little molecular orientation and at best only low levels of crystallinity or lateral order. Drawing must be performed above the glass-transition temperature in order that plastic-type molecular flow may take place. The effects of drawing on structure as revealed by X-ray diffraction are shown in Fig. 4. Other methods of measuring orientation are birefringence, infrared dichroism, magnetic anisotropy, sonic velocity, and mechanical response. The effects of

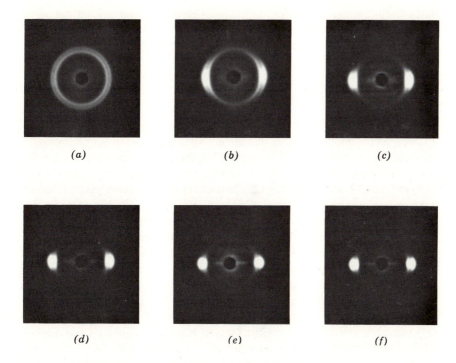

Fig. 4 X-ray diffraction patterns of nylon 66 with the following draw ratios: (*a*) 1.00, (*b*) 2.02, (*c*) 2.99, (*d*) 3.99, (*e*) 5.02, and (*f*) 5.99.

Table 5 Effects of Drawing on the Properties of Nylon 66 Fiber[a]

Draw Ratio	Birefringence	Elastic Modulus[b] [(dynes/cm^2) $\times 10^{10}$]	Tenacity at Break (g/tex)
1	0.0083	1.97	10.1
2	0.0330	2.74	15.6
3	0.0552	3.70	28.1
4	0.0590	4.59	38.4
5	0.0638	5.77	56.8
6	0.0690	6.74	68.2

[a] Data from Sakuma and Rebenfeld (21).
[b] Pulse-propagation technique.

drawing on fiber properties, to be associated with the development of an oriented fiber structure, are illustrated in Table 5. A direct outgrowth of orientation is the high level of anisotropy in nearly all fiber characteristics, particularly swelling behavior, mechanical response, and optical properties (20–22). Statton (23) has recently reviewed various concepts and methods of measurement of the orientation of molecules in fibers.

6. CONTRACTION AND ANNEALING

When exposed to high temperatures or to certain chemical environments, an oriented thermoplastic fiber will undergo a contraction in length unless it is mechanically restrained from doing so. The thermal or chemical contraction is a manifestation of molecular processes, and it has been shown in the case of nylon 66 (24) and polyester (25) that chain folding takes place as an important part of those processes. Orientation is retained unless temperatures near the melting point are reached. Chemically induced contraction is dependent on the nature of the chemical system and temperature. One of the best studied systems is nylon 66 in phenolic solutions (21, 26). The sorption of phenol and related compounds on the amide site disrupts intermolecular hydrogen bonds, thereby permitting molecular motion and the transformation of polymer chains into more stable, presumably folded, configurations.

The kinetics of the contraction process has been studied by Jacobs et al. (27), who presented a mathematical model based on sequential bond-breakdown and bond-re-formation reactions. By systematic variation of the phenolic reagents, Jacobs (28) postulated the existence of hydrophobic bonds as important intermolecular forces contributing to the overall cohesive structure of nylon 66.

Prevorsek and Tobolsky (29) have studied the nonflow shrinkage of fibers as a function of temperature and degree of orientation, and have interpreted their results in terms of a theory based on molecular dimensions. A microscopical method for the measurement of thermal shrinkage under stressfree conditions has been developed (30) and makes it possible to observe the course of the thermal shrinkage process. The shrinkage and melting of a γ-irradiated (slightly crosslinked) polyethylene fiber are shown in Fig. 5.

Ribnick (31), who studied the kinetics of the thermal contraction of nylon 66 and polyester, points out the similarity in form of the kinetics of fiber contraction and polymer crystallization. This fact was taken as a confirmation of the previously presented view (24) that shrinkage is a recrystallization process, presumably from a predominantly extended chain configuration to a predominantly folded one.

In many commercial operations thermoplastic fibers are subjected to either thermal or chemical treatments while under mechanical constraint to prevent shrinkage. Such annealing treatments involve molecular motions and processes similar to those occurring during fiber contraction. The modified or

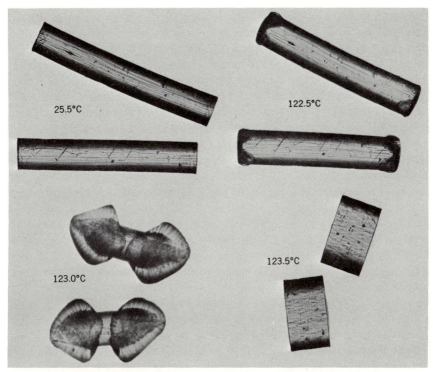

Fig. 5   Thermal shrinkage of γ-irradiated polyethylene filaments (29).

improved fiber structure resulting from annealing is reflected in modified chemical and physical properties. The heat-setting of thermoplastic fibers, frequently in a nonlinear fiber configuration, is a type of annealing where polymer recrystallization takes place. The crimping and bulking of polyamide, polyester, and acrylic yarn are examples of important industrial processes based on these molecular transformations.

7. SPANDEX ELASTIC FIBERS

In many textile applications it is desirable to have a fiber that will recover completely from long-range deformations immediately on removal of the deforming force. Fibers made from natural and synthetic latex rubbers have for years been the only fibers from which elastic fabrics could be manufactured. Although the elasticity of these fibers is high, the force of recovery from deformation and ultimate strength are not quite adequate. Their susceptibility to oxidative chemical degradation and particularly their poor dyeability are further serious drawbacks.

According to the classical theory of rubberlike elasticity, in an elastic fiber polymer chains must be flexible and easily extensible to an oriented configuration under the application of a tensile load to the fiber, but at the same time they must be able to return spontaneously to their disordered, more nearly random-coil, state on removal of the load. Ideally a synthetic elastic fiber should contain soft, extensible polymer segments and hard tie regions that bind the chains together to provide the retractive forces. These structural requirements were found in polyurethane chemistry (32, 33), and commercial Spandex (elastic) fibers are those in which the fiber-forming substance is a long-chain synthetic polymer composed at least 85% by weight of a segmented polyurethane. The term "segmented" refers to alternating soft and hard regions in the polymer structure. The formation of the segmented polyurethane structure first involves the formation of a polyester or polyether flexible, linear polyglycol, referred to as a macroglycol, and its subsequent reaction with excess diisocyanate, usually an aromatic one, resulting in the formation of an isocyanate-terminated soft-segment prepolymer. In the next step the hard segments are formed by reacting the isocyanate-terminated prepolymer with low-molecular-weight glycols or diamines. This results in a polymer with hydrogen-bonding sites through either urethane or urea groups, which provide the tie points in the segmented polyurethane structure responsible for the long-range elasticity.

Spandex fibers are usually formed as continuous filaments by traditional dry- and wet-spinning processes, although some use has been made of reaction spinning (34, 35). In reaction spinning the diisocyanate-terminated soft-segment prepolymer is extruded into an environment containing the glycol or diamine, and the reaction wherein the urethane or urea groupings are created takes place after fiber formation.

Spandex fibers have high extensibilities, low elastic moduli, and approximately 100% elastic recoveries from large deformations. These fibers are not only resistant to chemical degradation, light, and ultraviolet radiation but also adequate in thermal stability, their softening temperatures being higher than 200°C.

Spandex fibers are usually processed into fabrics as covered yarns. Covering the elastic fiber with either staple or continuous-filament hard fibers, such as polyester, polyamide, cotton, or wool, protects the elastic fiber and modifies the physical and chemical properties of the composite. Core spinning is a particular means of forming a composite yarn wherein partially extended Spandex continuous filament is fed into a spinning frame together with staple hard fibers. In a core-spun composite yarn the Spandex filaments form an inner core, which is surrounded with a sheath of staple fibers. Due to the partial extension of the Spandex filaments before core spinning, a fabric woven or knitted from these yarns shrinks during wet finishing. The final fabrics have high stretch and, more particularly, high recoverability.

8. DYEING

The sorption of molecular species from solution or from the vapor phase is an important chemical property of textile fibers. The ability to absorb dyestuffs is a requirement for all textile fibers, and no new fiber can look toward ready acceptance if its chemical structure does not allow the absorption of at least one of the common classes of dyes.

Dyes are complex organic molecules that can be classified according to their ionic character as cationic, anionic, and nonionic, although classifications according to organochemical structure and dye-application techniques are frequently more useful and common. Table 6 lists the common classes of dyestuffs and the types of fiber for which each is used.

Table 6   Classification and Application of Dyes

Dye Class	Description	Fiber Application
Basic	Salts of organic bases (cationic)	Silk, wool, acrylic
Acid	Sulfonic or carboxy acid salts (anionic)	Wool, silk, nylon, acrylic, polypropylene, Spandex
Direct	Contain azo group; sulfonic or carboxy acid salts	Cotton, rayon, and other cellulosic fibers
Mordant	Anthracene derivatives; no affinity; fibers must be pretreated with metal oxides	Cellulosic and proteinaceous
Sulfur	Contain sulfur; applied in reduced form and oxidized on fiber	Cotton, rayon
Azoic	Insoluble pigments formed within the fiber	Cotton, rayon, nylon, polyester
Vat	Insoluble; applied in reduced "leuco" form and oxidized on fiber	Cotton, rayon, polyester
Disperse	Nonionic, limited solubility; applied from dispersions	Acetate, triacetate, polyester, nylon, Spandex, acrylic
Reactive	Forms covalent bonds with fiber	Cotton, rayon, wool

There are three aspects of dyeing that must be considered: dye affinity, rate of dyeing, and properties of the fiber–dye complex. The affinities of various dyes for textile fibers are thermodynamic quantities that are determined only by the chemical structures of the fiber and the dye. Unless there is some affinity, thermodynamically defined as the difference in the chemical potential of the dye in the fiber and in the external liquid or vapor phase, there can be no useful fiber–dye complex formation.

Of equal importance, particularly from a practical point of view, is the rate of dyeing. Since dyeing is normally carried out from solutions, one can visualize the following important steps in the process:

1. Diffusion of dye in the liquid phase.
2. The sorption of dye at the fiber surface.
3. Dye transport at the interface.
4. Dye desorption at the surface.
5. Dye diffusion through the fiber bulk.

The rate-controlling step in this sequence is invariably the diffusion of dye through the fiber. The driving force for this diffusion is the concentration gradient, and in general the kinetics of dyeing is influenced by dye concentration, degree of dye aggregation in both the liquid and solid phases, pH, temperature, and presence of ionic species and other dyeing additives. The rate of dyeing is also markedly influenced by the fiber's supramolecular organization, with orientation, crystallinity, and lateral order exerting important effects.

Finally consideration must be given to the properties of the fiber–dye complex. Stability of these complexes, referred to as fastness, is a prime requirement. The fiber–dye complex must be colored to an acceptable depth and must be capable of withstanding exposure to light, actinic rays, repeated washing, dry cleaning, and other degrading influences. Washfastness and lightfastness are of particular importance, and it should be noted that these and other fastness characteristics are properties of the fiber–dye complex, not of the dyestuff alone. An excellent review of the physicochemical aspects of dyeing has recently been presented by Valko (36).

9. GROWTH AND MORPHOLOGY OF COTTON FIBERS

The cotton fiber is a single biological cell that grows as a seed hair on the plant belonging to the genus *Gossypium* of the tribe Hibisceae. Since the cotton fiber is essentially cellulosic in nature, its organochemical structure is best described as that of cellulose poly(1,4-β-D-anhydroglucopyranose). The degree of polymerization of the cellulose obtained from purified cotton fibers that have not been subjected to any degradation other than the incidental effects of the purification process is estimated to be approximately 3000. This value may be an order of magnitude higher in the native cotton fiber, but some degradation during isolation and purification cannot be avoided. The chains are closely packed, and, in view of strong intermolecular hydrogen bonds, cellulose in its native state from cotton is highly crystalline.

The fiber grows in several distinct stages, each contributing to its complex, heterogeneous structure. Growth proceeds as a lengthening of an epidermal cell for about 13 to 20 days, in the form of a thin membrane (0.2 micron) called the primary wall. Fiber thickening or development takes place after full length has been achieved and proceeds via the deposition of cellulose on the inner surface of the primary wall over a period of 25 to 40 days. This material, referred to as the secondary wall, never quite fills the cell but leaves a center canal called the lumen. When the cotton boll opens, drying and loss of water result in a collapse of the lumen, and the lint hair, or cotton fiber, assumes a characteristic flat, ribbonlike shape, with frequent twists or convolutions (37).

The deposition of cellulose during the development of the secondary wall is the crucial step in the growth of the cotton fiber. The cellulose fibrils are deposited in a characteristic helical fashion, the angle of the helix with respect to the fiber axis being constant through the cross section and along the length of the fiber. The helical angle was thought to vary for different types, varieties, and strains of cotton; however, recent studies have shown that this angle is the same for all cottons and that apparent variations are due to the superposition of the convolution and the helical angles (38).

The fibrillar nature of the cellulose in the secondary wall of the cotton fiber has been unequivocally established by electron microscopy (39). The fibrils are further associated or aggregated in laminar sheets, or growth layers, which can be readily seen in a cross section of a highly swollen fiber under ordinary high-magnification microscopy, as shown in Fig. 6. The growth layers are in some way associated with the day–night cycle and the corresponding variations in temperature, absolute moisture content, relative humidity, and light intensity. The exact manner in which these and possibly other variables affect the growth-layer characteristics of cotton fibers has been the subject of several studies, which have not as yet fully clarified the complexities of this problem.

The angle of the fibrils with respect to the fiber axis, frequently referred to as fibrillar orientation, is one of the cotton fiber's most important structural features and is readily evaluated by X-ray diffraction techniques. These methods are based on an azimuthal examination of the cellulose 002 diffraction arc and are expressed in angular measures of that arc (40). Many variations in cotton-fiber properties, particularly those describing response to mechanical deformation, can be related to the fibrillar orientation (41, 42). Cotton fibers with a high degree of fibrillar orientation, as characterized by a low X-ray angle, have high elastic moduli and low extensions at break, whereas fibers with low degrees of orientation, or a large X-ray angle, have

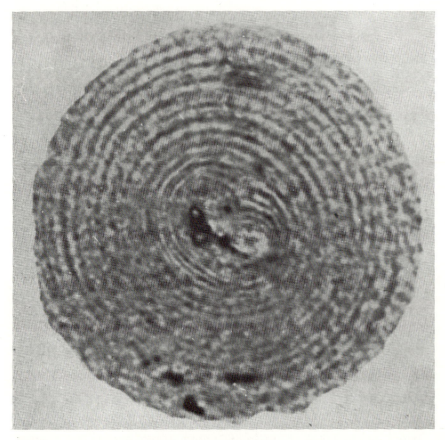

Fig. 6 Cross section of a highly swollen cotton fiber, showing growth layers. [Photomicrograph by the Southern Utilization Research and Development Division of the U.S. Department of Agriculture, New Orleans, Louisiana (39).

low moduli and high extensions at break. Other mechanical properties—such as strength, resilience, work recovery, and some chemical and physical characteristics—are also correlated with the X-ray angle. Some of these relationships are shown in Fig. 7.

In general the correlations between cotton-fiber properties and fibrillar orientation are the same as those for man-made fibers, where orientation with respect to the fiber axis is achieved by drawing operations. In making this comparison cotton would have to be considered as a highly crystalline fiber with almost the ultimate orientation achievable by the usual drawing operations. Other naturally occurring vegetable fibers, such as jute, ramie, linen, and hemp, have similar fibrillar structures, with well-defined helical configurations and characteristic helical angles. These bast fibers, derived from the

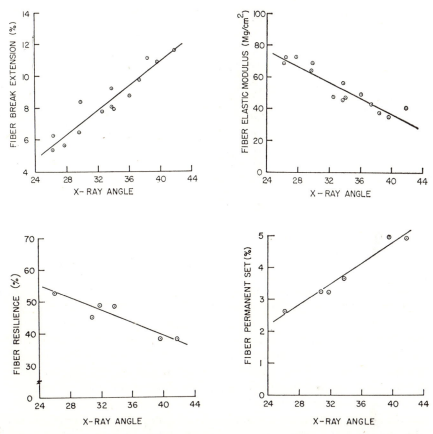

Fig. 7 Relationships between the mechanical properties of single cotton fibers and the fibrillar orientation as estimated by the X-ray angle.

stem of the plant, have angles ranging from 5 to 15°. Thus they have a higher degree of fibrillar orientation with respect to the fiber axis than cotton and correspondingly lower extensibilities and higher elastic moduli.

Although the helical angle of the fibrils with respect to the fiber axis is constant over the full length of the fiber, the sense of the helix reverses many times, as shown schematically in Fig. 8. These points of structural reversal—a unique feature of cotton fibers, not found in other native cellulosic fibers—occur anywhere from 30 to 100 times and can be seen under elliptically polarized illumination in light microscopy as complementary color bands. The reversal frequency is primarily a varietal characteristic, but it is also dependent on environmental growth conditions. The distances between points of reversal appear to be random, but it has been established that the reversal frequency is higher toward the tip of the fiber (43).

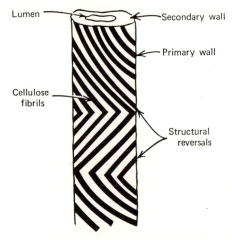

Fig. 8   Schematic diagram of a cotton fiber.

The reversal in the sense of the helix can be considered a weak spot in the cotton fiber and therefore a structural imperfection. When single cotton fibers are subjected to tensile deformations leading to rupture, approximately 45% of the fibers fail at a reversal point. Considering the total length of the fibers involved in regions of reversal, it would be expected that approximately 15% of the fibers would break at reversals and 85% would break between reversals if the rupture were random. The fact that 45% of the fibers break at reversals indicates that rupture is not random and that the helical reversals are points of weakness. This preferential breakage at helical reversals suggests that the fibrillar structure at these points is under strain and that the additional strain imposed in the tensile test causes fiber rupture to take place preferentially at these reversals due to stress concentrations.

Studies by Raes and co-workers (44) of fiber strength and of the distribution of the frequency of structural reversals have led to the conclusion that there must be other structural discontinuities between reversals similar in properties to those of the visible reversals. These studies have provided the basis of a fiber-growth model based on the random deposition of fibrils during the development of the secondary wall.

10. CROSSLINKING OF CELLULOSIC FIBERS

Textile fibers in final fabric form are subjected to various chemical treatments, known as wet-finishing procedures. These range anywhere from simple washing procedures to complex and highly sophisticated chemical modification reactions. In the case of cellulosic fibers, both natural and regenerated, chemical modification reactions are of particular importance as a means of imparting certain desirable properties to the fabric. The reaction of cellulosic fibers with bifunctional reagents to introduce covalent crosslinks (reaction I) forms the basis of industrial processes by which shrink-resistance, wash-and-wear, and durable-press properties are imparted to fabrics. The most commonly used bifunctional reagents are dimethylolurea, the dimethylol derivative of cyclic alkyleneureas, triazones, triazines, and carbamates.

$$R_{cell}-OH + OHCH_2-R-CH_2OH \rightarrow R_{cell}-OCH_2-R-CH_2O-R_{cell} \qquad (I)$$

Definitive chemical evidence for the existence of covalent cross links has not been obtained, but the properties of the product are consistent with the view that at least some crosslinking is achieved, although a large portion of the reagent may be present in the form of bulk-deposited polymer or may have reacted with the cellulose monofunctionally. Since the crosslinking reactions are invariably carried out on fabrics and not on bulk or loosely assembled fibers, the results of the reactions are normally evaluated in terms of such commercially important fabric characteristics as wrinkle recovery, abrasion resistance, tensile strength, and extensibility. In the case of cotton the crosslinking reactions increase the fabric's resilience, which can be estimated by wrinkle recovery, and decrease its durability, as estimated by abrasion resistance and energy to rupture (45). Functional relationships between fabric properties and the extent of crosslinking, as determined by the add-on of bifunctional reagent, are shown in Fig. 9.

Although the crosslinking reactions are normally carried out on fabrics, the properties of the individual fibers are modified in such a way as to indicate that crosslinking has indeed taken place. The stress-versus-strain curves for untreated and crosslinked rayon and cotton fibers are shown in Fig. 10. The difference in the response of cotton and rayon fibers to the crosslinking treatment is a reflection of the difference in their structures (46, 47). Cotton, being a highly crystalline fiber, is embrittled by covalent crosslinks, but the less ordered rayon is reinforced by them.

Several attempts have been made to characterize more definitively the chemical structure of crosslinked cellulosic fibers. However, the location of the crosslinks within the fiber's structure, along the cellulose chain, and within the monomeric anhydroglucose unit has not been established.

Several important improvements in crosslinking procedures, particularly for cotton, are based on the disruption of the intermolecular organization of the fiber prior to crosslinking by swelling treatments (48) and the deposition of polymer (49). Preferential crosslinking in the core of the fiber has been shown to be a useful method of minimizing the deleterious effects of crosslinking on fabric properties (50, 51).

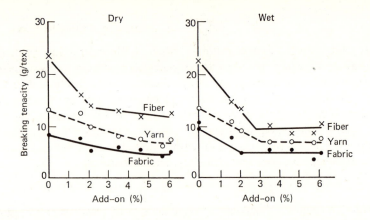

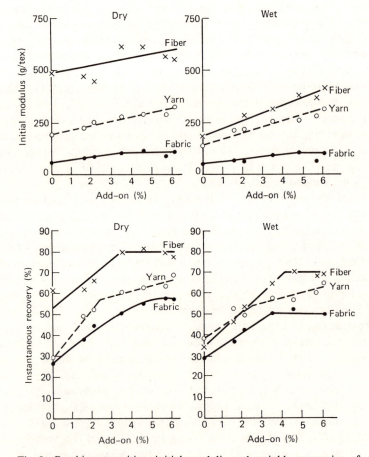

Fig. 9 Breaking tenacities, initial moduli, and wrinkle recoveries of cotton fibers, yarns, and fabrics as functions of the extent of crosslinking.

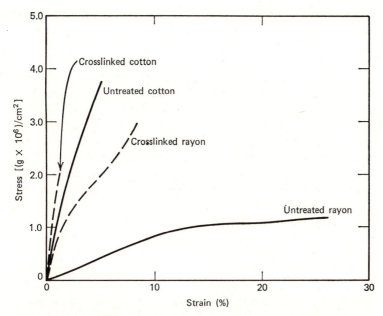

Fig. 10 Stress-versus-strain curves for crosslinked and untreated cotton and rayon fibers in the wet state.

11. MORPHOLOGY OF KERATIN FIBERS

Keratin fibers, of which wool and human hair are the best known examples, are undoubtedly the most complex fibrous structures that have received scientific attention and industrial acceptance. The keratin protein is a mixture of polypeptide chains composed of at least 18 amino acids (52). Approximately 33% (mole %) of the amino acids in wool keratin are neutral ones, including glycine, alanine, phenylalanine, leucine, isoleucine, and valine; 23% are hydroxylated amino acids, including serine, threonine, and tyrosine; 18% are acidic amino acids, including aspartic and glutamic acids; 11% are basic amino acids, including lysine, arginine, and histidine; and the remainder are the important amino acids cystine, methionine, proline, and tryptophan.

The amino acid composition varies somewhat with the source of the keratin. One of the most significant differences lies in the cystine content, which, for example, is considerably higher in human hair than it is in wool.

At the macromolecular level of organization the polypeptide backbone is in a helical configuration in the crystalline regions of α-keratin, that is, in an unextended keratin fiber (53). The helix has 3.7 amino acid residues per turn and is intramolecularly and intermolecularly stabilized by forces arising from the peptide group in the backbone chain and the various functional side groups. The stabilizing disulfide bonds are a unique structural feature of α-keratin, and therefore the cystine content, which determines the degree of covalent-bond stabilization or crosslinking, is of paramount importance. The α-helices are further coiled together in a helical configuration, forming a multistranded rope, referred to as a protofibril.

The helical configuration of keratin may be disrupted by extending the fiber and essentially pulling out the α-helix into relatively straight polypeptide chains, which are referred to as the β-configuration (54). The α-β transformation can be readily observed by X-ray techniques and is completely reversible unless specific setting reactions are introduced. In general these setting reactions require a stabilization of the extended configuration involving some operation on the disulfide crosslink.

When a wool fiber is subjected to tensile deformation, the stress-versus-strain curve, shown schematically in Fig. 11, reflects the response of the various structural elements in the fiber. The stress-versus-strain curve reveals three distinctly different regions, which may be associated with various molecular motions and transformations. In the initial stages of fiber extension the behavior is essentially Hookean and only bond-angle deformations are involved. At a well-defined strain, usually 2%, the elastic limit is reached, and the fiber's stress-versus-strain curve enters the yield region. This region of the stress-versus-strain curve represents the unfolding of the α-helices to the β-configuration. The process takes place against very little resistance, and the yield region of the stress-versus-strain curve is therefore essentially flat. At approximately 30% extension the stress-versus-strain curve reaches the post-yield region, in which there is a sudden increase in the stress with increasing

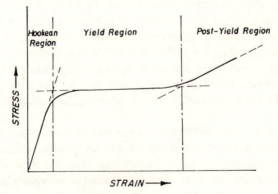

Fig. 11   Typical stress-versus-strain curve for wool fiber, showing three distinct regions of viscoelastic behavior.

strain, resulting in fiber rupture at approximately 50% strain, before completion of the $\alpha-\beta$ transformation. The postyield region represents the involvement of highly crosslinked domains in the fiber, which interrupts the $\alpha-\beta$ transformation taking place under the applied strain. Molecular interpretations of the stress-versus-strain curves of keratin fibers have been the subject of numerous investigations and reviews (55–57).

Microscopic observation of a wool fiber in cross section reveals the existence of many individual cells in the cortex of the fiber. This cellular nature of a wool fiber is consistent with the fact that keratin was once living tissue. The cortical cells are spindle shaped, with pointed ends, and are approximately 100 microns in length and 5 microns in diameter. The existence of intercellular material has not been established. A more detailed examination of the cortical cells reveals that they are composed of discrete microfibrils embedded in a matrix (58). In this visualization of keratin-fiber structure the microfibrils are presumably fully crystalline arrangements of the previously mentioned protofibrils. The nature of the matricullar material between the microfibrils has not been fully established, but presumably the polypeptide chains in the matrix are not crystalline and represent regions of relatively high sulfur content. Filshie and Rogers (59) postulated that the microfibrils have a further substructure in that the protofibrils are arranged in a characteristic "9 + 2" pattern. In this pattern nine protofibrils are located in the periphery and two in the center of each microfibril. This "9 + 2" arrangement has come under severe attack from a number of investigators (60, 61), although most recent evidence suggests the existence of some discrete annular and core structure in the microfibrils (62).

A unique structural feature is the fact that the individual cortical cells in certain keratin fibers can be classified into two distinct categories on the basis of their response to chemical treatments (63, 64). The portion of the cortex that is more receptive to dyestuffs, and more readily attacked by enzymes and other chemical reagents, is referred to as the orthocortex; the less dye-receptive and less reactive portion is called the paracortex. The paracortex is a higher sulfur-containing segment than the orthocortex, and since the sulfur arises principally from cystine, it can be assumed that the paracortex is more highly crosslinked. The fibrillar structure is essentially the same for the cells in the orthocortex and the paracortex, and it is probable that the different chemical structures implied by the higher sulfur content or higher disulfide-crosslink density in the paracortex is to be associated with the matrix.

The orthocortex and the paracortex of fine wool fibers are disposed along the fiber's length in such a way that they exist as hemicylinders wound around each other or twisted on themselves, as shown schematically in Fig. 12. The intertwining of the two cortical components takes place in phase with the natural crimp of the wool fiber, so that the orthocortex is always on the

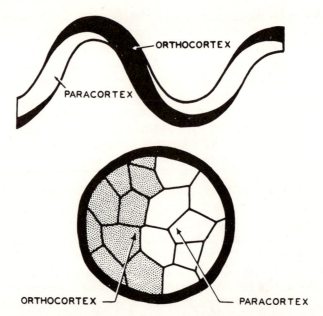

Fig. 12   Schematic representation of bilateral asymmetry and crimp in a wool fiber.

outside of a crimp curvature (65, 66). It is in fact this bilateral structure that is responsible for the crimp. During the growth of the fiber, at the critical stage of keratinization, the wool fiber acts as a twisted bimetallic strip with differential expansion properties. The result, after keratinization, is a structurally crimped wool fiber whose equilibrium state is the crimped configuration.

It should be noted that it is the natural crimp of wool fibers that is responsible for the bulky character of wool fabrics and that the production of highly crimped and bulked synthetic fibers is based on the bilateral principle of keratin fibers. By extruding, as a single filament, two polymers with different response characteristics to heat or moisture, a potentially bilateral fiber can be formed. Exposure of such a fiber to moisture at elevated temperatures will cause one of the polymer components to shrink to a greater extent than the other, resulting in a highly twisted and curled fiber (67).

Not all keratin fibers exhibit cortical asymmetry; for example, most human hair is cortically homogeneous and, by analogy with asymmetrical keratin fibers, is classified as paracortical in structure. It is difficult to isolate keratin fibers that are essentially all orthocortical in view of their relatively high reactivity. Cortically symmetrical fibers are not crimped, and this provides further evidence that the crimp of keratin fibers is based on the fiber's internal structure. Cortically asymmetrical fibers need not be bilaterally asymmetrical, and keratin fibers in which one component is disposed as an inner core surrounded by a layer of the other cortical component are quite common.

Under certain conditions of temperature and chemical environment, keratin fibers undergo a decrease in length, a phenomenon known as super-contraction. The process is equivalent to denaturation in globular proteins, and it should be noted that the longitudinal contraction of ordered, oriented fibers is by no means restricted to keratin. As already noted, when exposed to aqueous phenol solutions, a nylon fiber undergoes major contractions in length, the magnitude depending on phenol concentration, temperature, and extent of drawing. Supercontraction occurs in hydrogen-bond-breaking systems, such as concentrated solutions of alkali-metal halides, urea, and phenolic compounds. The effects of keratin-fiber structure, particularly the concentration of free sulfhydryl groups and the disulfide-bond density, on the supercontraction process have recently been reviewed by Rebenfeld(68).

REFERENCES

1. *Textile Organon* **40**, 97 (1969).

2. D. J. Montgomery and W. T. Milloway, *Textile Res. J.* **22**, 729 (1952).

3. H. D. Smith, *ASTM. Proc.* **44,** 542 (1944).

4. H. Wakeham and W. P. Virgin, *Textile Res. J.* **26**, 177 (1956).

5. M. M. Platt, *Textile Res. J.* **20**, 1, 519 (1950).

6. S. C. Scheier and W. J. Lyons, *Textile Res. J.* **34**, 410 (1964).

7. F. L. Scardino and W. J. Lyons, *Textile Res. J.* **37**, 874, 982, 1005 (1967).

8. W. J. Hamburger, *Textile Res. J.* **18**, 102 (1948).

9. D. Prevorsek and W. J. Lyons, *J. Appl. Phys.* **35**, 3152 (1964).

10. P. H. Geil, *Polymer Single Crystals*, Interscience, New York, 1963.

11. R. H. Marchessault and J. A. Howsmon, *Textile Res. J.* **27**, 30 (1957).

12. A. Koshimo and T. Kakishita, *J. Appl. Polymer Sci.* **9**, 91 (1965).

13. A. Koshimo and T. Tagawa, *J. Appl. Polymer Sci.* **9**, 117 (1965).

14. A. Keller, *Phil. Mag.* **7**, 1171 (1957).

15. P. H. Lindenmeyer, in P. H. Lindenmeyer, Ed., *Supramolecular Structure in Fibers* (*J. Polymer Sci.* C, 20), Interscience, New York, 1967, p. 145.

16. P. Predecki and W. O. Statton, *J. Appl. Phys.* **37**, 4053 (1966).

17. J. L. Koenig and M. J. Hannon, *J. Macromol. Sci.-Phys.* **B1**, 119 (1967).

18. J. L. Koenig and M. C. Agboatwalla, *J. Macromol. Sci.-Phys.* **B2**, 391 (1968).

19 A. Peterlin, in H. F. Mark, S. M. Atlas, and E. Cernia, Eds., *Man-Made Fibers — Science and Technology*, Vol. I, Interscience, New York, 1967, p. 283.

20. G. Caroti and J. H. Dusenbury, *J. Polymer Sci.* **22**, 399 (1956).

21. Y. Sakuma and L. Rebenfeld, *J. Appl. Polymer Sci.* **10**, 637 (1966).

22. J. H. Wakelin, E. T. L. Voong, D. J. Montgomery, and J. H. Dusenbury *J. Appl. Phys.* **26**, 786 (1955).

23. W. O. Statton, in P. H. Lindenmeyer, Ed., *Supramolecular Structure in Fibers* (*J. Polymer Sci.* C, 20), Interscience, New York, 1967, p. 117.

24. P. F. Dismore and W. O. Statton, in R. H. Marchessault, Ed., *Small-Angle Scattering from Fibrous and Partially Ordered Systems* (*J. Polymer Sci.* C, 13), Interscience, New York, 1966, p. 133.

25. J. H. Dumbleton, *J. Polymer Sci.* A-2, **7**, 667 (1969).

26. M. V. Forward and H. J. Palmer, *J. Textile Inst.* **41**, T267 (1950).

27. M. I. Jacobs, L. Rebenfeld, and H. S. Taylor, *J. Appl. Polymer Sci.* **13**, 427 (1969)

28. M. I. Jacobs, *J. Appl. Polymer Sci.* **13**, 1191 (1969).

29. D. Prevorsek and A. V. Tobolsky, *Textile Res. J.* **33**, 795 (1963).

30. D. C. Prevorsek, A. B. Coe, and A. V. Tobolsky, *Textile Res. J.* **32**, 960 (1962).

31. A. Ribnick, *Textile Res. J.* **39**, 428, 742 (1969).

32. E. M. Hicks, Jr., A. J. Ultee, and J. Drougas, *Science* **147**, 373 (1965).

33. H. Rinke, *Angew. Chem. Intern. Ed.* **1**, 419 (1962).

34. H. A. Pohl, *Textile Res. J.* **28**, 473 (1958).

35. S. Y. Fok, R. A. Mickles, and R. G. Griskey, *Textile Res. J.* **36**, 131 (1966).

36. E. I. Valko, in H. F. Mark, S. M. Atlas, and E. Cernia, Eds., *Man-Made Fibers—Science and Technology*, Vol. III, Interscience, New York, 1968, p. 533.

37. M. L. Rollins, in D. S. Hamby, Ed., *The American Cotton Handbook*, 3rd ed., Vol. II, Interscience, New York, 1965.

38. N. Morosoff and P. Ingram, *Textile Res. J.*, **40**, 250 (1970).

39. M. L. Rollins and V. W. Tripp, *Textile Res. J.* **24**, 345 (1954).

40. L. B. DeLuca and R. S. Orr, *J. Polymer Sci.* **54**, 457 (1961).

41. R. Meredith, *J. Textile Inst.* **42**, T291 (1951).

42. L. Rebenfeld and W. P. Virgin, *Textile Res. J.* **27**, 286 (1957).

43. H. Wakeham and N. Spicer, *Textile Res. J.* **25**, 585 (1955).

44. G. Raes, T. Fransen, and L. Verschraege, *Textile Res. J.* **38**, 182 (1968).

45. H. -D. Weigmann, M. G. Scott, and L. Rebenfeld, *Textile Res. J.* **39**, 460 (1969).

46. J. L. Gardon and R. Steele, *Textile Res. J.* **31**, 160 (1961).

47. L. Rebenfeld and H. -D. Weigmann, in *Proceedings of the First International Cotton Research Conference, Paris*, 1969, p. 595.

48. C. J. Gogek, W. F. Olds, E. V. Valko, and E. S. Shanley, *Textile Res. J.* **39**, 543 (1969).

49. N. R. S. Hollies and N. F. Getchell, *Textile Res. J.* **37**, 70 (1967).

50. J. J. Willard, G. C. Tesoro, and E. I. Valko, *Textile Res. J.* **39**, 413 (1969).

51. A. E. Lauchenauer, H. H. Bauer, P. F. Matzner, G. W. Toma, and J. B. Zurcher, *Textile Res. J.* **39**, 585 (1969).

52. W. G. Crewther, R. D. B. Fraser, F. G. Lennox, and H. Lindley, *Adv. Protein. Chem.* **20**, 191 (1965).

53. L. Pauling and R. B. Corey, *Nature* **171**, 59 (1943).

54. L. Pauling and R. B. Corey, *Proc. Natl. Acad. Sci. U.S.* **37**, 251, 261 (1951).

55. M. Feughelman and A. R. Haly, *Kolloid-Z.* **168**, 107 (1960).

56. H. -D. Weigmann and L. Rebenfeld, in A. V. Tobolsky, Ed., *The Chemistry of Sulfides*, Interscience, New York, 1968.

57. B. M. Chapman, *J. Textile Inst.* **60**, 181 (1969).

58. G. E. Rogers, *Ann. N.Y. Acad. Sci.* **83**, 378 (1959).

59. B. K. Filshie and G. E. Rogers, *J. Mol. Biol.* **3**, 784 (1961).

60. G. R. Millward, *J. Cell Biol.* **42**, 317 (1969).

61. D. J. Johnson and J. Sikorski, in *Proceedings of the International Wool Textile Research Conference, Paris*, 1965, Vol. I, p. 147.

62. R. D. B. Fraser, T. P. MacRae, and G. R. Millward, *J. Textile Inst.* **60**, 343 (1969).

63. M. Horio and T. Kondo, *Textile Res. J.* **23**, 37 (1953).

64. E. H. Mercer, *Textile Res. J.* **23**, 388 (1953).

65. J. Menkart and A. B. Coe, *Textile Res. J.* **28**, 218 (1958).

66. J. H. Dusenbury and A. B. Coe, *Textile Res. J.* **25**, 354 (1955).

67. E. M. Hicks, E. A. Tippetts, J. V. Hewett, and R. H. Brand, in H. F. Mark, S. M. Atlas, and E. Cernia, Eds., *Man-Made Fibers—Science and Technology*, Vol. I, Interscience, New York, 1967, p. 375.

68. L. Rebenfeld, in A. V. Tobolsky, Ed., *Structure and Properties of Fibers* (*J. Polymer Sci.* C, 9), Interscience, New York, 1965, p. 91.

Equilibrium Polymerization

16

A. V. Tobolsky

1. STATISTICAL THERMODYNAMIC THEORIES

The theory of polymer solutions permits one to predict an equilibrium distribution of sizes in a linear reversibly associating polymer system (1, 2). If there are N_0 solvent molecules, N_1 monomers, N_2 dimers, N_3 trimers, etc., and N_x x-mers, the number of configurations, based on the lattice model, is

$$\Omega = \frac{(N_0 + \sum N_x)!}{N_0! \, \Pi N_x!} \left(\frac{z - 1}{N_0 + \sum N_x} \right)^{\Sigma(x-1)N_x} \tag{1}$$

where z is the coordination number of the lattice.

If the standard free energy for the formation of one bond in the polymer chain is ΔF°, the standard free energy of bond formation for the system as a whole is

$$\Delta F_T^\circ = \sum (x - 1)N_x \, \Delta F^\circ \tag{2}$$

The complete expression for the free energy includes ΔF_T° and configuration free energy $-kT \ln \Omega$, where Ω is given by Eq. 1.

$$F = \sum (x - 1)N_x \, \Delta F^\circ - kT \ln \Omega \tag{3}$$

The equilibrium distribution of sizes is obtained by minimizing the free energy F with respect to the variables N_x, subject to the restriction that the total number of polymer segments is constant (1).

$$\sum x N_x = N_T \tag{4}$$

383

The results of this calculation give the following expression for the equilibrium weight fraction of x-mer (for simplicity we set N_0 equal to zero):

$$W(x) = xp^{x-1}(1 - p)^2 \tag{5}$$

where p is defined by the equation

$$\frac{p}{(1 - p)^2} = (z - 1)e^{-\Delta F^\circ/kT} = K \tag{6}$$

Equation 5 is the random or most probable distribution. The quantity p appearing in Eqs. 5 and 6 is related to the number-average and the weight-average degrees of polymerization as follows:

$$P_n = \frac{1}{1 - p}; \qquad P_w = \frac{1 + p}{1 - p} \tag{7}$$

The quantity K in Eq. 6 can be regarded as an equilibrium constant for the reaction (2) in terms of volume fractions v_x and v_y.

$$x\text{-mer} + y\text{-mer} \rightleftarrows (x + y)\text{-mer} \tag{8}$$

$$\frac{xy}{x + y} \frac{v_{x+y}}{v_x v_y} = K = (z - 1)e^{-\Delta F^\circ/kT}$$

Alternatively Eq. 8 can be rewritten so that the activities can be expressed in concentration units of moles per unit volume, c_{x+y}, c_x, and c_y. In this case

$$\frac{c_{x+y}}{c_x c_y} = K' \tag{8a}$$

$$\ln K' = -\frac{\Delta F^\circ}{kT} + \ln(z - 1)V_s$$

where V_s is the molar volume per structural unit (2).

2. TWO-CONSTANT THEORIES

Suppose we had a system that can undergo reversible polymerization. To a good approximation this can be treated by a two-constant theory (3). Such a system is exemplified by sulfur, which can exist as S_8 rings or diradical chains. The equilibrium can be written as follows (4):

$$M \overset{K}{\rightleftarrows} M^* \qquad\qquad K = \frac{M^*}{M}$$

$$(S_8 \rightleftarrows \cdot SSSSSSSS \cdot) \qquad M^* = KM$$

$$M^* + M \overset{K_3}{\rightleftarrows} M_2^* \qquad\qquad K_3 = \frac{M_2^*}{M^* \cdot M}$$

$$M_2^* = K_3 M^* M \qquad M_2^* = KK_3 M^2 \qquad (9)$$

$$M_2^* + M \overset{K_3}{\rightleftarrows} M_s^* \qquad\qquad K_3 = \frac{M_3^*}{M_2^* M}$$

$$M_3^* = K_3 M_2^* M \qquad M_3^* = KK_3{}^2 M^3$$

$$\cdot \; \cdot \; \cdot \; \cdot \; \cdot \; \cdot \; \cdot \; \cdot \; \cdot \; \cdot$$

$$M_{n-1}^* + M \overset{K_3}{\rightleftarrows} M_n \qquad M_n^* = KK_3^{n-1} M^n$$

In the above equation M represents S_8 rings, M^* represents the S_8 diradical, etc. We note first that two equilibrium constants are assigned, one for the initiation reaction and one for all the growth reactions. Second, all the activities appearing in the equilibrium equations are defined in moles per kilogram (more convenient than moles per liter) in accord with the theory of polymer solutions. Third, equilibrium depends only on the initial and final state, not on the reaction path or reaction mechanism. Equations 9 are a complete characterization of initial and final products. We need not be concerned here whether reactions of the type $M_n^* + M_p^* \rightleftarrows M_{n+p}^*$ occur or not, provided we are interested only in equilibrium.

Let N be the total equilibrium concentration of polymer molecules:

$$N = \sum_1^\infty M_n^* = \sum_1^\infty KK_3^{n-1} M^n = \frac{KM}{1 - K_3 M} \tag{10}$$

Let $W = $ total equilibrium concentration of monomer segments incorporated into polymer:

$$W = \sum_1^\infty nM_n^* = \sum nKK_3^{n-1} M^n$$

$$= KM[1 + 2K_3 M + 3(K_3 M)^2 + 4(K_3 M)^3 + \cdots] \tag{11}$$

$$= \frac{KM}{(1 - K_3 M)^2}$$

The number-average degree of polymerization P is given by the following:

$$P = \frac{W}{N} = \frac{1}{1 - K_3 M} \tag{12}$$

The total concentration of S_8 units, whether in ring form or as segments of a polymer chain, is designated by M_0. This is a constant, equal in this case to 3.90 moles/kg. It is clear that

$$M_0 = M + W = M + \frac{KM}{(1 - K_3 M)^2} = M(1 + KP^2) \tag{13}$$

Equations 12 and 13 provide a complete characterization of the equilibrium between S_8 rings and chains of all sizes, as discussed in Section 4.

3. EQUILIBRIUM POLYMERIZATION OF CAPROLACTAM

The overall reaction of ε-caprolactam with small amounts of water to an equilibrium mixture of caprolactam, water, and nylon 6 polymer can be written as follows:

$$
\begin{array}{c}
\text{CO}\text{————}\text{NH} \\
\mid \qquad\qquad \mid \quad + \text{H}_2\text{O} \;\; \overset{K}{\rightleftarrows} \;\; \text{HOOC(CH}_2)_5\text{NH}_2 \\
\text{CH}_2\text{—(CH}_2)_3\text{—CH}_2
\end{array}
$$

$$\text{HOOC(CH}_2)_5\text{NH}_2 + \text{M} \;\; \overset{K_3}{\rightleftarrows} \;\; \text{HOOC(CH}_2)_5\text{NHCO(CH}_2)_5\text{NH}_2\text{, etc.}$$

Symbolically we can write the above equations as follows:

$$
\begin{aligned}
M + X &\overset{K}{\rightleftarrows} M_1 \\
M_1 + M &\overset{K_3}{\rightleftarrows} M_2 \\
M_2 + M &\overset{K_3}{\rightleftarrows} M_3 \\
M_{n-1} + M &\overset{K_3}{\rightleftarrows} M_n
\end{aligned}
\tag{14}
$$

In the above equations M denotes caprolactam concentration, X denotes water concentration, and M_1, M_2, M_3, M_n denote the various polymeric species. Proceeding as in Section 2, we define $N = \sum M_n$, $W = \sum n M_n$, and $P = W/N$; M_0 and X_0 are the initial (unreacted) concentrations of caprolactam and water. The final results are (5)

$$P = \frac{1}{(1 - K_3 M)} = \frac{M_0 - M}{X_0 - X} \tag{15}$$
$$M_0 = M(1 + KXP^2); \qquad X_0 = X(1 + KMP)$$

Once again we note that equilibria depend *only* on the initial and final states, not on the reaction path or mechanism. Although reactions of the type $M_n + M_p \rightleftarrows M_{n+p} + \text{H}_2\text{O}$ do occur, they need not be included in Eqs. 14, which provide a full representation of the initial and final states.

The equilibrium distribution is a random or most probable distribution, the mole fraction of n-mer being given by

$$X(n) = p^{n-1}(1 - p) \tag{16}$$

with $p = K_3 M$. Equation 16 also holds for the results derived in Section 2.

In order to apply the theoretical results summarized in Eqs. 15 it is necessary to compute K and K_3 from limited experimental data and then see how well Eqs. 15 describe the entire range of experimental results. At a given temperature and for a given value of M_0 and X_0 data for P and M are available in the literature. The constant K_3 is immediately obtained from $P = 1/(1 - K_3 M)$, and X is then obtained from $P = (M_0 - M)/(X_0 - X)$. Finally K is obtained from $X_0 = X(1 + KMP)$. The concentration units are all expressed in moles per kilogram.

The consistency of the treatment was checked by using values of K and K_3 to predict the values of P and M for other values of M_0 and X_0. The results were very successful.

The constants K and K_3 were determined at several temperatures. Application of the van't Hoff equation to the values of K and K_3 gave for the polymerization of caprolactam

$$K = \exp\left(\frac{\Delta S^\circ}{R}\right)\exp\left(-\frac{\Delta H^\circ}{RT}\right) \tag{17}$$

$$K_3 = \exp\left(\frac{\Delta S_3^\circ}{R}\right)\exp\left(-\frac{\Delta H_3^\circ}{RT}\right)$$

$$\Delta S^\circ = -6.8 \text{ e.u.}; \qquad \Delta S_3^\circ = -7.0 \text{ e.u.};$$

$$\Delta H^\circ = 2240 \text{ cal/mole}; \qquad \Delta H_3^\circ = -4030 \text{ cal/mole}$$

The standard state is 1 mole/kg.

Since $P = 1/(1 - K_3 M)$, it may be pointed out that for large values of P, $M \approx 1/K_3$; that is, the equilibrium concentration of caprolactam is a function of temperature only, predictable from Eq. 17.

4. EQUILIBRIUM POLYMERIZATION OF SULFUR

In Section 2 the mathematical treatment of the equilibrium polymerization of sulfur was presented, leading to the following results (Eqs. 12 and 13):

$$P = \frac{1}{1(1 - K_3 M)}; \qquad M_0 = M(1 + KP^2)$$

These two relationships are capable of explaining the remarkable effects that occur when elemental rhombic sulfur is heated in a sealed evacuated tube.

At first the sulfur melts at 113°C into a pale-yellow liquid consisting primarily of S_8 rings. This is a liquid of normal low viscosity, which decreases slightly with increasing temperature up to about 159°C. At this point there is a fairly abrupt and very large increase in viscosity, followed by a gradual decrease at still higher temperatures, as shown in Fig. 1. The temperature of 159°C is regarded as a transition or floor temperature for reasons that will soon be clarified.

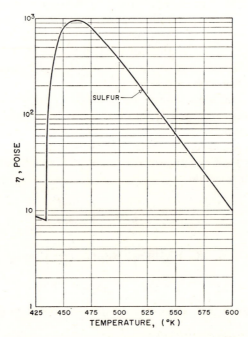

Fig. 1   Viscosity versus temperature for molten sulfur.

The phenomena described above have long been regarded as due to an equilibrium between sulfur in ring form and sulfur in chain form. Preliminary thermodynamic treatments and collation of data were made by Powell and Eyring (6) and by Gee (7). A complete theory based on the equations of Section 2 was presented by Tobolsky and Eisenberg (4).

Experimental points for the average degree of polymerization P and the equilibrium monomer concentration (S_8 rings) are shown in Figs. 2 and 3. It is clear that a transition occurs in the neighborhood of 159°C. At this temperature the P-versus-T curve is in the neighborhood of its maximum value, and M begins to depart from the value of M_0 (3.90 moles/kg). This transition is called a floor temperature because *below* this temperature the sulfur exists only as monomer. In other equilibrium polymerizations we

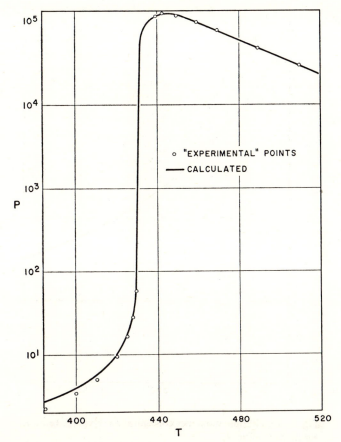

Fig. 2 Degree of polymerization versus temperature for molten sulfur (4).

have ceiling temperatures where only monomer exists *above* the ceiling temperature.

From the equations $P = 1/(1 - K_3 M)$ and $M_0 = M(1 + KP^2)$ one can calculate the values of K and K_3 at a given temperature from the experimental values of M and P. This was done for two temperatures above 159°C, and by use of the van't Hoff equation the following relations were obtained for the equilibrium polymerization of sulfur:

$$\ln K = -\frac{\Delta H^\circ}{RT} + \frac{\Delta S^\circ}{R} \tag{18}$$

$$\ln K_3 = -\frac{\Delta H_3^\circ}{RT} + \frac{\Delta S_3^\circ}{R}$$

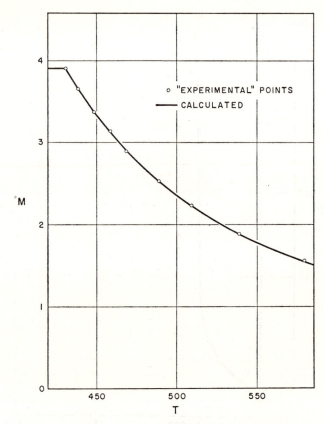

Fig. 3 Equilibrium monomer concentration versus temperature for molten sulfur (4).

$$\Delta H^\circ = 32{,}800 \text{ cal/mole}; \qquad \Delta S^\circ = 23 \text{ cal/}^\circ\text{C-mole}$$
$$\Delta H_3^\circ = 3170 \text{ cal/mole}; \qquad \Delta S_3^\circ = 4.63 \text{ cal/}^\circ\text{C-mole}$$

From Eqs. 18 one can now compute M and P (from Eqs. 12 and 13) at all temperatures. Remarkably this led to a complete prediction of the P-versus-T curve and M-versus-T curve, as shown in Figs. 2 and 3, including the prediction of the sharp transition.

Further mathematical analysis shows that the transition temperature T_t (in this case a floor temperature) is defined by the following equation:

$$T_t = \left| \frac{\Delta H_3^\circ}{\Delta S_3^\circ} \right| \qquad (19)$$

If ΔH_3° is positive, as is true in sulfur, the transition temperature is a floor temperature. If ΔH_3° is negative, the transition temperature is a ceiling temperature.

The sharpness of the transition depends on the ratio $K/K_3 M_0$. For a very sharp transition it is necessary that

$$\frac{K}{K_3 M_0} \ll 1 \tag{20}$$

The concept of floor and ceiling temperatures and Eq. 18 were first presented by Dainton and Ivin (8). It should be pointed out, however, that if only one equilibrium constant is involved, as Dainton assumed and as presented in Section 1, the "transition" is so diffuse that it would be impossible to discern on a P-versus-T or M-versus-T curve. It is only when two equilibrium constants are involved, and a relation like Eq. 20 obtains, that one can really speak of a sharp transition.

The sulfur equilibria as treated in this section assume only S_8 rings and chains. It is known that S_6 rings are also formed, but this has not been explicitly treated. If macrorings, such as S_{16} or S_{24}, were formed in appreciable amounts, the equilibria would not be as simple as presented here. The excellent agreement between theory and experiment shown in this section seems to indicate that macrorings are not a major factor in this case.

Macrorings do, however, play a major role in the polymerization of siloxanes or the polymerization of cyclooctene via an olefin metathesis reaction. Macroring–chain equilibria can be treated by the theory of Jacobsen and Stockmayer (9), later modified by Flory and Semlyen (10). The basis of these theories is that ΔS° for macroring formation is dependent on the size of the ring. This is related to the probability of the end-to-end distance of a random chain being zero, as discussed in Chapter 3.

The effect of the nature of the solvent on equilibrium polymerization was noted by Bywater (11). Ivin and Leonard have observed an effect of polymer concentration on the equilibrium concentration of monomer (12).

5. EQUILIBRIUM COPOLYMERIZATION

By using the equilibrium constants for the homopolymerization of sulfur and the homopolymerization of selenium, it has been possible to predict the copolymerization of sulfur and selenium without introducing any other mathematical constants (13).

Another treatment for equilibrium copolymerization is based on an Ising–model treatment suggested by Alfrey and Tobolsky (14). Consider an equilibrium copolymer composed of N_A monomeric A units and N_B monomeric

B units. At equilibrium the number M_{AB} of AB pairs along the chain is approximately given by

$$\frac{(N_A - M_{AB})(N_B - M_{BA})}{M_{AB}{}^2} = K = \exp\left(\frac{\Delta F_{AB}}{RT}\right) \tag{21}$$

where ΔF_{AB} is $2F_{AB} - F_{AA} - F_{BB}$, and F_{AB} is the molar free energy of formation of an A—B bond, etc. An elegant treatment of this type of Ising problem is given by Rushbrooke (15).

Recent advances in the theory of equilibrium copolymerization are given in reference 16.

REFERENCES

1. A. V. Tobolsky, *J. Chem. Phys.* **12**, 402 (1944).

2. P. J. Flory, *J. Chem. Phys.* **12**, 425 (1944).

3. A. V. Tobolsky, *J. Polymer Sci.* **31**, 126 (1958).

4. A. V. Tobolsky and A. Eisenberg, *J. Am. Chem. Soc.* **81**, 780 (1959).

5. A. V. Tobolsky and A. Eisenberg, *J. Am. Chem. Soc.* **81**, 2302 (1959).

6. R. Powell and H. Eyring, *J. Am. Chem. Soc.* **65**, 648 (1943).

7. G. Gee, *Trans. Faraday Soc.* **48**, 515 (1952).

8. F. S. Dainton and K. J. Ivin, *Quart. Rev.* **12**, 67 (1958).

9. H. Jacobsen and W. H. Stockmayer, *J. Chem. Phys.* **18**, 1600 (1950).

10. P. J. Flory and J. A. Semlyen, *J. Am. Chem. Soc.* **88**, 3209 (1966).

11. S. Bywater, *Makromol. Chem.* **B14**, 52, 120 (1962).

12. J. Leonard, *Macromolecules* **2**, 661 (1969).

13. A. V. Tobolsky and G. D. T. Owen, *J. Polymer Sci.* **59**, 319 (1962).

14. T. Alfrey and A. V. Tobolsky, *J. Polymer Sci.* **38**, 269 (1959).

15. G. S. Rushbrooke, *Introduction to Statistical Mechanics*, Oxford, 1949, pp. 296–300.

16. *Polymer Preprints* (American Chemical Society) **11**, No. 1, February 1970. (Symposium, chaired by K. F. O'Driscoll).

Index